PRINCIPLES OF APPLIED ENGINEERING

Kenneth Reid
Virginia Tech

David Reeping
Ohio Northern University

Elizabeth A. Stephan
Clemson University

Michael Hacker
Hofstra University

David Burghardt
Hofstra University

Richard M. Leuptow
Northwestern University

David R. Bowman
JobScope

William J. Park
Clemson University

Benjamin L. Sill
Clemson University

Matthew W. Ohland
Purdue University

We would like to express special thanks to Elizabeth Spingola, a graduate student at Virginia Tech, for her review of this text.

Boston • Columbus • Indianapolis • New York • San Francisco
Amsterdam • Cape Town • Dubai • London • Madrid • Milan • Munich • Paris • Montréal • Toronto
Delhi • Mexico City • São Paulo • Sydney • Hong Kong • Seoul • Singapore • Taipei • Tokyo

Pearson

Cover Art: Courtesy of Brandstock/Shutterstock

Pearson

330 Hudson Street, New York, NY 10013

Hardcover ISBN 10: 0-13-442801-3

Hardcover ISBN 13: 978-0-13-442801-7

2 17

BRIEF CONTENTS

CONTENTS

ABOUT THE AUTHORS

Kenneth Reid is the Assistant Department Head for Undergraduate Programs in Engineering Education at Virginia Tech. He earned his PhD in Engineering Education in 2009, only the seventh person to earn this type of degree in the nation. He is active in engineering within K-12, serving on the TSA Board of Directors. In 2014, Reid and his coauthors were awarded the William Elgin Wickenden Award, recognizing the best paper in the Journal of Engineering Education, as well as the best paper award for the Educational Research and Methods (ERM) Division of ASEE. He was awarded an IEEE-USA Professional Achievement Award in 2013 for designing the nation's first BS degree in Engineering Education. Reid was named NETI Faculty Fellow for 2013–2014, and the Herbert F. Alter Chair of Engineering (Ohio Northern University) in 2010. His research interests include success in first-year engineering, engineering in K-12, introducing entrepreneurship into engineering, and international service and engineering. He has written two texts in Digital Electronics, has taken multiple teams of students to the Dominican Republic through Solid Rock International, and has received a number of teaching awards.

David Reeping is an undergraduate student at Ohio Northern University, majoring in Engineering Education. He is currently pursuing admission to a graduate program in the same discipline. He was a Choose Ohio First scholar inducted during the 2012–2013 school year and was the recipient of the Remsburg Creativity Award for 2013. In addition, Reeping received The DeBow Freed Award for outstanding leadership as an undergraduate student (sophomore) in 2014. He is a member of the mathematics, education, and engineering honor societies: Kappa Mu Epsilon, Kappa Delta Pi, and Tau Beta Pi, respectively. Reeping has extensive experience in curriculum development for K-12 and develops material for the Technology Student Association's annual TEAMS competition. As an undergraduate research assistant, he has studied the analysis and refinement of the first year engineering experience, authentic projects and assessments, and P-12 engineering.

Elizabeth A. Stephan is the Director of the General Engineering Program at Clemson University. She earned a BS in Chemical Engineering from The University of Akron. During her undergraduate work, she completed a cooperative education experience with Dow Chemical in Midland, MI, conducted research on coal purification methods, and was named the College of Engineering Outstanding Senior. After graduation, she was employed by Boride, a wholly owned subsidiary of Dow Chemical in Traverse City, MI, specializing in high-performance ceramics. She returned to The

University of Akron on a College of Engineering Fellowship, earning her PhD in Chemical Engineering focusing on multiphase transport processes. She has taught at The University of Akron and Wayne College, and served in several post-doctoral positions. She joined the faculty at Clemson in January, 2002 in the General Engineering Program, assuming the role of Director in 2007. Beth has served as a national official as a district director in Tau Beta Pi, the engineering honor society, since 1996. She is the chief advisor for the South Carolina Alpha Chapter of Tau Beta Pi, and an advisor for the Clemson chapter of Alpha Omega Epsilon, a professional sorority.

Michael Hacker serves as the Co-Director of the Center for Technological Literacy at Hofstra University. For over 45 years, engineering and technology education have been at the core of his professional life. He has been a classroom teacher, department supervisor, and university professor. He rose to local, state, and national prominence as an advocate for contemporary engineering and technology education programs in the nation's schools. As the former New York State Education Department Supervisor for Technology Education, Hacker has managed the development and implementation of innovative middle- and high-school curricula that have served as national models. He was named State Supervisor of the Year in 1993 by the ITEEA, received the ITEEA Award of Distinction in 1995, and was inducted into the prestigious ITEEA Academy of Fellows in 2004. He was honored with the Epsilon Pi Tau Distinguished Service Award in 2001. Hacker is the author of five secondary-school textbooks and dozens of journal articles. He served as a co-editor of three sets of NATO proceedings, as well as a major international publication focused on the role of engineering and technology education in fostering human development. He has served nationally as a consultant to the American Association for the Advancement of Science, as a National Science Foundation expert reviewer, and as a writer for the National Standards for Technological Literacy.

David Burghardt is a Distinguished Professor of Engineering, Co-Director of the Center for Technological Literacy, and the former chair of Computer Science at Hofstra University. He is the author of ten texts in engineering and technology as well as many articles in engineering and technology education. He is a professional engineer in New York State, a chartered engineer in the United Kingdom, a member of the IEEE committee on linking colleges of engineering and education, an NSF reviewer, a principal investigator on NSF and NASA projects, and a writer for the New York State math, science, and technology learning standards. At the Center for Technological Literacy, Burghardt has helped create a graduate program in integrated math, science, and technology for elementary and middle-school teachers. In addition, he designed a course that places informed design at the center of learning.

David R. Bowman has been teaching in the General Engineering Program at Clemson University since January, 2006. He earned his degrees from Clemson University, including a BS and MS in Computer Engineering and is currently pursuing a PhD. A member of ASEE, David has experience in the design and development of software tools for engineering education research and pedagogy. During his undergraduate and graduate work, David hosted *All Screams Considered*, an award winning radio show on WSBF-FM, whose name apes the popular NPR program *All Things Considered*. In addition to broadcasting, David enjoys performing music on acoustic, electric, and bass guitars.

William J. Park is currently an associate professor in the Engineering and Science Education Department at Clemson University. Following a few years as a cattle farmer, he completed three degrees at Clemson University: a BS in Ornamental Horticulture with a particular emphasis on xerophytic plants, an MS in Electrical Engineering focusing on electronic music synthesis, and a PhD in Electrical Engineering conducting research in electronic counter-counter measures. Bill is currently faculty advisor for a student team renovating a very large 1970's vintage electronic organ, and is a moderately accomplished pianist.

Benjamin L. Sill is Alumni Professor Emeritus of Civil Engineering, having retired in 2008 after 32 years at Clemson University. He earned a BS and MS from N.C. State University in Aerospace Engineering and a PhD from Virginia Tech in Aerospace and Ocean Engineering. Before he joined Clemson, Ben was employed by the Naval Ordnance Station, Indian Head, MD, and by Duke Power Company, Charlotte, NC. At Clemson, he was a founder of Clemson's Wind Load Test Facility. Beginning in 1999 he served as the Director of Clemson's General Engineering Program. In 2007, he helped establish a new Engineering and Science Education Department at Clemson, and served as its chair until his retirement. He is the recipient of numerous teaching and research awards, including the prestigious Clemson Class of 1939 Award. Outside the university, he gives numerous presentations with topics ranging from humorous to educational – including talks on ancient coins, old maps, wildflowers, houseplants, snakes, birds, and hurricanes. Ben has authored three bird books, has published technical articles on snakes, frogs, fish, volleyball, and bromeliads and has created and registered many new bromeliad hybrids.

Matthew W. Ohland is currently an associate professor of Engineering Education at Purdue University, West Lafayette, IN. He earned a BS in Engineering and a BA in Religion from Swarthmore College, MS degrees in both Mechanical Engineering and Materials Engineering from Rensselaer Polytechnic Institute, and a PhD in Civil Engineering from the University of Florida. Matt was an NSF postdoctoral fellow for science, mathematics, engineering, and technology education and joined the faculty of General Engineering at Clemson University in 2001. In 2006, he joined the faculty at Purdue University. He was the 2002–2006 National President of Tau Beta Pi, the engineering honor society. He currently serves as the Chair of the Educational Research and Methods division and an ABET Program Evaluator for the American Society of Engineering Education, on the Administrative Committee of the IEEE Education Society, and as the Chair of the Steering Committee of the IEEE Transactions on Learning Technology.

ACKNOWLEDGEMENTS

I have to begin by thanking my colleagues who continuously offer support to pursue everything that is interesting, valuable, or fun. The most supportive person has always been and continues to be my beautiful wife, Jenny, and I can't thank her enough. And a special thanks to Bryan and Chrissie, who heard quite a bit about writing a book through the process.
-Ken Reid

I would like to thank everyone who has offered their words of encouragement through the process of developing this text. My parents deserve an additional and special thanks for their undying support of my work, this text, and beyond - I appreciate it more than they could possibly imagine.
-David Reeping

And from both of us, thanks to the team at Pearson and Emergent Learning.

CHAPTER 0
A SCENARIO IN THE FUTURE

The year is 2025. Your relatively new engineering career is going great! You graduated with a degree in mechanical engineering a few years ago. Now you have landed a job you love with a multinational engineering firm, complete with the opportunity to travel and meet new people along the way. You get along well with the team at work; in fact, you even meet up outside of work from time to time. While you had the chance to work with teams throughout college, it became clear that engineering is often more about teamwork than the formulas etched into your brain.

Figure 0-1

Teamwork
Design

One aspect of your job that you enjoy is the chance to think outside the box – to be creative. Your designs have been successful (most of the time), but failure is a natural part of design. In the same way that your first draft is not your final draft, your first iteration of a product will not be the design you take to market. Through constant improvement, your contributions have always met the requirements of the project; you have met the criteria and designed within all constraints.

Criteria
Constraints

0.1 HAPPY NATURE NATIONAL PARK

We all remember the former manufacturing site that was restored a few years ago (keep in mind – it is 2025). When the government came in and cleaned out the site, they discovered beautiful scenery like rock formations, streams, lakes, and small waterfalls. Since it was so beautiful, they decided to make it a national park. After a huge contest, the new name was selected: Happy Nature National Park!

Figure 0-2 This family couldn't be any happier as they visit Happy Nature National Park!

Figure 0-3 A member of your team fills her water bottle at the clean and clear stream.

Everyone agrees: The park is a magical place where the skies are always sunny, the grass is always green, and the families are always happy. Further, everyone agrees that it is hard to picture this as a formerly contaminated area.

Since you and your team all have vacation time coming up, you decide to explore Happy Nature. "Maybe we should rent a cabin and stay for a few days!" Your team agrees and the plans are set.

Your team of three engineers packs up and hits the road. You make the drive to Happy Nature. You are all engineers, but you decide to leave work behind and just visit nature.

When you arrive at the cabin, a note on the door assures you that this park is SO CLEAN that there is no need for separate drinking water! A few empty bottles and a map to the local stream where you can fill them were left on the counter. Fortunately, your cabin is also equipped with a beautiful, flat, study table (this will be important later).

The water is clear and seems perfectly safe! One member of your team likes to fish, one likes to hike, one prefers to swim – and there is time for all of it.

One day, as your team is hiking, you see a sign warning of radiation in the area. You are sure that this must be left over from the former manufacturing site, so you aren't concerned with this now. The hike continues.

The days are filled with a mix of relaxation and adventure.

Figure 0-4

0.2 TROUBLE IN PARADISE?

You begin to notice that a few things seem a little out of place…

Later on in the evening, you hear some other hikers walking by, so you give them a friendly wave "hello!" While they do wave back and respond with a friendly "hello," their apparel seems strange for hikers. Your team talks with these friendly strangers for a little while. However, when you ask them why they are dressed that way, they brush your concern aside. "You never can be too careful, right?" they say, laughing. While it does seem strange, and you aren't completely satisfied with their explanation, you decide that maybe they just don't understand what they should wear on a hike.

Figure 0-5 Not typical apparel for hiking.

One teammate heads off for an afternoon of fishing and brings back a bounty of fish. The team decides that a fish dinner sounds like a great idea!

Figure 0-6 Radioactive ... fish?

When your teammate pulls the fish from the cooler, however, one looks ... strange.

Since this is 2025, we know the government recently found a way to label animals (including wildlife) if they happen to be contaminated. It seems that your teammate missed a pretty obviously marked fish among his haul. Yes, one is clearly marked as radioactive. You haven't seen a radioactive fish or animal yet on this trip, so maybe this was brought here? Maybe it swam to the lake from a stream? You decide to investigate and head to the lake, where you see that there is no stream feeding the lake. Now that you are paying attention, you notice that there are quite a few of these radioactive fish!

Finally, you see one more indication that things may not be quite as safe as you thought: The lake water in the water bottles gives off a slight glow when the lights are off. Obviously, this doesn't seem right.

0.3 PROBLEM SOLVING

To summarize:

a. The water labeled as safe to drink may be radioactive.
b. There is a full bottle sitting exactly 18" from the edge of the table: At this point, you don't want to touch the bottle.
c. You are a member of an engineering team that is capable of solving problems creatively.

Your team decides to meet to determine what to do. In a strange, but fortunate, series of events, one of your teammates happened to pack clothing and three items: a toothbrush, a Geiger counter, and a radiation-proof container that can hold one water bottle. The Geiger counter confirms that the water in the bottle is radioactive.

Your task: The team must design and build a device to lift the radioactive water bottle into a radiation-proof container.

Figure 0-7 Geiger counter.

The team decides to build this device. The team members scour the cabin in search of supplies. You find a toolbox and assorted other material, including paper clips, a spool of wire, tape, rubber bands, and more.

They also find a toolbox with a note saying:

> You can use these tools, but you may NOT include any of these tools in a design you build.

0.4 YOUR TASK

In teams of 3 or 4, gather the supplies listed in our story.

Materials list:

(Note: Your teacher may modify the first part of this list if necessary; as long as each team has the same set of supplies, the design goes on.)

- 5 cardboard strips: 12" long, 2.5" wide
- 5 rubber bands
- 8 clothespins
- 25 paper clips
- 20 craft sticks
- 6 pencils
- 8 binder clips
- 3 feet masking tape
- 3 feet wire
- 3 feet string or twine

Also (not to be incorporated into the radioactive water-lifter):

- Scissors
- Wire clippers
- Ruler
- Water bottle (at least ¾ full)
- Bucket (preferably smaller, but large enough to hold a water bottle)

You have 15 minutes to build a radioactive water-lifter able to lift a full water bottle into the air and place it in a bucket. The operator may not get any part of his or her body within 18" of the water bottle. You may test while you build.

After the activity, discuss as a class:

1. Was this activity "engineering"? Why or why not?

2. Did the teams exhibit effective teamwork?

3. If you had time to plan, would your design have improved?

CHAPTER 1
ENGINEERING AS A DISCIPLINE

1.1 DEFINE ENGINEERING

engineering

The field of **engineering** operates under a slew of misconceptions held by the public. A modern day engineer does not operate trains; the scope of work within engineering has broadened since the early days of the hostler. Professionals in the field will likely not be found underneath our cars changing oil. Instead, engineers can be found behind the scenes of technological innovations – from the devices in our pockets capable of connecting us with others around the world to the composition of the rubber soles in shoes.

accrediting
ABET

We hear the word engineering in advertisements, tossed around as a potential career for those gifted in math and science (though one can have average math skills and still succeed), but what do we mean by *engineering*? The organization responsible for **accrediting** (or evaluating) programs in the field, the Accreditation Board for Engineering and Technology (**ABET**), provides a definition to encompass the primary aspects of what it means to be an engineer.

> *Engineering is defined to be a profession that draws upon knowledge of mathematics and science to create ways to economically use available materials and the forces of nature to benefit the human race through the process of design.*

Scattered through the definition are various key words and phrases which characterize the profession: knowledge of mathematics and science, economic use of materials and natural forces, benefiting the human race, and finally – design. Now, what do all of these phrases mean and how do they capture what engineering is all about?

Knowledge of Mathematics and Science: Engineers are heavily trained to ensure they are proficient in mathematics and science. The typical coursework an undergraduate engineering student will face includes study in the realms of calculus and physics. The mathematical concepts and methods we learn help us solve physical problems. Rising to the call of being an engineer goes beyond mere textbook exercises; instead, it requires us to solve problems on a grander scale. As eloquently stated by Galileo Galilei, "The laws of Nature are written in the language of mathematics." Considering that our ability to harness the natural forces around us is dependent on being able to use mathematics in a practical sense, we are trained to speak the language fluently.

criteria
constraints

Economic Use of Materials and Natural Forces: Although we can theoretically sketch out plans for an extravagantly costly and wasteful project, engineers must weigh their options and work within specifications, often called **criteria**. Beyond the requirements set by a customer, societal or political **constraints** could affect the creation and implementation of the project. Practically, the common limitation that engineers face is the fact that material and energy cannot be created or destroyed; therefore, we need to be mindful of our consumption of natural resources.

Benefiting the Human Race: Through an engineer's work, he or she is meeting a particular need. Whether the need in question is relieving poverty in a disadvantaged region of the nation or delivering a quality surround-sound experience in a family's home theater, the process is fundamentally the same. Our modern lives can be attributed to the work of engineers through the centuries.

design

Design Process: Of all the terms in the definition, **design** lies at the heart of the engineering profession. In order to benefit the human race in a tangible way, we need to undergo a process that takes an idea and implements it physically. The process in question is called the **design process**, an *iterative* series of steps engineers use to arrive at the solution to a problem.

design process

Throughout this text, we will delve into the field of engineering and the manner in which an engineer goes about meeting the technological challenges of this century – design.

1.2 THE HISTORICAL DEVELOPMENT OF ENGINEERING

When did technology actually begin? People often think of our times as technological times and earlier eras as "before technology." In fact, people have always used technology. Recall that technology is simply using the knowledge we have to create things that people need. Because our knowledge has grown, technology has grown with it—from the simple spears and grinding stones of our earliest ancestors to the complex machinery and systems of today.

Much of our history has been defined by the materials that people used in a particular era. They used these materials to make tools that helped them carry out everyday tasks. The names of these materials have given us a way to classify historical periods. Three important periods of early history are the Stone Age, the Bronze Age, and the Iron Age.

Stone Age

Stone Age

The **Stone Age**, the period during which people used stone to make tools, lasted for almost 1 million years—from about 1 million B.C. to about 3000 B.C. (Other common materials used during this time were bones and wood.)

During this period, people lived very different lives from ours. Early humans used stone, bone, and wood to make axes, spears, scrapers, and even tools for starting a fire. They used the bow for hunting animals. They made needles from bone and used them to sew clothing from animal skins.

During the Stone Age, there were few villages. Most people lived nomadic lives, wandering from place to place. They hunted animals and gathered plants, fruits, seeds, and roots for food. They used fire for cooking meat and for protection from wild animals. When they had used up food sources in an area, they moved to a new location. The pace of technological change was very slow.

Bronze Age

Bronze Age

During the **Bronze Age**, which lasted from about 3000 B.C. to about 1200 B.C., people began to craft tools and weapons from bronze rather than from stone. The Bronze Age actually began with the discovery of copper. Copper is a soft metal, so it was not practical for tools. During the Bronze Age, however, people learned that copper ore could be heated using charcoal to yield pure copper. They discovered that by melting other ores with copper, they could produce a stronger metal. Bronze is a mixture of the metals copper and tin. A mixture of two or more metals is called an **alloy.**

alloy

Bronze is lighter than stone, and bronze tools and weapons can be made sharper than those made of stone. Using bronze, early humans created new tools such as knives, hooks, and pins. They could also craft large urns and vessels to hold water and food (Figure 1-1).

Figure 1-1 Because bronze is lighter than stone, it was used to make urns, pots, and tools.

Iron Age

Iron Age

The **Iron Age,** the era when iron came into common use, followed the Bronze Age at different times around the world. It began around 1200 B.C. in the Middle East and about 450 B.C. in Great Britain. It was shorter than either of the two previous periods, lasting only about 1,000 years.

smelting

Iron was made from iron ore through a process called **smelting,** in which the ore was melted to take out the impurities. The early iron-smelting furnace was a clay-lined hole in the ground. Iron ore and charcoal were placed in the hole, and then air was pumped in with a bellows. Air from the bellows made the charcoal burn hot enough to soften the ore into a mass of iron. This mass was hammered into the desired shape while it was still red hot (Figure 1-2).

Figure 1-2 Iron-smelting furnaces were used to soften iron core. The blacksmith could then hammer the iron into different shapes.

Iron eventually replaced bronze as the primary metal used in toolmaking because it was available in more areas and was cheaper to process. It could also be crafted into sharper, longer-wearing tools, such as chisels and saws. As with bronze, the techniques used to make iron varied in different parts of the world. People experimented with temperatures and with the amounts of the various ores they used. They often developed these early technologies by trial and error, without an understanding of the underlying scientific principles.

1.2.1 KEY INVENTIONS OF EARLY HISTORY

During early history, people in different parts of the world used stone, bronze, and iron for many of their needs. Tools gradually became more sophisticated and easier to use. People in colder climates had a greater need for warm clothing and shelter. People who lived far from a river or lake wanted to be able to store water for later use. Over many thousands of years, important inventions were developed to meet these different needs.

Although many tools were used in similar ways throughout the world, the inventions in a particular area often reflected the specific needs of those people. For example, when horses were used as a form of transportation, the saddle was then invented so riders could ride more comfortably.

Agriculture

About 10,000 years ago—in 8000 B.C.—the nomadic, hunter-gatherer way of life changed. People learned how to grow their own food. They cultivated different types of grains for food, such as wheat and barley. They became farmers who grew flax, which provided fibers to make linen cloth. They invented the wheel and then vehicles with wheels. They built roads and stone houses.

The plow was first developed approximately 4000 B.C. This simple tool (Figure 1-3) allowed farmers to turn over soil for planting more easily. It made farming much more efficient. Soon, strong animals such as oxen were used to pull the plow. Farmers could

Agricultural Era

now plow more land in less time. This period in history is described as the **Agricultural Era,** a time when most people lived off the land. Many tools and discoveries had to do with growing and harvesting crops.

Irrigation, Sewage Systems, and Aqueducts

With farming and the rise of towns and cities came the need for water systems. People developed irrigation systems to bring water to dry soil and increase plant production. Sewage systems helped safeguard health.

aqueducts

The ancient Romans built aqueducts to carry water from the nearby hills to the city. An **aqueduct** (Figure 1-4) is a channel built to carry water. The underground ones were channels carved into rock. Some were raised on arches to pass over rivers or valleys. All were built with a slight slope to allow for gravitational flow.

Figure 1-3 The plow allowed most of the heavy work of turning the soil to be done by strong animals such as oxen or horses.

Figure 1-4 The Pont du Gard is the largest and best preserved of the Roman aqueducts.

Figure 1-5 Water wheel.

Water Wheels and Machinery

An important early invention was the water wheel (Figure 1-5). In the earliest water-wheel systems, water from a river or stream turned a large wheel. This wheel turned a mill for grinding grain into flour. It replaced human muscle power with water power. Before the water wheel, grain was ground by hand using two large stones. This was such hard work that people were sometimes made to do it as punishment.

The water wheel introduced the machine age. It provided power not only for grinding grain but also for pumping water, making cloth, and heating iron ore for smelting. The use of machinery such as the water wheel was perhaps the most important technological development of the Middle Ages.

Writing and Printing

Writing enabled people to record their ideas and make them permanent. It increased the amount of knowledge that could be transmitted from one generation to the next and from one culture to another. People first wrote on stone or clay tablets. Years later, paper made from wood pulp was invented.

The printing press further transformed the way information was spread. Before the printing press, books were copied by hand, a process that often took weeks or even months. In the mid-1400s, a German goldsmith, Johann Gutenberg, devised a printing press that used small metal molds for all the letters of the alphabet. Once the letters were set on the press, text could be printed in a matter of minutes.

1.2.2 INDUSTRIAL REVOLUTION

Industrial Revolution

The **Industrial Revolution** was a period in which human and animal muscle power was replaced by machines. The Industrial Revolution began in Britain during the late 1700s and lasted until the mid-1900s. During this time, people invented machines that could spin wool, weave fabrics, and pump water.

An English metalworker named Abraham Darby was able to make iron that was very pure and extremely strong. He did this by heating it with coke. Coke is a carbon-like substance that comes from coal. Darby's process led to a much better quality of iron. Iron could now be used to build large structures and new types of machines for industry. Many of these machines were powered by steam (Figure 1-6). This new technology provided a great leap forward in the manufacturing process.

A New Way of Making Things

factory system

With the invention of these new machines came the factory system of work. In the **factory system,** people made goods using machines rather than by hand. Before this innovation, craftspeople had used their own tools and workshops to make things. Shoes, cloth, and all sorts of tools were crafted individually. Each product was slightly different. With machines, however, products such as fabric were no longer created by an individual. Instead, massive looms manufactured yards of fabric at a time. Many products could be made faster and cheaper.

Richard Arkwright, a British inventor and cotton manufacturer, developed the first manufacturing system. He linked together machines of his own design. Using a

SETTING THE SCENE: Technology, Then and Now

Technology has led to dramatic changes in the way we grow food, treat disease, manufacture complex machinery, commute to school and work, and even heat our homes. The pictures in this box tell the story of one of the most exciting changes in technology—how we communicate.

- The telegraph, invented in 1837, allowed people to send short written messages. Words were translated into a special code that was sent by electric pulses across a telegraph wire.
- In 1876, Alexander Graham Bell invented the telephone. For the first time ever, the sound of the human voice was transmitted by electrical signal.
- About 100 years later, the first cellular telephone was invented. Clunky and expensive, it was still innovative.

Today, telephone communication is inexpensive, and calls can be made from anywhere to anywhere. Wall jacks and traditional phone lines may soon be a thing of the past, as more devices are combining phones with computers and many people are replacing land lines with cell phones.

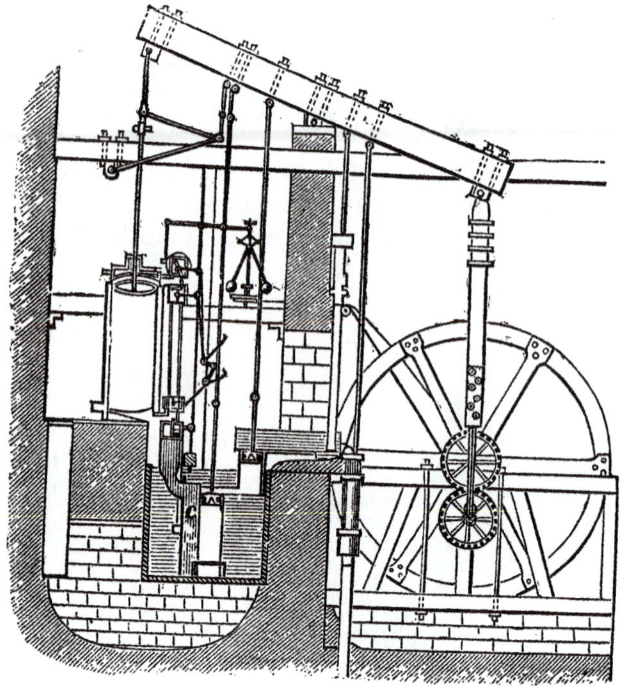

Figure 1-6 During the Industrial Revolution, many machines, such as this early engine, were powered by steam.

sequence of steps, each machine carried out one job and prepared the material to be processed by the next machine. The cotton thread machine changed raw cotton into thread. Arkwright built many mills that produced cloth cheaper and faster than it could be produced by hand. Because of this new technology, fabric making was transformed from a cottage craft into a factory job.

In England and the United States, the assembly line became important for manufacturing. Assembly lines were based on a system of interchangeable parts. In a car, for example, identical copies of each part were assembled in the same way to build many identical cars (Figure 1-7).

Changes Produced by the Industrial Revolution

The factory system produced goods more cheaply and quickly than ever before. But it also created change and difficulty in people's lives. Craftspeople could not compete with machine manufacturing. Many had to give up their crafts and move to cities to find work. People became workers on factory assembly lines, using their employers' tools and machines instead of their own. They received wages for their work. Because parents were often gone from the home all day, schools were built to educate and care for children during the workday.

The technology of the Industrial Revolution also affected politics—how laws were made and civic affairs were governed. Men, women, and children often worked 14 hours a day in crowded and unsafe conditions. The mistreatment of employees led people to form labor unions. New laws addressed employees' rights, safety issues, working conditions, and fair wages.

Patent laws were also developed during the Industrial Revolution. People who invented a new device or improved on an old one could patent their idea. With patents, no one could legally use another's idea or invention without paying for it. In addition, patents gave official credit where and when it was due, because one's creativity was central to the idea. Because people could now make money from their inventions, there was an incentive to invent and improve products and systems. Existing ideas were used as building blocks for newer and more powerful technologies (Figure 1-8).

Figure 1-7 Assembly lines modernized the assembly of automobiles.

Combining Technologies				
Existing Technologies				**New Technology**
Lenses	+	Chemistry	→	Photography
Shipbuilding	+	Steam engine	→	Steamship
Telegraph	+	Printing press	→	Newspaper
Internal combustion engine	+	Wagons & carriages	→	Automobile
Lightweight gasoline engine	+	Kite and glider technologies	→	Airplane
Materials technology (e.g., titanium)	+	Jet engine	→	Spy aircraft
Photography	+	Satellites	→	Exploration of the Earth

Figure 1-8 By combining two simple technologies, people were able to create newer and more powerful technologies.

Information Age

Information Age

The changes caused by machinery and factory production gradually gave way to a new type of technology: information. The last few decades—a tiny fraction of human history—have been marked by extremely rapid change. This period, which continues today, is called the **Information Age.** In 1947, the first transistor was invented using a piece of semiconducting material and some wires. Transistors were soon combined with other components into integrated circuits. These were used in televisions and radios (Figure 1-9) and later in computers and electronic devices.

Inventions became increasingly focused on electronics and the computer. The first successful personal computer—the Apple II—was invented in 1978. Computers have since become much smaller, faster, and cheaper, and many of today's inventions are now based on computer technology. Computers also have changed how we work. Jobs depend on workers being well-educated and staying informed about changes in technology.

(a) (b)

Figure 1-9 Information Age Inventions (a) The transistors used in this early radio were an important invention of the Information Age. (b) Another important invention of the Information Age has been the Internet.

1.2.3 THE RATE OF TECHNOLOGICAL CHANGE

The rate of technological change has increased over the course of history. During prehistoric times, it took many thousands of years to progress from stone tools to metal tools. As time passed, however, knowledge increased. People lived closer together in cities. They could publish and share their ideas in books and newspapers.

exponential rate of change

Technological change is faster now than it has ever been. Some people say that the world's body of knowledge is now doubling every four years. Today, technology undergoes an **exponential rate of change,** with extraordinarily rapid developments.

When something changes at an increasing rate, we say that it changes exponentially. This is true of technological knowledge. It took about a million years to go from using stone tools to using tools made of bronze, but only about 5,000 years to go from using bronze tools to using machines.

Think about the changes that have taken place in the short time since the first personal computer was sold in 1981. Today's PCs have about 50,000 times as much memory and are a great deal faster than the early versions. The rate of technological change continues to increase (Figure 1-10).

Moore's law

The rate of technological change is often theorized to follow **Moore's Law**. Moore's law states that the number of transistors within a computer processor doubles every two years. As the advance of technology continues to increase, Moore's Law is sometimes said to double the number of processors every 18 months. It is sometimes applied to other metrics of computers: the amount of storage (hard drive storage and Random Access Memory, or RAM) doubles every 18 months; the cost of a given system is cut in half every 18 months; the speed of the processor doubles, the bus speed within the computer, etc. Moore's law has largely held true over the past few decades.

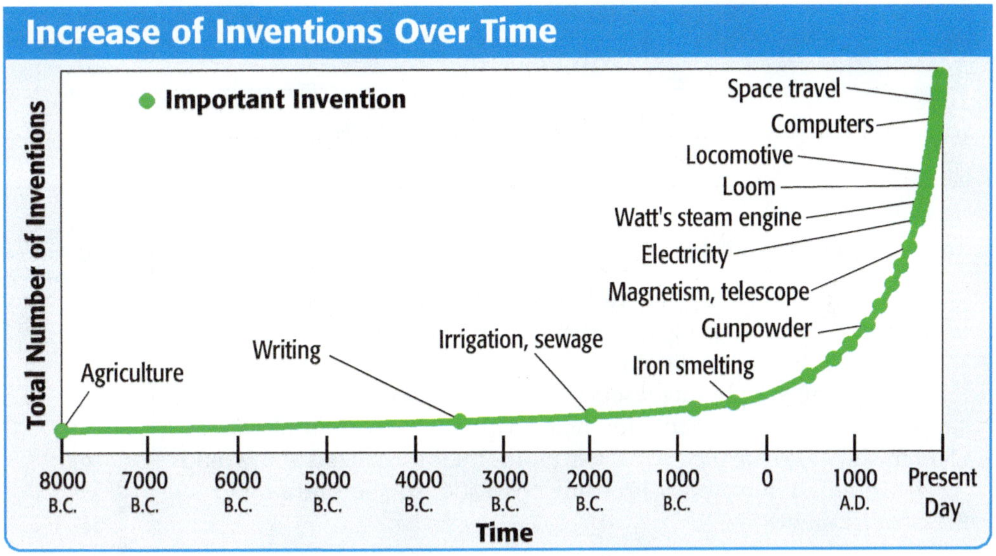

Figure 1-10 The pace of invention has continually increased over time.

CONNECTING TO STEM

STEM = Science, Technology, Engineering, and Mathematics

Exponential Rate of Change

What does "exponential rate of change" mean? Let's look at an example. Sandra's mother agreed to let her download ten songs from iTunes a month. Her brother, Todd, wanted to download some, too. He got his mother to agree to this plan: He would download two songs the first month and then double the number he downloaded each month thereafter. Who do you think would have more songs after six months?

Sandra: Linear Rate of Change

At the end of the first month, Sandra had ten songs. Each month, she downloaded another ten. After six months, she had 60 songs, as shown in the graph. Because her supply increased by the same rate each month, the number of songs showed a **linear rate of change**. On a graph, this rate looks like a straight line.

Monthly Song Downloads

Month	Sandra's Purchases (Number of CDs)	Todd's Purchases (Number of CDs)
1	10	2
2	10	4
3	10	8
4	10	16
5	10	32
6	10	64

Todd: Exponential Rate of Change

Todd started out by buying two songs. He doubled his purchases each month thereafter. In the second month, he bought four, and in the third month, he bought eight.

Todd's song supply increased faster each month. As a result, his supply increased exponentially. In a graph, an **exponential rate of change** shows a sharp curve. Their mother soon realized she would be spending all her money on songs for Todd, so she put an end to the arrangement.

Total Number of CDs Purchased

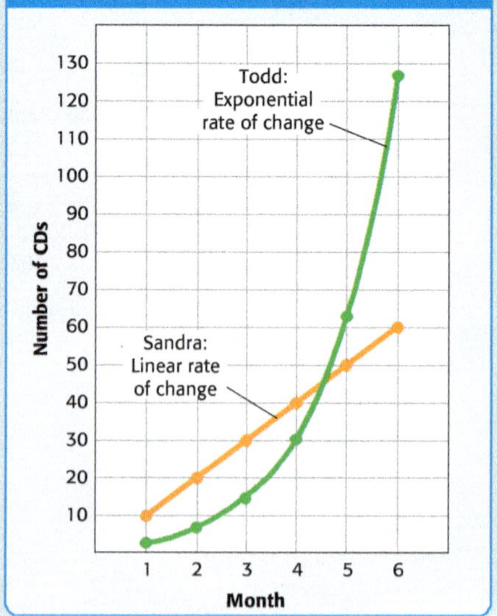

1.3 ENGINEERING IN TODAY'S SOCIETY

Technology always causes change. Change can be good for some and bad for others. It can create jobs and restructure whole communities. It can also inflict economic hardship on people who are forced to move to find new jobs. People who live in different parts of the world have different technological choices and opportunities because of such factors as differences in economic resources, geographic location, and cultural values.

Technology Affects People's Jobs

In the late 1800s, the steam locomotive powered the westward expansion along the railroads. People moved west, and goods and services could now be exchanged easily between the east and west coasts. New towns were established along the rail lines, where machinists, boilermakers, and repair people maintained the equipment. Soon, diesel power replaced steam. Diesel engines were more fuel efficient, so they required fewer stops en route. The train communities withered as people moved in search of new jobs.

Similar changes have resulted from the new technologies of the Information Age. In the 1960s, transistors replaced vacuum tubes. Semiconductor manufacturers prospered and hired people, but vacuum tube manufacturers went out of business.

More recently, the Internet created a new industry almost overnight. The dot.com industry, made up of companies doing business only on the Web, created many new jobs, but even this high-tech industry has had many businesses fail. This pattern of technological advancement, resulting in both winners and losers, is observed time and again (Figure 1-11).

Technological systems, processes, and products evolve and change over time in response to demands from people. These demands can be economically driven in the desire for less expensive goods and services; politically, culturally, or socially driven because of societal needs, such as clean drinking water; or technologically driven because of discoveries that allow new designs.

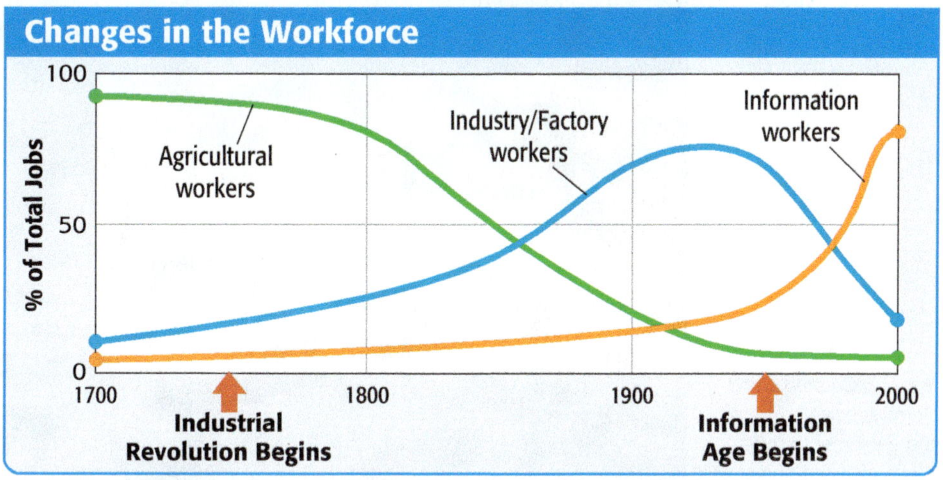

Figure 1-11 Changes in technology have led to the development of different types of jobs in different eras.

Technology Increases Comfort, Safety, and Accessibility

Using technology, products have been designed to make our lives more comfortable. Chairs are shaped for back support. Drinking glasses for small children are made so that they are easy to hold. Tables and desks are at a convenient height. Comfortable auto seats make driving more enjoyable. Designers match products to people's needs.

Products are also designed for safety and for accessibility to all types of people. The blades of kitchen machines such as food processors are protected to avoid the possibility of people accidentally cutting themselves. Automobiles are designed to be as safe as possible in the event of an accident. Special bicycles, cars, and skis are made so that people with disabilities can remain active and independent (Figure 1-12). Special buses allow those in wheelchairs to use mass transportation. Technology has done much to make public service accessible to all members of our society.

Technology Changes Our Landscape

Technology has also changed our surroundings. Since the Industrial Revolution, cities have grown in number and in size. They have become more and more crowded. In the early twentieth century, the development of a mass transportation technology—the electric trolley car—allowed suburbs to flourish. The trolley was later replaced by the automobile. As roads become more congested, mass transportation becomes more important. Today, technological research is focused on exploring new types of transportation.

1.3.1 TECHNOLOGY AND THE ENVIRONMENT

Technology is often considered to have a negative effect on the environment. In fact, the conflict between technology and the environment has existed for thousands of years. In the 1600s, forests were cut down for fuel. Garbage was burned in open fires. Human waste was poured into rivers. However, all this occurred on a small scale.

Figure 1-12 Created using technology, specially designed equipment allows people with all types of disabilities to enjoy an active lifestyle.

Technology Has Changed Our Landscape

Today, technology has the potential to harm the environment on a worldwide scale. Pollution is a worldwide issue. Cars and trucks give off dangerous gases that pollute our air. Wastes from some factories pollute our rivers and oceans. We burn more than seven billion tons of coal worldwide each year to provide energy. Burning coal produces pollutants such as sulfur dioxide. In the United States, vehicles add huge amounts of pollutants to the air.

The chemicals given off by cars and industrial plants can produce acid rain. Acid rain has killed forests and fish in the United States and other countries (Figure 1-13). Chemicals are also used increasingly in farming to control pests and disease, but they can remain in the soil and air. They can get into the food we eat and the water we drink.

Should We Go Back to Nature?

If we went back to a simpler life, would problems like pollution disappear? Probably not. Pollution has been a problem since people first began to live in cities. The rivers around Rome were so full of waste that people were forbidden to bathe there. One hundred years ago, before the automobile, horses produced tons of manure in the streets of New York and other cities. We need to create technologies that will protect the environment as well as serve our needs. Some technological solutions to environmental problems, such as dredging, involve trade-offs between cost and societal benefit. Others, such as clean drinking water, have positive economic and environmental effects.

Figure 1-13 Pollution is a worldwide issue. Burning coal and other fuels can create acid rain, which destroyed these trees.

LIVING GREEN

Did you know that the average family household with two cars can contribute up to 20,000 pounds of carbon dioxide emissions in just one year? In families that drive in urban areas, where traffic congestion is a problem, this number may be much higher. Automobiles emit carbon dioxide (CO_2). Carbon dioxide is a greenhouse gas, meaning it traps heat inside Earth's atmosphere and over time can contribute to global warming.

Make a Change

Taking several small trips around town can really add to the amount of pollution your family car gives off. Encourage your family to combine errands, such as grocery shopping and trips to the post office, and to carpool to and from school events and sports practices.

Using Technology to Solve Environmental Problems

Hybrid electric vehicles

Advances in technology have helped solve pollution problems. Pollution control devices on cars and scrubbers on factory smokestacks have reduced pollution. Automobiles powered by nonpolluting energy sources are now available. **Hybrid electric vehicles** (HEVs) combine the engine of a conventional vehicle with the battery and electric motor of an electric vehicle (Figure 1-14). These cars achieve twice the fuel economy of conventional vehicles (about 50 miles per gallon).

Advances in the technology of waste management have also helped protect the environment. Recycling items such as glass, paper, and aluminum has decreased the waste sent to landfills. In turn, this decreases the need for more land to be allotted to

Figure 1-14 This hybrid vehicle uses both electricity and gas. It captures energy from braking and uses it to recharge its batteries.

landfills. Wastewater treatment plants help keep our rivers and oceans clean by controlling the levels of pollutants released into them.

Many companies are trying to limit the harm caused by chemicals in the environment. They conduct technology assessment studies to determine the effects that chemicals will have on people. Such studies can save lives and prevent future health problems.

Our environment must be treated with care. We must use resources wisely and look for alternatives for scarce resources. We must develop technologies that do not cause environmental harm. For example, solar energy systems can provide for some of our energy needs without harming the environment.

1.3.2 TECHNOLOGY AND ETHICS

ethical
ethical dilemma

The development and use of technology can also pose a number of ethical questions. Sometimes, new technologies or new uses for a technology cause concern about whether using that technology is right or wrong. Society is often faced with these ethical dilemmas. The word **ethical** means "having to do with right and wrong" and the word *dilemma* means "a difficult decision." An **ethical dilemma** is a difficult decision that a person has to make about whether something is right or wrong.

The field of biotechnology, in particular, has created moral dilemmas for society. Scientists have successfully cloned sheep, mice, and cows. To clone the sheep, Dolly, cells were taken from an adult female sheep. Dolly became an exact duplicate of her biological mother. An ethical dilemma was posed: If cloning sheep and other animals is possible, what about cloning humans? Should society allow research into human cloning? Some say absolutely not because it is immoral. Others support cloning research because they believe we might learn how to avoid genetic diseases.

Another ethical dilemma concerns who will receive high-cost medical assistance. If a high-cost medical technology will keep a person alive, but with a low quality of life, should that technology be used? When is the use of certain technologies such as life support justified, given the financial and emotional costs? Should these technologies be available only to people who can afford to pay for them? All these questions must be considered as new technologies are developed.

Regulations

Government has an important role in ensuring the safety and well-being of its citizens, hence the need for laws and regulations. In democracies, the legislative branch of government passes laws; regulations are developed by the executive branch most often as a way to implement laws. This is also done internationally, with regulations affecting commerce between countries, by land, sea, and air, and hence affecting technologies in these and other areas. Locally, there may be ordinances regulating proper waste disposal, whereas nationally there can be regulations dealing with interstate transportation.

TECHNOLOGY IN THE REAL WORLD

Replacement Hearts

Creative medical technology has allowed doctors to replace hands, arms, or even entire legs to provide new mobility and independence with human-made prosthetic devices. But it is a bit trickier to replace organs that lie inside the body. Every year, people with various organ failures wait for human donors to provide healthy replacements of a kidney, liver, or heart. The list of donors, however, is much shorter than the need for transplants. Even after an organ has been transplanted successfully, patients must take dozens of medications to ward off their body's response to the new organ, and it is still sometimes rejected.

Instead of Muscle, Titanium

A new direction in medical technology is to replace the diseased organ with a completely human-designed organ. A medical technology company, Abiocor, has done just that with its new replacement heart.

In Abiocor's heart, form is entirely functional. This technological masterpiece is a two-pound mass of clear plastic and titanium. Unlike previous technologies, the Abiocor heart is powered by a wireless system that requires no direct opening to the outside body. This design drastically reduces the chance of infection. The Abiocor heart also responds to the body like a natural heart: Its pump speeds up and slows down according to the body's demands.

Making a Real Difference

Every year, more than 300,000 people in the United States die from congestive heart failure. Although the Abiocor heart is a new, experimental technology, it holds the promise of changing the lives of thousands of people who suffer from heart disease. As medical technology grows, the list of human-made organs may grow as well.

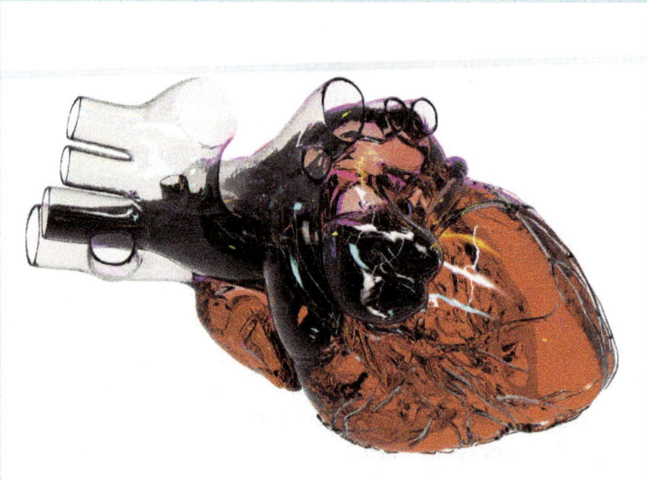

A replacement heart's design mimics the valves and arteries of a real heart.

1.4 NAE GRAND CHALLENGES FOR ENGINEERING

Grand Challenges

History (and prehistory) is replete with examples of technological innovations that forever changed the course of human society: the mastery of fire, the development of agriculture, the wheel, metallurgy, mathematics of many flavors, the printing press, the harnessing of electricity, powered flight, nuclear power, and many others. The NAE has established a list of 14 challenges for the twenty-first century, each of which has the potential to transform the way we live, work, and play. Your interest in one or more of the **Grand Challenges** for Engineering may help you select your major in college. For more information, visit the NAE website at http://www.engineeringchallenges.org/. In case this address changes after we go to press, you can also type "NAE Grand Challenges for Engineering" into your favorite search engine.

A burgeoning planetary population and the technological advances of the last century are exacerbating many current problems as well as engendering a variety of new ones, for example:

- Relatively inexpensive and rapid global travel make it possible for diseases to quickly span the globe whereas a century ago, they could spread, but much more slowly.
- The reliance of the developed world on computers and the Internet makes the fabric of commerce and government vulnerable to cyber terrorism.
- Increased demand for limited resources not only drives up prices for those commodities, but also fosters strain among the nations competing for them.

These same factors can also be a force for positive change in the world.

- Relatively inexpensive and rapid global travel allows even people of modest means to experience different cultures and hopefully promote a more tolerant attitude toward those who live by different sets of social norms.
- Modern communications systems—cell phones, the Internet, etc.—make it essentially impossible for a government to control the flow of information to isolate the members of a population or to isolate that population from the political realities in other parts of the world. An excellent example was the rapid spread of rebellion in the Middle East and Africa in early 2011 against autocratic leaders who had been in power for decades.
- Increased demand for, and rising prices of limited resources is driving increased innovation in alternatives, particularly in meeting the world's energy needs.

As should be obvious from these few examples, technology not only solves problems, but also creates them. A significant portion of the difficulty in the challenges put forth by the NAE to solve critical problems in the world lies in finding solutions that do not create other problems. Let us consider a couple of the stated challenges in a little more detail. You probably already have some familiarity with several of them, such as "make solar energy economical," "provide energy from fusion," "secure cyberspace," and "enhance virtual reality," so we will begin with one of the NAE Grand Challenges for Engineering that is perhaps less well known.

1.4.1 THE NITROGEN CYCLE

Nitrogen is an element required for all known forms of life, being part of every one of the 20 amino acids that are combined in various ways to form proteins, all five bases used to construct RNA and DNA, and numerous other common biological molecules such as chlorophyll and hemoglobin. Fortunately, the supply of nitrogen is—for all practical purposes—inexhaustible, constituting over 75% of the Earth's atmosphere. However, nitrogen is mostly in the molecular form N_2, which is chemically unavailable for uptake in biological systems since the two nitrogen atoms are held together by a very strong triple bond.

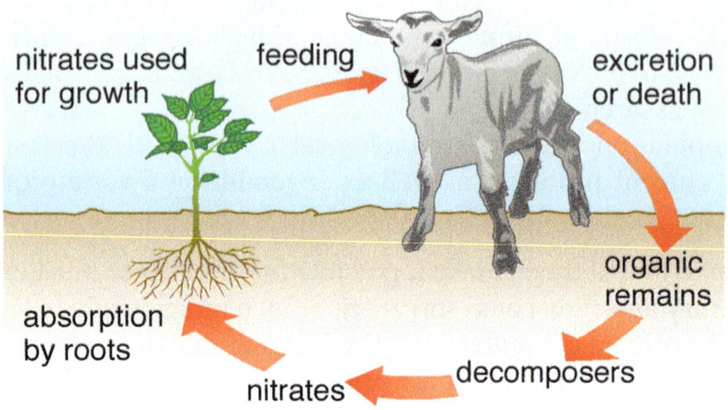

Figure 1-15 Illustration of the Nitrogen cycle.

For atmospheric nitrogen to be available to biological organisms, it must be converted, or "fixed," by the addition of hydrogen, into ammonia, NH_3, that may then be used directly or converted by other microorganisms into other reactive nitrogen compounds for uptake by microorganisms and plants. The term nitrogen fixation includes conversion of N_2 into both ammonia and these other reactive compounds, such as the many oxides of nitrogen. Eventually the cycle is completed when these more readily available forms of nitrogen are converted back to N_2 by microorganisms, a process called denitrification (Figure 1-15).

Prior to the development of human technology, essentially all nitrogen fixation was performed by bacteria possessing an enzyme capable of splitting N_2 and adding hydrogen to form ammonia, although small amounts of fixed nitrogen are produced by lightning and other high-energy processes. In the early twentieth century, a process called the Haber-Bosch process was developed that would allow conversion of atmospheric nitrogen into ammonia and related compounds on an industrial scale. Today, slightly more than a century later, approximately one-third of all fixed nitrogen is produced using this process.

The ready availability of relatively inexpensive nitrogen fertilizers has revolutionized agriculture, allowing people to increase yields dramatically and to grow crops on previously unproductive lands. However, the widespread use of synthetic nitrogen has caused numerous problems, including water pollution, air pollution, numerous human health problems, and disruption of marine and terrestrial ecosystems to the extent that entire populations of some organisms have died off.

Deliberate nitrogen fixation is only one part of the nitrogen cycle problem, however. Many human activities, especially those involving the combustion of fossil fuels, pump huge quantities of various nitrogen compounds into the atmosphere. Nitrous oxide (N_2O), also of some notoriety as the dissociative anesthetic commonly known as "laughing gas," is particularly problematic since it is about 200 times more effective than carbon dioxide as a greenhouse gas, and persists in the atmosphere for over a century.

Altogether, human-caused conversion of nitrogen into more reactive forms now accounts for about half of all nitrogen fixation, meaning that there is twice as much nitrogen fixed today than there was a little more than a century ago. However, we have done little to augment the natural denitrification process, so the deleterious effects of excessive fixed nitrogen continue to increase. We have overwhelmed the natural nitrogen cycle. If we are to continue along this path, we must learn to manage the use of these products more efficiently and plan strategies for denitrification to bring the cycle back into balance.

1.4.2 REVERSE-ENGINEERING THE BRAIN

The development of true artificial intelligence (AI) holds possibly the most overall potential for positive change in the human race as well as the most horrendous possible negative effects. This is reflected in science fiction, where the concept of thinking machines is a common plot device, ranging from Isaac Asimov's benevolent R. Daneel Olivaw to the malevolent Skynet in the Terminator movies. If history is any guide, however, the potential for disastrous consequences seldom deters technological advances, so let us consider what is involved in the development of AI.

Although great strides have been made in creating machines that seem to possess "intelligence," almost all such systems that have come to the public notice either rely on brute-force calculations, such as the chess-playing computer, Deep Blue, that defeated world champion Garry Kasparov in 1997, or reliance on incredibly fast access to massive databases, such as the Jeopardy-playing computer, Watson, that defeated both the highest money winner, Brad Rutter, and the record holder for longest winning streak, Ken Jennings, in 2011. Perhaps needless to say, these are oversimplifications, and there are many more aspects to both of these systems. However, one would be hard-pressed to argue that these computers are truly intelligent—that they are self-aware and contain the unexplainable spark of creativity, which is the hallmark of humans, and arguably other highly intelligent creatures on Earth.

Today's robots perform many routine tasks, from welding and painting vehicles to vacuuming our homes and cutting our grass. However, all of these systems are programmed to perform within certain restrictions and have serious limitations when confronted with unexpected situations. For example, if your school utilized vacuuming robots to clean the floors in the classrooms, it would probably be unable to handle the situation effectively if someone became nauseous and regurgitated on the carpet. If we could endow such robots with more human-like intelligence, the range of tasks that they could successfully accomplish would increase by orders of magnitude, thus increasing their utility tremendously.

To date, we have almost exclusively attempted merely to construct intelligent systems that mimic behavior and thought, not design systems that actually store and process information in a manner analogous to that of a biologically based computer (a brain). The human brain utilizes a network of interconnections between specialized subsections that makes even the most advanced computers look like a set of children's building blocks. Although some understanding has been gained, the means of encoding information and its transfer in the brain is almost completely a mystery.

Gaining even a basic understanding of brain function might allow us to develop prosthetic limbs that actually function as well as the originals, restore sight to the blind, repair brain damage, or even enhance human intelligence.

1.5 ENGINEERING DISCIPLINES

The following paragraphs briefly introduce several different types of engineering majors. By no means is this list completely inclusive.

Bioengineering or Biomedical Engineering

Bioengineering (BioE) and biomedical engineering (BME) apply engineering principles to the understanding and solution of medical problems. Bioengineers are involved in research and development in all areas of medicine, from investigating the physiological behavior of single cells to designing implants for the replacement of diseased or traumatized body tissues. Bioengineers design new instruments, devices, and software, assemble knowledge from many scientific sources to develop new procedures, and conduct research to solve medical problems.

Typical bioengineers work in such areas as artificial organs, automated patient monitoring, blood chemistry sensors, advanced therapeutic and surgical devices, clinical laboratory design, medical imaging systems, biomaterials, and sports medicine.

Bioengineers are employed in universities, industry, hospitals, research facilities, and government. In industry, they may be part of a team serving as a liaison between engineers and clinicians. In hospitals, they select appropriate equipment and supervise equipment performance, testing, and maintenance. In government agencies, they are involved in safety standards and testing.

Biosystems Engineering

Biosystems engineering (BE) is the field of engineering most closely allied with advances in biology. BE emphasizes two main areas: (1) bioprocess engineering, with its basis in microbiology, and (2) ecological engineering, with its basis in ecology. The field focuses on the sustainable production of biorefinery compounds (biofuels, bioactive molecules, and biomaterials) using metabolic pathways found in nature and green processing technologies. Further, BE encompasses the design of sustainable communities utilizing low-impact development strategies (bioretention basins, rainwater harvesting) for stormwater retention and treatment—and ecologically sound food and energy-crop production. Scientific emphasis is shifting toward the biosciences. Biosystems engineers apply engineering design and analysis to biological systems and incorporate fundamental biological principles to engineering designs to achieve ecological balance.

Here are some activities of biosystems engineers:

- Design bioprocesses and systems for biofuels (biodiesel, hydrogen, ethanol), biopharmaceutical, bioplastics, and food processing industries
- Develop ecological designs (permeable pavement, bioswales, green infrastructure) to integrate water management into the landscape
- Integrate biological sustainability concepts into energy, water, and food systems
- Provide engineering expertise for agriculture, food processing, and manufacturing
- Pursue medical or veterinary school or graduate school in the fields of BE, BME, or ecological engineering

Chemical Engineering

Chemical engineering (ChE) incorporates strong emphasis on three sciences: chemistry, physics, and mathematics. Chemical engineers are involved in the research and development, manufacture, sales, and use of chemicals, pharmaceuticals, electronic components, food and consumer goods, petroleum products, synthetic fibers and films, pulp and paper, and many other products. They work on environmental remediation and pollution prevention, as well as in medical and health-related fields. Chemical engineers:

- Conduct research and develop new products
- Develop and design new manufacturing processes
- Earn additional degrees to practice medicine or patent, environmental, or corporate law
- Sell and provide technical support for sophisticated chemical products to customers

- Solve environmental problems; work in biotechnology
- Troubleshoot and solve problems in chemical manufacturing facilities

Descriptions of Engineering

My current responsibilities include:
- Analysis of traffic signal operations and safety for municipal and private clients;
- Preparation of traffic impact studies;
- Review of plans and traffic studies for municipalities and counties;
- Design of traffic signal installations and traffic signing projects.

C. Hill, CE

I develop, manage, and support all software systems. I also deal with system scalability, customer satisfaction, and data management.

J. Comardelle, CpE

I am a digital designer and work on the read channel for hard disk drives. The read channel is the portion of the controller SOC (system on a chip) that decodes the analog signal read from the hard disk and converts it to digital data.

I am responsible for writing Verilog RTL code, verification, synthesis into gates, and meeting timing requirements of my blocks.

E. D'Avignon, CpE

Civil Engineering

Civil Engineering (CE) involves the planning, design, construction, maintenance, and operation of facilities and systems to control and improve the environment for modern civilizations. This includes projects of major importance such as bridges, transportation systems, buildings, ports, water distribution systems, and disaster planning.

Here are just a few of many opportunities available for civil engineers:

- Design and analyze structures ranging from small buildings to skyscrapers to offshore oil platforms
- Design dams and building foundations
- Develop new materials for pavements, buildings, and bridges
- Design improved transportation systems
- Design water distribution and removal systems
- Develop new methods to improve safety, reduce cost, speed construction, and reduce environmental impact
- Provide construction and project management services for large engineered projects throughout the world

I am responsible for assisting in the management of commercial and healthcare projects for Brasfield & Gorrie. Working closely with the owner and architect, I maintain open lines of communication and aim to provide exceptional service to the entire project team from the preconstruction phase of the project through construction. I assist in establishing and monitoring procedures for controlling the cost, schedule, and quality of the work in accordance with the construction contract.

L. Edwards, CE

Figure 1-16

Computer Engineering

Computer engineering (CpE) spans the fields of computer science and engineering, giving a balanced view of hardware, software, hardware-software trade-offs, and basic modeling techniques that represent the computing process involving the following technologies:

- Communication system design
- Computer interface design
- Computer networking
- Digital signal processing applications
- Digital system design
- Embedded computer design
- Process instrumentation and control
- Software design

As a Radiation Effects Engineer I test the performance of electronic components in a specific application exposed to different types of radiation. Responsibilities include interfacing with design and system engineers, creating test plans, performing testing and data analysis, and authoring test reports.

A. Passman, EE

Figure 1-17

Electrical Engineering

Electrical engineering (EE) ranges from the generation and delivery of electrical power to the use of electricity in integrated circuits. The rapid development of technology, based on integrated circuit devices, has enabled the pervasive use of computers in command, control, communication, and computer-aided design. Some systems electrical engineers work on include the following:

- Communication system design
- Control systems—from aircraft to automotive
- Electrical power generation and distribution
- Electromagnetic waves
- Integrated circuit design
- Process instrumentation and control
- Robotic systems design
- Telecommunications

Figure 1-18

My group supports [a major automotive manufacturer's] decisions pertaining to where to put new plants around the world, what products to build in them, and at what volumes.

In particular, my work involves understanding what the other auto manufacturers are planning for the future (footprint, capacity, technology, processes, etc.), so that information can be used to affect decisions about how to compete around the globe.

M. Peterson, EE

Environmental Engineering

Environmental engineering (EnvE) is an interdisciplinary field of engineering that is focused on cleaning up environmental contamination, as well as designing sustainable approaches to prevent future contamination. Environmental engineers apply concepts from basic sciences (including chemistry, biology, mathematics, and physics) to develop engineered solutions to complex environmental problems.

Environmental engineers design, operate, and manage both engineered and natural systems to protect the public from exposure to environmental contamination and to develop a more sustainable use of our natural resources. These activities include the following:

- Production of safe, potable drinking water
- Treatment of wastewater so that it is safe to discharge to surface water or reuse in applications such as landscape irrigation

- Treatment of air pollutants from mobile (e.g., automobiles) and stationary (e.g., power plants) sources
- Characterization and remediation of sites contaminated with hazardous wastes (e.g., polychlorinated biphenyls, or PCBs)
- Disposal of municipal solid wastes
- Management of radioactive wastes, including characterization of how radioactive materials move through the environment and the risks they pose to human health
- Evaluation of methods to minimize or prevent waste production and inefficient use of energy by manufacturing facilities
- Reduce human health risks by tracking contaminants as they move through the environment
- Design a more sustainable future by understanding our use of resources

With a BS degree in EnvE, students will find employment with consulting engineering firms, government agencies involved in environmental protection, and manufacturing industries.

I work with scientists and engineers to protect their innovations by writing patent applications describing their inventions and presenting the applications before the United States Patent & Trademark Office. I also assist clients in determining whether another party is infringing their patents and help my clients to avoid infringing other's patents.

M. Lauer, EnvE

Industrial Engineering

Industrial engineering (IE) deals with the design and improvement of systems, rather than with the objects and artifacts that other engineers design. A second aspect of IE is the involvement of people in these systems—from the people involved in the design and production to the people who are ultimate end users. A common theme is the testing and evaluation of alternatives that may depend on random events. Industrial engineers use mathematical, physical, social sciences, and engineering combined with the analytical and design methods to design, install, and improve complex systems that provide goods and services to our society. Industrial engineers are called upon to:

- Analyze and model complex work processes to evaluate potential system improvements
- Analyze how combinations of people and machines work together
- Analyze how the surroundings affect the worker, and design to reduce the negative effects of this environment
- Develop mathematical and computer models of how systems operate and interact
- Improve production and service processes from the perspectives of quality, productivity, and cost
- Work on teams with other professionals in manufacturing, service industries, government agencies

Figure 1-19

My primary job responsibilities include maintaining, upgrading, and designing all the computer systems and IT infrastructure for the Vermont Railroad. I handle all the servers and take care of network equipment. When needed, I also program customized applications and websites for customers or our own internal use. I also serve as a spare conductor and locomotive engineer when business needs demand.

S. Houghton, CpE

As the Business Leader for Central Florida at a major power company, I develop and manage a $42 million budget. I ensure that our engineering project schedule and budget match, and report on variances monthly. I also conduct internal audits and coach employees on Sarbanes–Oxley compliance requirements.

R. Holcomb, IE

Currently I am working on a project to determine patient priorities for evacuations from healthcare facilities during emergencies. The assumption of an evacuation is that there will be enough time to transfer all of the patients, but in the event of limited resources, there may not be enough time to move all of the patients to safety. Further—and depending on the emergency type—it may be an increased risk to transport some patient types. Based on certain objectives, we are developing guidelines to most ethically determine a schedule for choosing patients for emergency evacuations.

A. Childers, IE

Materials Engineering

Materials engineering or metallurgical engineering focuses on the properties and production of materials. Nature supplied only 92 naturally occurring elements to serve as building blocks to construct all modern conveniences. A materials engineer works to unlock the relationship between atomic, molecular, and larger-scale structures and the resultant properties. This category includes majors such as ceramic engineering, metallurgical engineering, and polymer science and engineering.

Here is a partial list of products designed and manufactured by Material Engineers:

- Brick, tile, and whitewares research and manufacturing for the home and workplace
- Ceramic spark plugs, oxygen sensors, and catalytic converters that optimize engine performance
- Metal and ceramic materials that enable biomedical implants and prosthetics
- Microwave responsive ceramics that stabilize and filter cellular phone reception
- Nanotechnology, including silver nanoparticles used as antibacterial agents in socks and t-shirts and carbon nanotubes used to reinforce the fork of racing bicycles
- Plastics found in bulletproof vests, replacement heart valves, and high tension wires on bridges

- Superconducting metals that are used in medical imaging devices like magnetic resonance imaging (MRI)
- Ultrapure glass optical fibers that carry telephone conversations and Internet communications

As a Metallurgical Engineer, my duties include

- Consulting firm management/administration
- Failure analysis
- Subcontracted metals testing services
- Metallurgical quality systems design/auditing
- Metallurgical expert in litigation cases
- Materials selection and design consultant, in-process and final inspection and testing services

E. Basta,
Materials Engineer

Mechanical Engineering

Mechanical engineering (ME) involves areas related to machine design, manufacturing, energy production and control, materials, and transportation. Areas supported by mechanical engineers include:

- Construction
- Energy production and control
- Environmental systems
- Food production
- Management
- Materials processing
- Medicine
- Military service
- Propulsion and transportation systems
- Technical sales

In my job as a management consultant, I address CEO-level management decisions as part of a project team by helping clients identify, analyze, and solve business-related problems. My responsibilities include generating hypotheses, gathering, and analyzing data, conducting benchmarking and best practices assessments, recommending actions, and working with clients to develop implementation plans.

M. Ciuca, ME

As Plant Engineering Manager I report directly to the Plant Manager. My primary responsibilities are managing all the capital investments; providing technical support and expertise to the plant leadership team; and mentoring and developing the plant's engineering staff and technical resources.

J. Huggins, ME

I am a salesman, so at the end of the day I'm looking to grow my market share while trying to protect the market share I already have. I help companies maintain a safe, reliable, and efficient steam and condensate system by utilizing the many products and services that we have to offer. This is mostly done by designing and installing upgrades and improving my customer's existing steam systems.

T. Burns, ME

WISE WORDS: WHAT DID YOU DO YESTERDAY MORNING AT WORK?

I worked on completing a failure analysis report for an industrial client.

E. Basta, Materials Engineer

I reviewed the results of the overnight simulation runs. There were several failures, so I analyzed the failures and devised fixes for the problems.

E. D'Avignon, CpE

On any given day, my morning might be spent this way: in meetings, at my computer (e-mail, drafting documents/reports), making phone calls, talking to other project members, running a test on the manufacturing lines. Not glamorous, but necessary to solve problems and keep the project moving forward.

S. Forkner, ChE

I continued to design a warehouse/office building on a nuclear expansion project.

T. Hill, CE

Yesterday I designed a spreadsheet to assist in more precisely forecasting monthly expenditures.

R. Holcomb, IE

I attended the Plant Morning Meeting and the Boardmill Leadership Team Meeting, followed by the Plant Budget Meeting. In between meetings I returned e-mails and project-related phone calls. Typically I spend about 50% of my time in meetings. I use the information I receive at these meetings to direct and focus the efforts of the engineering staff.

J. Huggins, ME

Testing some failed prototype biostimulators returned by a trial user, to determine why and how they failed and how to prevent it from happening in coming production versions.

J. Kronberg, EE

Yesterday, I worked on a patent infringement opinion involving agricultural seeding implements, a Chinese patent office response for a component placement and inspection machine used in circuit board manufacturing, and a U.S. patent office response for database navigation software.

M. Lauer, EnvE

In my current position, I spend much of my time reading technical manuals and interface control documents. I attended a meeting detailing lightning protection for the Ares rocket.

E. Styles, EE

1.6 THE FUTURE OF ENGINEERING

Existing ideas are used as building blocks for newer and more powerful technologies. Future technologies will become more and more complex. Space shuttles, satellites, robotic factories, and maglev trains all use ideas from a variety of technological systems—electrical, transportation, manufacturing, and scientific. This confluence, or flowing together, of systems means that new technologies will not be separate and distinct. Many systems will be combined to produce new technologies.

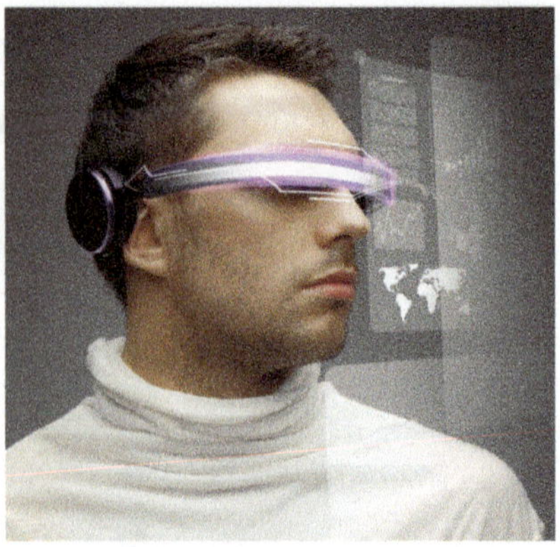

Figure 1-20 Virtual reality can change the way we perceive the world around us.

Consumers of the Future

What effects will technology have on your life in the future? What amazing new inventions will you see in the next few years, or by the end of your lifetime? Here are some possibilities for the near future.

- Computer screens will be lightweight, flexible, and rolled into a metal tube for storage.
- Chemists will combine elements to create new foods that are as healthful and tasty as the foods we eat today.
- Holographic telephones will project life-size images of the caller and the person being called.
- Fossil fuels will run out. Renewable energy sources will fill most of our energy needs.
- People will manufacture items they need in their homes.

As technology changes and becomes more complex, these possibilities may become a reality. Part of preparing for the future will be deciding how we want technology to change.

Because of the rapid improvements in computers and electronics, new communication technologies are emerging every day. What will technology be like in the future? Faster, more powerful chips and new software will enable computers to recognize and converse with people. When people look back on the computers of today, they'll say: "What did they do? They couldn't see; they couldn't listen; they couldn't speak." Computers will continue to get smaller and become even more powerful.

1.6.1 NANOTECHNOLOGY AND COMPUTERS

nanotechnology

Linking all of these wireless networks will be faster—and much smaller—computers. Computers of the future will be based on nanotechnology. **Nanotechnology** is the creation of materials, devices, and systems using individual atoms and molecules. The word comes from the prefix *nano-*, which means one billionth. Nanotechnology uses particles that are 1/80,000 the diameter of a human hair. At such a small scale, new physical, chemical, and biological properties become evident.

Instead of using transistors and silicon chips, quantum computers will follow the laws of physics that govern subatomic particles. For example, electrons will be able to "tunnel" through a barrier that would normally block their path. Quantum computers may be able to solve extremely complex problems, such as breaking codes for government work.

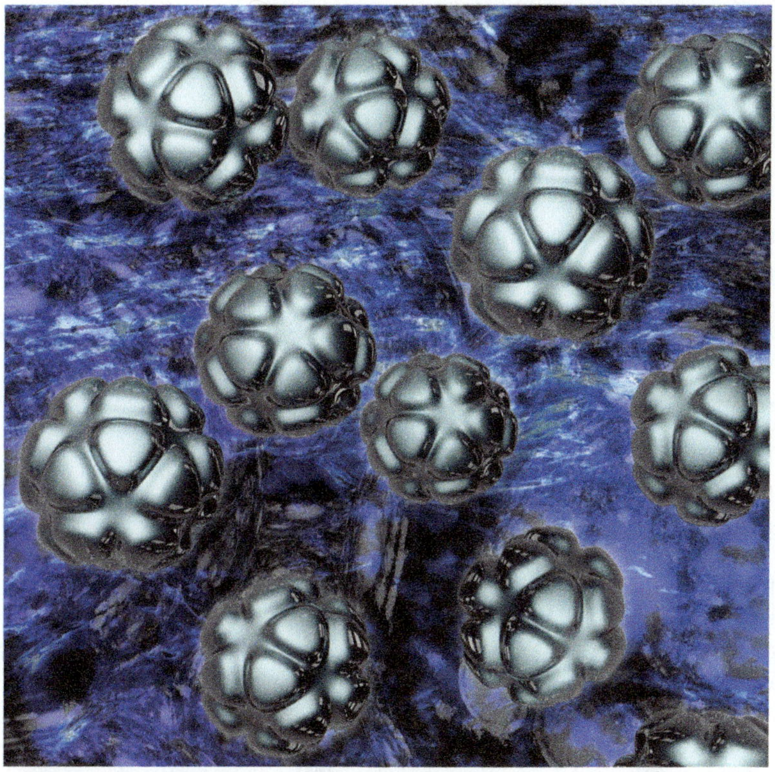

Figure 1-21 These interlocking gears were assembled using many individual carbon atoms.

Like today's computers, electronic nanocomputers will represent information as the storage and movement of electrons. Future transistors will be much smaller, allowing up to 10,000 transistors to fit in the space taken up by one current transistor. Mechanical nanocomputers will calculate by using moving molecule-size rods and rotating gears that spin on shafts and bearings (Figure 1-21). These tiny machines and computers will be assembled by mechanically positioning atoms one at a time.

1.6.2 ARTIFICIAL INTELLIGENCE

artificial intelligence

Artificial intelligence (AI) is any use of machines to imitate human intelligence. AI can be used to recognize and re-create speech, to make humanlike robots, and to create complete virtual reality environments.

Voice Recognition Technologies

As it continues to improve, voice recognition will go hand in hand with a wireless environment. New voice recognition technologies will allow speech recognition (converting speech into text) and speech synthesis (converting text into speech).

speech recognition

In **speech recognition** applications, the sound waves of speech are digitized and matched to coded waves. These matches are then converted to text, as in typed language. Speech recognition technology allows you to speak into devices when it is not practical to enter data using a keyboard, such as when using handheld computers and car phones. Also, computer games can become more friendly and realistic if players can speak to the characters.

TECHNOLOGY IN THE REAL WORLD

From the Film Studio to the Living Room

Motion-sensing, or motion capture, technology is any technology that translates motion into a digital signal that can be interpreted by a computer. If you have seen almost any movie with special effects in the last few years, you have seen one of the things that can be created with motion capture technology.

In film, motion capture technology enables an actor's movements to be tracked by computers. The actor wears an outfit that has reflective dots stuck to it. These dots are tracked by cameras. The cameras communicate the motions to the computer, which can then create an animated character that moves just like the actor. This results in more lifelike animated characters and saves the film studio a great deal of time and money. They don't have to draw or have a computer create every single element of a motion. Gollum in *The Lord of the Rings* movies, the Na'vi in *Avatar*, and King Kong in the 2005 version of the movie are animated characters that were created using motion capture technology.

Another recent use of motion-sensing technology that many people are familiar with is Nintendo's Wii, a gaming console with a remote, motion-sensing controller. This controller lets players interact with their games by moving the controller through space, swinging it like a bat, pointing it like a gun, or turning it like a steering wheel, for example.

The Wii controller uses infrared (IR), just like a regular television remote control, as well as Bluetooth wireless technology to communicate with a sensor bar that must be placed near the television. The controller can track its own distance and angle from the sensor bar. There is also a device inside the controller, called the accelerometer, that can sense how fast the controller is moving through space, and another one, called a gyrometer, that can sense how the controller is being tilted. Finally, it all shows up on the television because the controller uses Bluetooth to communicate all that location and motion information to the game console, which interprets it as shooting the bad guy or hitting the baseball.

Manufacturers are using motion-sensing technology to make product design and testing more efficient. It allows designers to interact with computer simulations and models.

Marketers are using it to create interactive advertising, such as billboards that react to people walking close by or moving their hands. Corporations are also using motion-sensing computer controls to run dramatic sales or information presentations. Instead of clicking and dragging with a mouse, special glove and camera systems can allow people to move things around on a computer screen using hand gestures. This can add an exciting new element of showmanship to presentations.

Companies are finding new ways to use motion-sensing technology every day. Many of those discoveries will be finding their ways into homes very soon, just like the Wii already has.

speech synthesis

Speech synthesis applications convert text into speech. This involves breaking down the written words into the basic sounds of a language and then converting the sounds into digital audio. Because the sound is generated electronically, it does not sound as natural as a real human voice. One important use of speech synthesis is to help people who are visually impaired get information from text materials.

Robotics

robotics

Robotics is the application of AI in the design and construction of robots. Several different types of problems must be addressed when designing a robot. In many cases, the robot must be programmed to "think" so it can develop a plan for doing a job. Mechanical problems might include the need to design a robot's arm so that it can move strongly in all directions. Sensory problems include the need to design a robot so that it can pick up fragile objects without breaking them.

Another design challenge is robotic vision (Figure 1-22). Vision systems must be developed to allow robots to recognize objects and interpret visual images. When you look around, recognizing desks or chairs is easy, but it is very difficult for a robot. Another problem that is easy for you but hard for a robot is finding a path from one place to another. Knowing how to avoid bumping into objects must be programmed into the robot.

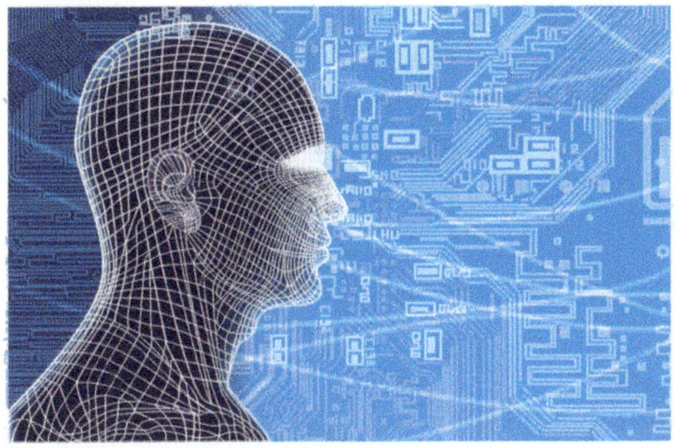

Figure 1-22 Artificial intelligence is being used to make robots communicate more like people.

Virtual Reality

virtual reality

Virtual reality uses AI to create a three-dimensional computer-generated environment into which a person is projected using special eyewear or a glove. In virtual reality, you can go where you want to go, experience what you want to experience, and be who you want to be (Figure 1-23). Soon, people will be able to "virtually" touch each other. People will be using virtual reality glasses to put images directly on their retinas. Virtual reality will allow you to drive your dream car and look under its hood. It will take you on tours through foreign countries. It will allow you to view the inner workings of molecules.

Figure 1-23 Virtual reality games are one application of artificial intelligence.

"Smart" Houses and Cars

In the future, "smart" houses will have built-in computers that can control systems. These houses will have bathrooms in which you can choose a shower, a spring rain, or a warm breeze. Bedrooms will have voice-activated radio and television systems.

"Smart" cars will be able to travel on autopilot. Communication systems will allow cars to exchange information about speed, direction, and location with an automated highway control network and with other vehicles. Superhighways will send signals to computers in cars, keeping them on the road until the correct exit is reached. Magnetic markers will be installed along the center or edges of highways. The markers will contain digital information to warn drivers of exits or curves.

TECHNOLOGY IN THE REAL WORLD

Nanotechnology in Medicine

You don't need to look into the future to see technology in use in the human body. Many electronic devices are already being used in the human body to help it function better. Heart pacemakers keep the heart beating regularly. Hearing aids are implanted in the ear to aid in hearing. What will be different in the future is that these devices will become much smaller.

Technology as Small as a Pill

With nanotechnology, new medical sensors will improve hearing, smell, and even the sense of taste. They will connect directly to the nervous system and deliver information directly to the human brain.

Other types of sensors could provide people with valuable information about their environments. Small sensors on the body might detect information about the chemical makeup of your surroundings. This type of information could provide a type of security system and also help people avoid unhealthy or toxic environments. Some years from now, doctors may provide you with swallowable capsules containing millions of sensors that have been programmed to find and kill cancer cells.

Nanobots

In the future, nanobots—robots of microscopic size built using nanotechnology—will travel through the bloodstream to provide a picture of the way the brain's circuitry works.

To make examinations of the stomach and intestinal area easier, patients will be able to swallow a pill-sized capsule that moves smoothly and painlessly through the body. As the capsule passes through the stomach and intestinal area, it will transmit signals to an array of antennas placed on the outside of the patient's body. Data will be stored in a data recorder worn on the person's belt. A computer will process the data and produce a short video clip of the results.

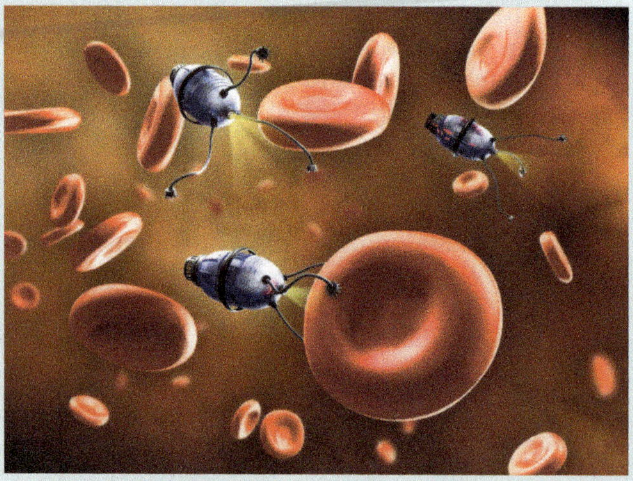

An artist's rendering of nanobots in the blood stream.

1.6.3 PERSONAL MANUFACTURING

Imagine that you need a new cell phone. Now you can purchase one online or go to a nearby store and select a model. What if you built your own? What if there were a whole variety of cell phone designs, from simple to complex, you downloaded the design, sent it to your personal fabricator and in a couple of hours you had your own custom-designed, and personally manufactured, cell phone? Not really so far-fetched. 3-D printers are machines that quickly create objects from specifications in a CAD file.

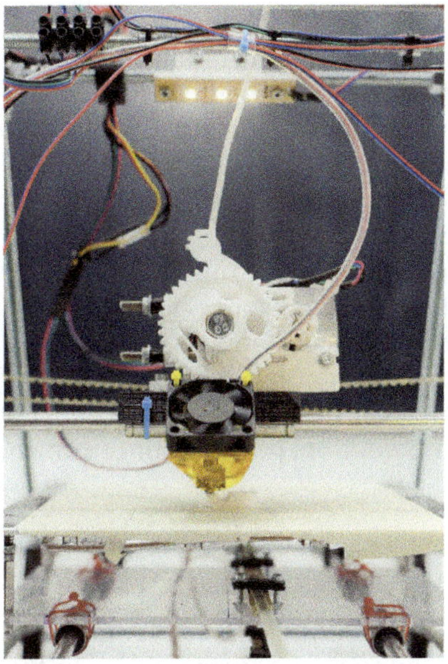

Figure 1-24 A 3-D printer can be used to manufacture objects that are personalized by and for you.

How Fabrication Works

Figure 1-24 illustrates a 3-D printer. In this system there is a syringe with a plastic or gel substance that is accurately extruded to form the object from the CAD files, which are converted to stereolithographic files. The object, such as the propeller, or gear, shown in the picture, is formed on the plate and the gear mechanisms raise the extruder as the object is built up. Thus, the extruder needs to move only vertically in the Z direction, but in a variety of patterns in the X-Y directions. Other designs used an extruder that is essentially like a hot glue gun, where a material is fed continuously through the nozzle and heated, so it melts, and forms the object. There are situations where the table base moves in the Z direction, and the nozzle only moves in the X-Y directions.

Biological Designs

One of the difficulties facing surgeons is replacing bones that have degenerated or been badly fractured. What if an image of the bone was sent to the 3-D printer and a custom-designed replacement was created with the precise size and shape, tailored to the patient? Teams are working on bones and cartilage that were created from the patient's own cells, thus avoiding the problem of rejection by the body. Cells are taken from the patient and grown in laboratories, so they multiply millions of times, and then these are used to form the bone or cartilage (Figure 1-25).

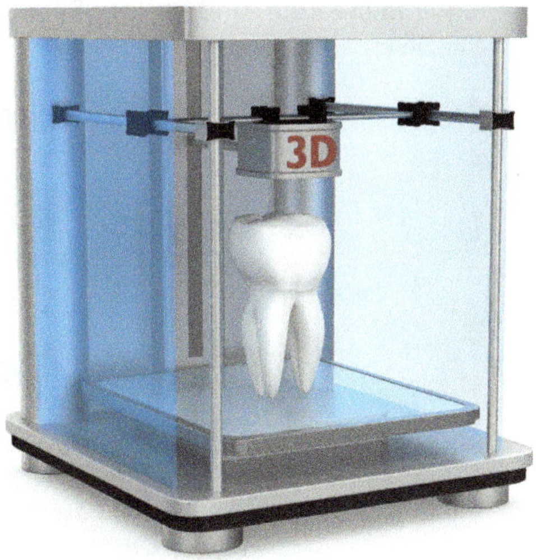

Figure 1-25 A 3-D printer concept printing a tooth for artificial dental replacement.

Bone and cartilage do not require a blood supply, but tissue does. There is great interest in creating organs, such as a kidney or the liver. The concept involves using a bio-ink, a mixture created from a person's cells. An organ is created as the bio-ink forms a three-dimensional structure with the cells self-assembling to form a functioning organ. This will become more of a medical issue as the population ages and there are fewer and fewer organ donations available. A significant hurdle is that the tissue requires blood vessels so that it can receive vital nutrients for its survival. How to form these and provide the nutritive blood has not been solved yet.

Self-Replicating Robots

One of the challenges in space exploration is to have devices that can repair themselves and their environment automatically, independent of human intervention. Thus, thinking robots that can self-replicate, can repair parts damaged by radiation and the hostile environment, are being developed. Researchers are developing electro-mechanical systems that can self-reproduce, analogously as biological systems reproduce.

Current efforts, such as creating a flashlight, which needs structural and mechanical elements, indicate that advanced 3-D printers could be used for this type of manufacturing. Other work has demonstrated that the artificial intelligence used in robots allows for self-replication. Other studies have indicated that robots can design and produce devices to meet specifications independent of human intervention. Many of the elements needed for self-replication exist. However, there may be dangers involved—because the robots do not need human intervention, people cannot control the outcomes.

3-D Printing and Art

Technology and art have been intertwined from the beginnings of humankind. There are many types of 3-D printing devices that can be used to create large objects. In this type of printer, the CAD image is again sliced into different layers. The printer itself puts down a thin layer of plastic. Another layer of plastic is put in place, and the object is built (Figure 1-26).

The renderings can be quite complex and interesting. People are using this technique in creating food objects. Sugar is a popular ingredient and elaborate renderings can be created.

Figure 1-26 A 3-D printer that deposits layers as a means to build an object.

CONNECTING TO STEM

Energy from Space: Using Helium-3 from the Moon

As fossil fuels become more and more depleted, we need to explore alternative sources of energy. One new source of energy may be nuclear fusion. In this process, atoms are forced to combine, releasing huge amounts of energy.

Although these reactions produce a lot of energy, they also require huge amounts of energy to start. Researchers are trying to find ways to get at least as much energy *from* the reaction as is needed to start it. Once this has been achieved, the excess energy can be used to generate electricity.

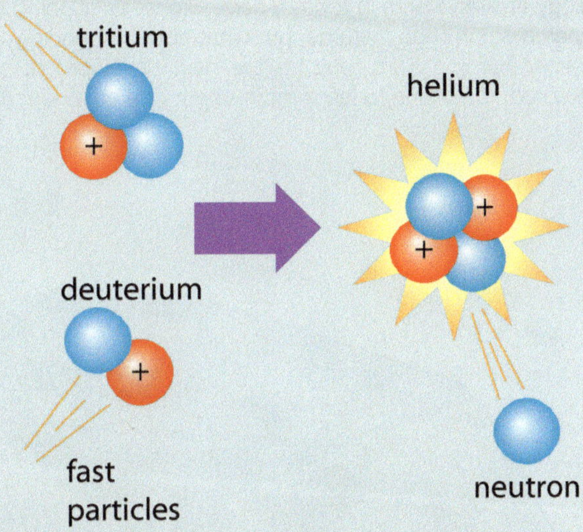

Fuel for the Future?

Because nuclear fusion is such a rich source of energy, some researchers predict that fusion reactors will be built by the year 2030. Energy from atomic fusion might well become a major energy source in the decades ahead. The simplest fuels used for fusion are two types of hydrogen atoms, deuterium and tritium.

Although deuterium and tritium produce a lot of energy, the perfect source of fuel for fusion may be helium-3, a variation of the helium used to blow up balloons. A fusion reactor using helium-3 and deuterium would produce very little radioactive waste. Therefore, you could safely build a helium-3 plant in the middle of a populated area.

An Abundant Resource

The problem is that helium-3 is very rare on Earth. However, it is abundant on the moon. Scientists estimate that there are about 1 million tons of helium-3 on the moon, enough to power the world for thousands of years. One Space Shuttle load of 25 tons could satisfy the energy needs of the United States for an entire year.

How Technology Works...

WHAT WILL YOU BE DRIVING? ALTERNATIVE ENGINES ONLINE

Technology is everywhere in today's cars. Global positioning systems, DVD players, and cellphones are common features, but what about technology under the hood?

Cleaner burning fuels may be the next development to affect the way cars are manufactured. Go online to learn more about one advance: the hydrogen engine.

Chapter 1 ASSESSMENT QUESTIONS

1. Write your own definition of engineering. What key concepts does your definition contain?
2. What are some of the earliest examples of engineering? List three, and explain why they should be considered engineering.
3. Explain how the invention of printing led to three of the greatest technological advances we have today.
4. Explain why the rate of technological advances has increased over time.
5. We know that the burning of fossil fuels causes pollution. Answer the following questions about the burning of fossil fuels in one paragraph each:
   (a) Why do we continue to burn fossil fuels?
   (b) What alternatives should be investigated or implemented?
   (c) Why are these alternatives not already in place?
6. Identify one innovation we use today that resulted from engineering. Investigate the history behind that innovation and write a 2-page report on the innovation. Be sure to include relevant illustrations, showing the history of this device. Finally, predict the future of this device: What will it look like in 10 years?

Note: Your teacher may have you work on this as a group assignment using a shared document editor such a Google doc.

7. List three technologies that pose ethical dilemmas.
8. Cell phones have had major impacts on how we communicate with each other. However, there are negative consequences from the technology, such as texting during class or while driving. Analyze the positive and negative impacts on society and include anticipated and unanticipated effects.
9. List 10 types of common transportation in chronological order. For each, identify why this was the preferred technology at that time, and identify one reason it was replaced. What will the most common type of transportation be in the year 2050?
10. Give at last three examples of complex technologies that were developed from simpler technologies. Do not use the examples given in the book.
11. Provide one example of a problem that technology has caused and one of the solutions that technology has provided.
12. In 2008, experts convened by the National Academy of Engineering (NAE) met and proposed a set of Grand Challenges for Engineering, a list of the 14 key goals for engineers to work toward during the 21st century. For any aspiring engineer, reading this list should feel like reading a description of the challenges you will face throughout your career. To read a description of each Challenge, as well as a description of some connected areas within each, visit the NAE's Grand Challenges for Engineering website at http://www.engineeringchallenges.org.

    After reading through the list, write a job description of your dream job. The job description should include the standard components of a job description: title, responsibilities (overall and specific), a description of the work hierarchy (whom you report to, whom you work with), as well as any necessary qualifications. Do not include any salary requirements, as this is a completely fictional position.

    After writing the description of your dream job, identify which of the 14 Grand Challenges for Engineering this position is instrumental in working toward solving. Cite specific examples of the types of projects your fictional job would require you to do, and discuss the impact of those projects on the Grand Challenges you have identified.

13. Consider the Engineering Grand Challenges.
    (a) Which do you think is the most difficult to solve in the next 10 years, and why?
    (b) Which do you think is the easiest to solve in the next 10 years, and why?
14. Identify one additional problem for humanity that you feel should be added to the Engineering Grand Challenges. Why do you feel it should be included?
15. A company is working on a design to prevent flood damage during or after a hurricane. Identify how engineers from each of these disciplines could be involved and what they might propose. Don't be afraid to think outside of the box; the solution could be wild or impractical – but it needs to be possible. (For example, your answer should not be, "Invent a time machine to go back in time before the hurricane.")
    (a) Electrical engineer
    (b) Mechanical engineer
    (c) Chemical engineer
    (d) Materials engineer
16. Describe 5 technological innovations we will see in the "smart house" 10 years from now. For each, describe why it doesn't yet exist and how it will change our everyday life.
17. Video game designs have gone from corded controllers to cordless, from cartridges with primitive graphics to disks and almost realistic graphics, from games you play at home to those that you can challenge people worldwide in game play. What will in-home video game systems feature in 5 years? Describe the system.
18. What is the difference between speech recognition and speech synthesis?
19. Investigation: Recent studies have shown that the use of 3-D printers may expose users to nanopollution. Investigate using the Internet and prepare a 1-page paper and 90 second verbal report on the potential danger and preventative measures needed for 3-D printing.

Note: Your teacher may assign groups for this problem.

20. A flight to Mars may take months or years.
    (a) Describe 5 recent technological innovations that could help make human travel to Mars a reality.
    (b) How will each of these innovations help a human survive a multi-year trip through space?
    (c) Can any of these be considered an emerging trend for mechanical engineering? Why or why not?
21. Investigation: Use the Internet to investigate fuel cells as a fuel source.
    (a) What is the fuel source for a fuel cell?
    (b) What is the waste product?
    (c) Why aren't we driving fuel-cell cars today?
    (d) Can this be considered an emerging trend in electrical engineering? Why or why not?
22. The invention and design of public sewage systems is widely regarded as an engineering feat that has been one of the most beneficial to humankind. Write a paragraph on how this invention has affected each of the following:
    (a) An individual – yourself, a family member, or a friend
    (b) Society within your hometown
    (c) Cultures within the United States vs. a developing country
    (d) The economy of the United States vs. a developing country
23. Write a paragraph on how a lack of public sewage systems would affect the environment.
24. Investigation: Define each of these engineering areas. Then, identify disciplines that are related to the field. Finally, identify career(s) that are related to that field.
    (a) Robotics
    (b) Process Control
    (c) Automation
    (d) Mechatronics

CHAPTER 2
THE PROFESSION OF ENGINEERING

What could we possibly mean by the *profession* of engineering? You may know someone who works as an engineer, but whether we would consider his/her status as an engineer to be a profession or a job can be subject to debate. To examine engineering in the sense of a profession versus a job, we pose the following question: if you decide to become an engineer, what are the implications for the general public, your employer, and the so-called profession of engineering?

A common definition of engineering is a profession for people "who are good at math and science" or people who "love math." Certainly, math and science are tools used by engineers, but engineering transcends the tedium of calculations and science facts. Engineering is a profession that involves helping people and helping society; a profession in which designs are created within a set of constraints - realistic boundaries and/or requirements.

The expansion of technology is visible every day and all around us. With the help of modern engineering achievements, we can swim underwater with scuba tanks, float through the air, and communicate worldwide. We can fly to the moon, circle it for months, and then fly back to Earth again. Our phones have evolved beyond their initial purpose – making voice calls – to incorporate functionality that enables us to take pictures, play our entire music library, send messages, and even make video calls. Engineering is the process of designing, creating, or modifying this technology.

The National Academy of Engineering describes engineering as the "application of science," because engineers build tangible products from abstract ideas. Engineers are people who create new products and "design under constraint," which means that a design has to conform to the requirements and should do exactly what it is expected to do without any unexpected consequences.

Engineering is constrained by the laws of nature; for instance, energy cannot be created or destroyed. Engineering is typically conducted by teams in which each member of the team contributes his or her specialized knowledge.

A profession is more than a job. People who belong to a profession are sufficiently competent in their abilities, have the necessary training to be successful, and achieve with the intent to help the general public. Members of a profession are typically paid with a salary instead of hourly and typically have a higher level of responsibility and must follow a code of ethics.

> What's the difference between a salaried and an hourly employee? An hourly employee gets paid for each hour that he or she works. Typically, if you don't work one day, you don't get paid. Salaried workers earn a steady salary, usually a given amount per year. To calculate an hourly rate for a salaried employee, dividing by 2080 gives a good estimate.
>
> For example, an engineer earning $60,000/year is earning roughly equivalent to $21.36 per hour.

2.1 WHAT IS A CAREER?

The most straightforward definition of a career is a job; however, there are qualities of a career that set it apart from a job. Typically, a career is a job that someone works to attain. Obtaining a career usually requires specialized knowledge and training; a person desiring a career has to invest in himself or herself prior to embarking on the career path. This may require college, other training, licensure, etc.

In today's society, the careers available to you upon your graduation are numerous and diverse. It is often difficult as a young adult to determine exactly what occupation you want to work at for the rest of your life because you have so many options. You cannot make a good decision without accurate information. No one can (or should) make the decision for you: not your relatives, teachers, advisors, or friends. You may not know all the answers to your questions right away. That means you will have to get them by gathering more information from outside resources and through your personal experience. Keep in mind that choosing your major and ultimately your career is a process. You constantly evaluate and reevaluate what you learn and experience. A key component is whether you feel challenged or overwhelmed. True success in a profession is not measured in monetary terms; it is measured in job satisfaction. As you find the answers, you can choose a major that leads you into a successful career path that you enjoy.

My team is responsible for implementing the engineered design in the field. We install, tune, test, and accept into operations all of the electronics that allow customers to use our state of the art fiber optic network to run voice, video, and data for their residential needs.

L. Gascoigne, CE

Figure 2-1

2.1.1 ENGINEERING CAREERS

Engineering careers are traditionally some of the highest-paying, especially for new college graduates. Numerous sites on the Internet show potential starting salaries for different engineering disciplines, especially compared to other starting salaries for recent college grads. Two of the more popular fields are electrical and mechanical engineering.

What does an electrical engineer actually do? If you read that electrical engineers might choose to go into solid state design, analog or digital communications, wireless technology, or power distribution, it is likely that you would understand very little of the differences between these areas. Similarly, if you see a list of possible mechanical engineering careers, including transportation systems, power production, or the production of energy or development of machines, it could look like there is a lot of overlap with electrical engineering. This is actually the case.

There is more to each engineering discipline than you might typically think: electrical engineers do more than just design electronics; mechanical engineers do more than just design robots. In fact, there are opportunities for both types of engineers (and other disciplines as well) in:

power grid

- *Power generation and distribution*: If you have seen wind turbines growing in popularity, notice that the mechanical action of the spinning blades generates electricity. That power needs to be transmitted throughout the community to reach each house and business. This transmission happens through the **power grid**, or just the grid. Both mechanical and electrical engineers play an active role in power and energy distribution.

Figure 2-2 Wind turbines generating electricity.

- *Wireless communication*: Look at cell phone technology of 10 years ago and notice the difference between that and today's technology. Your cell phone is now a music library, browsing device, game system, computer and, also, a phone. It contains remarkable power within a very small unit. Electrical engineers play a substantial role in developing this technology. In fact, the organization that specifies exactly how these devices communicate is the IEEE, the Electrical Engineering professional society.

Figure 2-3 Cell phones allow us to communicate using wireless technology.

- *Computers*: If you look inside a computer, you see many electronic components, as well as cooling units, which are crucial to the successful operation of your device. Electrical and mechanical engineers play a crucial role in the design and development of computers.
- *Other systems*: Robotics, space travel, transportation, amusement park rides, etc., all rely on engineering from different disciplines.

2.1.2 PROFESSIONAL CONDUCT

A professional is more than a person who works for a salary or someone who wears a suit to work. There are many aspects of a professional that are generally assumed to be true.

The professional is highly trained and often has a specialty. Teachers are excellent examples of trained professionals. They attend college for a degree in education and usually have to continue their education throughout their career. Teachers are known to take work home because it needs to be finished and often cannot be completed over the course of the work day. Teachers have areas of specialization: for example, your history teacher probably isn't qualified to be your math teacher.

A professional must be a good communicator: In almost every case, if not every case, a professional needs to communicate with his or her customers or clients. Professional managers – whether they are sports managers, personnel managers, or managers of companies – need to communicate with their employees as well as their managers or bosses. A manager who cannot communicate effectively is usually not a manager for long.

A professional typically does whatever it takes to get the job done: Doctors go through rigorous training, sometimes working 24 hour shifts (or worse). In times of tragedy, doctors are called upon to go to hospitals and clinics to help. The Hippocratic Oath, an oath taken by doctors as they enter their profession, requires them to treat the sick to the best of their ability, preserve patient privacy, and teach the secrets of medicine to the next generation, among other things. Doctors are expected to uphold this oath.

A professional works toward advancing his or her profession: One of the goals of a professional is typically to enhance that profession. Veterinarians, for example, have a responsibility toward their patients, but as they discover more effective techniques or new or unusual issues, they are generally expected to spread their expertise to others in their profession.

A professional maintains a professional attitude: However, the concept of a "professional attitude" is somewhat hard to define. If you work as a professional and you encounter someone who does not maintain a professional attitude, you will probably recognize it. This may mean anything from dressing professionally to writing a clear, articulate article. For example, you would rarely, if ever, see a law enforcement official wearing anything other than a uniform or suit. You would rarely see a politician without a suit, or a superintendent without a tie (or at least wearing khakis). Strive for your best work: Appear and act in a professional manner when in public or with clients, enhance others who work with your employer or in your profession, and uphold any oaths or standards of conduct or codes of ethics.

Figure 2-4 Professionalism.

2.1.3 RESUMES AND INTERVIEWS

When you apply for a professional position, your potential employer probably needs a more thorough picture of you as a potential employee. Most employers will require a cover letter, resume, and references (among other paperwork). They will use these to try to understand your experience, skills, and training, as well as aspects such as your work ethic and drive to excel. The potential employer usually considers hiring the potential employee as part of a long-term relationship, and expects the employee to make a solid contribution.

As part of a career search, you will need to prepare and organize the materials you will need.

Every job search requires the following:

- A *resume*, which is a written summary of your work-related skills, experience, and education. It introduces you to the prospective employer by presenting a snapshot of your qualifications. It should be brief and to the point, printed on white paper, true and accurate, and have no typographical, grammatical, or spelling errors. A good resume attracts the interest of the reader so he or she wants to learn more about you.
- A *cover letter*, or letter of interest, which is a letter of introduction that you send with a resume.
- A *list of references* that includes the names and contact information of people who know you and your qualifications and who are willing to speak about you to potential employers. Potential employers will likely call or email these references to obtain a richer understanding of you before deciding whether or not to offer you an interview, or the position.
- A *thank-you letter* that you send to the employer after an interview to show your interest in the position.

A cover letter is a strong and succinct way to introduce yourself to a potential client or employer. It also helps convince employers to review your resume by focusing attention on specific resume information, for example, credentials, skills, or specific projects in your portfolio of design work. The letter should be concise and well written, with no grammar or spelling mistakes. The cover letter is your first point of contact, so craft one that makes an excellent impression.

A resume, which summarizes your training and experience, is absolutely necessary when searching for a job, especially because employers will usually want to review all resumes before inviting an applicant for an interview. Employers often have to sort through large numbers of resumes, so make yours informative, easy to read, and concise. A resume should list your education, experience, and skills but should never exceed two pages. To make your resume stand out from the rest in a professional way, consider these tips:

- Make it easy to read. Leave space between lines so it is not crowded or overloaded.
- Use one, easy-to-read font, and apply different font styles and sizes for emphasis.
- Use bullets to make lines of text stand out.
- Use proper spelling, punctuation, grammar, and be careful to use or define any appropriate technical terms.
- Keep it to one page, if possible; two pages at the most. (If you use two pages, be sure to put your name in the header or footer on page 2, in case it becomes separated from page 1.)
- If you are printing the document, use plain white paper. Do not include a personal photograph.

In addition to a cover letter and resume, some potential employers may ask you to fill out an application form—sometimes online and sometimes on paper. Application forms require you to enter specific information about your education and past employment, including dates and locations. As with your resume, it is important to be truthful and accurate.

If an employer thinks you have the qualifications for the job, you will be invited for a job interview. A preliminary interview may be by phone or video conference, but almost all employers will expect a face-to-face meeting. The interview is an opportunity for you and the interviewer to ask questions and decide if the position is right for you. You should prepare for the interview by researching the company and the position and by practicing the answers to questions you think the employer might ask. You should also prepare a few questions that you can ask the employer. On the day of the interview, make sure your appearance is clean and neat. Dress professionally and arrive a few minutes early. Be prepared to shake hands and make eye contact with everyone you meet.

After the interview, you should write a thank-you note. A thank-you note reminds the interviewer that you are serious about wanting the job. Use the thank-you note to restate your interest in the job, highlight your key qualifications and to thank the interviewer for spending time with you. Refer to something specific that you discussed during the interview. Address the note to the person who interviewed you.

2.1.4 CLEAR AND CONCISE COMMUNICATION

What if you have a list of accomplishments that would push your resume to a third page, or you have an amazingly outgoing personality and desire for a position pushing your cover letter into two or three pages? Wouldn't your potential employer want to know each detail about your qualifications? Should you just ignore the fact that you assembled your own bicycle at the age of 8 years old, that you are able to recite the entire dialog of your favorite movie? Probably. Your writing in these instances and other, especially technical, writing should be concise.

concise

Concise writing is more than making sure the product is short: It is writing that is brief and comprehensive. Your cover letter, for example, should be one page, which may mean you need to leave out items you want to include, or may need to edit sections to use fewer, but possibly more descriptive words.

For example, a cover letter may describe an applicant's prior experience in extracurricular activities:

> From the age of 16 through graduation, I was very involved in many extracurricular activities. One of my favorite extracurricular activities was when I was part of a team that built a competitive robot for our school's FIRST Robotics team. I also was involved with three competitions in the Technology Student Association, including Video Game Design, The CO_2 racer design where I helped the team place third. I was also involved in a new competition for our school, the TEAMS competition, a competition involving extensive mathematics. Our team placed fourth in the state for TEAMS. I was also a member of the Spanish Club, where we had to only speak Spanish during each meeting and I was a member of the Student Engineering Club.

There is nothing in the previous 124-word paragraph that is negative, so there is nothing wrong with including any of that information. The mention of Spanish Club doesn't seem to add to the description: It's not bad, but all this description shows is that this applicant attended meetings and spoke some Spanish while there, so it could be deleted. It could be condensed and written in a more concise manner. For example,

> During High School, I was involved in a number of activities related to engineering, including the Engineering Club and competitions for the Technology Student Association (TSA) and FIRST Robotics, including a third place finish in TSA's CO_2 racer competition. I was in the first team to compete in the TEAMS competition, involving extensive mathematics.

This paragraph comes in at a concise 54 words and offers the same information, which can help keep that cover letter to one page. It is shorter, but also clear and effective: Each positive item from the original paragraph is emphasized.

Technical documentation is often concise; it is different than writing a book or story where rich, deep, and descriptive prose is appropriate.

2.2 ENGINEERING AS A PROFESSION

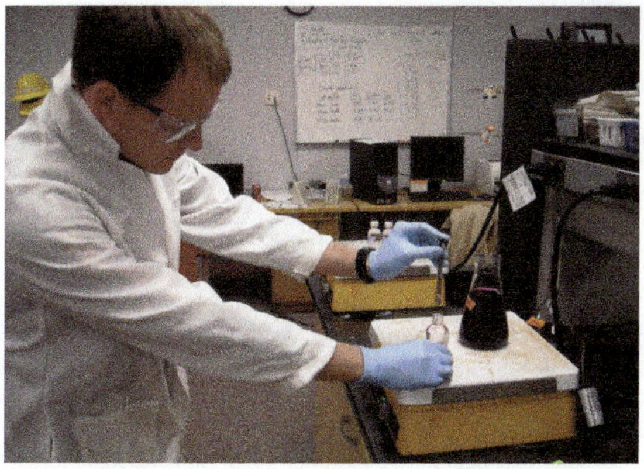

Figure 2-5

My research is part of a Water Research Foundation project, which is investigating the formation of emerging disinfection byproducts (DBPs) in drinking water treatment.

DBPs are undesirable, toxic compounds that are formed when water is chlorinated. I am investigating the effects of pH, bromide and iodide concentrations, and preoxidants on the formation of a specific family of DBPs.

D. Jones, BE

Other professions are related to engineering; perhaps the closest is the field of engineering technology. We described engineering as applying math and science toward problem solving intended to benefit the public. University programs in Engineering and Engineering Technology are assessed for their effectiveness by an organization called ABET, one organization that offers descriptions of both fields.

Engineering programs typically focus on theory and conceptual design and require higher-level mathematics, including multiple semesters of calculus and beyond. Engineering technology programs focus more on application and implementation of designs and systems, and have more focus on algebra, trigonometry, and applied calculus. Engineering is more typically associated with design, while engineering technology is more typically associated with implementation and manufacture.

While it is possible to earn an Associate's degree in Engineering Technology, it is clearer to compare the Bachelor's degrees in Engineering Technology and Engineering to observe the formal differences. The student outcomes required of accredited Engineering Technology programs are shown in the following table alongside those required of accredited Engineering programs. Note that the criteria do not appear in the same order in the two sets of criteria. In some cases, whereas the wording may be different between the two criteria, the intent is similar. For example, Engineering student outcome "(g), an ability to communicate effectively", is similar to Engineering Technology student outcome "g. an ability to apply written, oral, and graphical communication in both technical and nontechnical environments; and an ability to identify and use appropriate technical literature". Whereas the Engineering Technology outcome provides much greater detail, there is no reason to suggest that these are different outcomes.

Table 2-1 ABET Criteria for Engineering Technology and Engineering Programs

Engineering Technology (1)	Engineering (2)
a. an ability to select and apply the knowledge, techniques, skills, and modern tools of the discipline to broadly defined engineering technology activities;	(k) an ability to use the techniques, skills, and modern engineering tools necessary for engineering practice.
b. an ability to select and apply a knowledge of mathematics, science, engineering, and technology to engineering technology problems that require the application of principles and applied procedures or methodologies;	(a) an ability to apply knowledge of mathematics, science, and engineering
c. an ability to conduct standard tests and measurements; to conduct, analyze, and interpret experiments; and to apply experimental results to improve processes;	(b) an ability to design and conduct experiments, as well as to analyze and interpret data
d. an ability to design systems, components, or processes for broadly defined engineering technology problems appropriate to program educational objectives;	(c) an ability to design a system, component, or process to meet desired needs within realistic constraints such as economic, environmental, social, political, ethical, health and safety, manufacturability, and sustainability
e. an ability to function effectively as a member or leader on a technical team;	(d) an ability to function on multidisciplinary teams
f. an ability to identify, analyze, and solve broadly defined engineering technology problems;	(e) an ability to identify, formulate, and solve engineering problems
g. an ability to apply written, oral, and graphical communication in both technical and nontechnical environments; and an ability to identify and use appropriate technical literature;	(g) an ability to communicate effectively
j. a knowledge of the impact of engineering technology solutions in a societal and global context;	(h) the broad education necessary to understand the impact of engineering solutions in a global, economic, environmental, and societal context

Engineering Technology (1)	Engineering (2)
i. an understanding of and a commitment to address professional and ethical responsibilities including a respect for diversity;	(f) an understanding of professional and ethical responsibility
	(j) a knowledge of contemporary issues
h. an understanding of the need for and an ability to engage in self-directed continuing professional development;	(i) a recognition of the need for, and an ability to engage in lifelong learning
k. a commitment to quality, timeliness, and continuous improvement.	

(1) http://www.abet.org/DisplayTemplates/DocsHandbook.aspx?id=3150
(2) http://www.abet.org/DisplayTemplates/DocsHandbook.aspx?id=3149

Generally, Engineering program curricula are more academic, focusing on theory and concepts, whereas Engineering Technology program curricula are more practical, focusing on applications and skills. This difference can be seen in the preceding table whereas E(b) requires that engineering graduates are able to design and conduct experiments, ET(c.) does not require engineering technology graduates to be able to design experiments. Engineering outcome E(e) requires the more theory-oriented ability to "formulate" problems, an outcome that is missing from ET(f).

Engineering graduates have wide-ranging jobs ranging from design to analysis, office work to field work, from companies that make things to companies that design things that are made by others. Graduates of four-year engineering technology programs are more often found in jobs where things are made or sold, and are more often engaged in field work. Four-year technology graduates are called technologists—the term "technician" is appropriate for two-year Engineering Technology graduates. Thus, generally, engineering careers are more flexible whereas engineering technology careers tend to result in more tangible accomplishments (rather than accomplishments on paper).

I am an aerospace engineering manager responsible for developing unique astronaut tools and spacewalk procedures and for testing and training for NASA's Hubble Space Telescope servicing missions. My job ranges from tool and procedure design and development to underwater scuba testing to real-time, on-console support of Space Shuttle missions.

R. Werneth, ME

Figure 2-6

2.3 SAFETY

A product, system, or process must be designed to ensure that it is safe for consumers to use. For example, cooking utensils made of heat-conducting materials should have handles composed of materials that do not conduct heat. Similarly, electrical devices must be designed to shield the user from the internal electrical parts. Safety codes and regulations must be met. These are especially important for devices that can pose a risk to humans: components of devices that sustain life, such as medical devices, and large equipment, such as automobiles or amusement rides.

Everything has a life span; living organisms die and physical devices fail. One important responsibility of a designer is to ensure that a product's failure does not endanger people.

safety factor

Engineers are responsible to ensure their designs are safe. As engineers design, they may consider a **safety factor** in their design. A safety factor is a multiple of some criteria used in a design. For example, if a bridge is designed to hold 8,000 tons, an engineering firm may design to a safety factor of 1.5 and design the bridge to accommodate 12,000 tons. Suppose an engineer is designing an aquarium tank for a large aquatic center and calculates the thickness of the glass walls to be 2" thick. A safety factor of 3 would mean a thickness of 6". However, sometimes this will depend on where the safety factor is taken into consideration: if the aquarium specifications called for 20,000 liters in a tank giving a thickness of 6", we may multiply the volume of water by 3, and design the tank for 60,000 liters. This may or may not give a thickness of 6", but we can still claim a safety factor of 3.

Figure 2-7 Safety features are an important part of good design.

Considering safety is also important as we work with different tasks in engineering. For example, the manufacture of almost any electronic device involves designing and building circuit boards; if you have ever seen the inside of a computer, there is usually a large, green board inside with electronic components – the motherboard. Producing the printed circuit board involves chemically removing copper and epoxy (glue), which in turn produces waste. When the components are placed on the board, they are held in place by solder. Until a few years ago, almost all solder was a mixture of tin and lead. This means, as older computers are discarded, they contain copper, tin, lead, and a number of other chemicals that may or may not be hazardous.

Figure 2-8 Printed circuit board.

On a smaller scale, we produce waste every day. Many of us don't think twice about grabbing a bottle of water, and many of these bottles end up being recycled. However, many – approximately 60 million – end up in landfills or incinerators each year. Plastic bottles take up room in a landfill, but they are not considered to be hazardous waste.

hazardous waste

hazardous materials

What is the difference? **Hazardous waste** is classified as waste that is potentially harmful to health or the environment. **Hazardous materials** are similar to hazardous waste: These are materials that can be hazardous to health or the environment but are still in their original, usable form. Some household waste is considered hazardous (used cleaning products, broken compact fluorescent, or CFL, light bulbs), while a lot more is generated by industry. Almost every household has some hazardous material (cleaning products, gasoline, and paint thinners) as well. Your classroom probably has some material classified as hazardous.

Figure 2-9 Hazardous materials, hazardous waste, and even non-hazardous wastes should be disposed of properly.

The Environmental Protection Agency (EPA) is in charge of classifying which wastes are considered hazardous, and how they are to be disposed of. Fortunately, you will likely not come into contact with most of these, but some are common.

As a student, and perhaps someone who will design, build, manufacture, or recycle material in the future, you need to be familiar with the regulations in place at your school (and, eventually, your place of employment). The federal and state Environmental Protection agencies have specific rules and methods to follow to store and use hazardous materials, and to dispose of waste products. Common rules include:

- Store flammable hazardous materials in a well-marked, explosion-proof cabinet (if there is a significant amount of the material).
- Clearly post safety rules and guidelines in areas where these materials are stored and used.
- Require safety goggles when working with these materials.

Material Safety Data Sheet (MSDS)

Many hazardous materials come with paperwork explaining how they are to be used and stored: This paperwork is called a **Material Safety Data Sheet**, or **MSDS**. These sheets should be kept in a visible, obvious location in case of an accident, and in case they need to be reviewed quickly.

Figure 2-10 Material Safety Data Sheets (MSDSs).

Hazardous waste has rules and procedures for handling and disposing similar to those of hazardous materials. For example:

- Do not pour waste products down the drain, including products that may have resulted from a chemistry experiment.
- Do not dispose of known hazardous material in the normal trash can.

Your classroom and/or laboratory should have a set of lab safety guidelines. You are responsible to be familiar with these: It's not just a matter of your own safety, but the safety of those around you and your local environment.

EXAMPLE 2-1

Work in small groups to answer the following questions:

(1) Identify three important safety procedures you will need to consider in this class. You are allowed to ask for advice from your teacher on this question.

(2) Find a Material Safety Data sheet, either in your classroom or on the Internet. What is the chemical? How is it to be handled and/or disposed of?

(3) Identify a piece of laboratory equipment you may use this year in class. How is it maintained? What could go wrong with it, and what is a good procedure if something does go wrong with it?

2.4 ETHICS

Every day, we make numerous ethical decisions, although most are so minor that we do not even view them as such.

- You know that another student has plagiarized an assignment; do you rat him or her out?
- When going through a public door, do you make a habit of looking back to see if releasing the door will cause it to slam in someone's face?
- A cashier gives you too much change for a purchase. Do you correct the cashier?

On a grand scale, few of these decisions are particularly important, although some might lead to undesirable consequences. However, as an aspiring engineer, you may face numerous decisions in your career that could affect the lives and well-being of thousands of people. Just like almost everything else, practice makes perfect, or at least better. The more you practice analyzing day-to-day decisions from an ethical standpoint, the easier it will be for you to make good decisions when the results of a poor choice may be catastrophic.

Some ethical decisions are clear-cut. For example, almost everyone would agree that it is unethical to kill someone because you do not like his or her hat. Unfortunately, many real-world decisions that we must make are far from "black and white" issues, instead having many subtle nuances that must be considered to arrive at what one believes is the "best" decision.

There is no proven algorithm or set of rules that one can follow to guarantee that the most ethical decision possible is being made in any particular situation. However, numerous people have developed procedures that can guide us in considering questions with ethical ramifications. A four-step procedure is discussed here, although there are various other approaches.

Step 1: **Determine *What* the issues are and *Who* might be affected by the various alternative courses of action that might be implemented.**

stakeholders

We will refer to the *Who* as **stakeholders**. Note that at this point, we are not trying to determine how the stakeholders will be affected by any particular plan of action.

- The issues (What) can refer to a wide variety of things, including, for example, personal freedom, national security, quality of life, economic issues, fairness, and equality.
- The term stakeholders (Who) does not necessarily refer to people, but might be an individual, a group of people, an institution, or a natural system, among other things.

EXAMPLE 2-2

Consider the question of whether to allow further drilling for oil in the Alaska National Wildlife Refuge (ANWR). List several issues and stakeholders.

Issues:

- *Oil independence*
- *The price of gasoline*
- *Possible impacts on the ecosystem*

Stakeholders:

- *Oil companies*
- *The general population of the United States*
- *Other countries from whom we purchase oil*
- *The flora and fauna in ANWR*
- *The native people in Alaska*

Step 2: **Consider the effects of alternative courses of action from different perspectives.**

Here, we look at three perspectives: consequences, intent, and character.

Perspective 1: Consequences

When considering this perspective, ask how the various stakeholders will be affected by each alternative plan being contemplated. In addition, attempt to assign a relative level of importance (weight) to each effect on each stakeholder. For instance, an action that might affect millions of people adversely is probably more important than an action that would cause an equivalent level of harm to a dozen people.

Perspective 2: Intent

The intentions of the person doing the acting or deciding are considered in this perspective, sometimes called the "rights" perspective. Since actions based on good intentions can sometimes yield bad results, and vice versa, the intent perspective avoids this possible pitfall by not considering the outcome at all, only the intentions.

Ask yourself the following questions:

(a) Is the action I am taking something that I believe everyone should do?
(b) Do I believe that this sort of behavior should be codified in law?
(c) Would I like to be on the receiving end (the victim) of this action?

EXAMPLE 2-3

Should you download music illegally over the Internet?

Rephrasing this question using the suggestions above yields:

(a) Should everyone illegally download the music they want if it is there for the taking?
(b) Should the laws be changed so that anyone who obtains a song by any means can post it on the web for everyone to get for free?
(c) If you were a struggling musician trying to pay the bills, would you like your revenue stream to dry up because everyone who wanted your music got it for free?

Perspective 3: Character

Character is the inherent complex of attributes that determines a person's moral and ethical actions and reactions. This perspective considers the character of a person who takes the action under consideration. There are different ways of thinking about this. One is to simply ask: Would a person of good character do this? Another is to ask: If I do this, does it enhance or degrade my character? Yet another way is to ask yourself if a person you revere as a person of unimpeachable character (whoever that might be) would take this action.

Step 3: **Correlate perspectives.**

Now look back at the results of considering the issues from the three perspectives. In many cases, all three perspectives will lead to the same or a similar conclusion. When this occurs, you have a high level of confidence that the action is the best choice from an ethical standpoint.

If the three perspectives do not agree, you may wish to reconsider the question. It may be helpful to discuss the issue with people whom you have not previously consulted in this matter. Did you omit any factors? For complicated issues, it is difficult to make sure you have included all possible stakeholders and consequences. Did you properly assign weights to the various aspects? All three perspectives may converge.

If not, go with two out of three.

Step 4: **Act.**

This is often the hardest step of all to take, since ethical action often requires courage. The whistle-blower who risks losing his or her job, Harriet Tubman repeatedly risking her life to lead slaves to freedom via the Underground Railroad, the elected official standing up for what she knows to be right even though it will probably cost her the next election, or even something as mundane as risking the ridicule of your friends because you refuse to go along with whatever questionable activities they are engaging in for "fun." Ask yourself the question: "Do *I* have the courage to do what I know is right?"

● EXAMPLE 2-4

Your company has been granted a contract to develop the next generation of electronic cigarette, also known as a "nicotine delivery system," and you have been assigned to the design team. Can you in good conscience contribute your expertise to this project?

NOTE

In the interest of brevity, this is not an exhaustive analysis but shows the general procedure.

Step 1: *Identify the issues (What) and the stakeholders (Who).*

Issues:

- *Nicotine is poisonous and addictive.*
- *These devices eliminate many of the harmful components of tobacco smoke.*
- *Laws concerning these devices range from completely legal, to classification as a medical device, to banned, depending on country.*
- *There are claims that such devices can help wean tobacco addicts off nicotine.*
- *The World Health Organization does not consider this an effective means to stop smoking.*
- *Whether an individual chooses to use nicotine should be a personal decision, since its use does not generally degrade a person's function in society.*
- *The carrier of the nicotine (80–90% of the total inhaled product) is propylene glycol, which is relatively safe, but can cause skin and eye irritation, as well as other adverse effects in doses much larger than would be obtained from this device.*
- *A profit can be made from nicotine products or anti-smoking devices.*

Stakeholders:

- *You (your job and promotions)*
- *Your company and stockholders (profit)*
- *Cigarette manufacturers and their employees and stockholders (lost revenue)*
- *Tobacco farmers (less demand)*
- *The public (less second-hand smoke)*
- *The user (various health effects, possibly positive or negative)*

Step 2: *Analyze alternative courses of action from different perspectives.*

1. *Consequences*

- *You may lose your job or promotion if you refuse.*
- *If you convince management to abandon the project, the company may lose money.*
- *If you succeed brilliantly, your company may make money hand over fist, and you receive a promotion.*
- *If the project goes ahead, the possibility of future lawsuits exists.*
- *Users' health may be damaged.*
- *Users' dependence on nicotine may either increase or decrease.*

2. *Intent*

- *Should everyone use electronic cigarettes, or at least condone their use?*
- *Should use of electronic cigarettes be unrestricted by law?*
- *Would I like to risk nicotine addiction because of using these devices?*
- *Would I be able to kick my tobacco habit by using these devices?*

3. *Character*

- *Would a person of good character develop this device, use it, or condone its use?*
- *Would work on this project (thus implicitly condoning its use) or use of the device itself enhance or degrade my character?*
- *Would my personal spiritual leader, or other person I revere, condone development or use of this product?*

Step 3: *Correlate perspectives.*

Here we enter the realm of subjective judgment. The individual author responsible for this example has a definite personal answer, but it is in the nature of ethical decision making that different people will often arrive at different results in good conscience. You would have to weigh the various factors (including any that have been overlooked or knowingly omitted) to arrive at your own conclusion. We refuse to dictate a decision to you.

Step 4: *Act on your decision.*

If your decision was that working on this project poses no threat to your soul (if you happen to believe in such), probably little courage is required to follow through, since your career may blossom, or at least not be curtailed.

On the other hand, if you believe that the project is unethical, you need to have the intestinal fortitude to either attempt to change the minds of management or refuse to work on the project, both of which may put your career at risk.

2.4.1 ETHICS VS. MORALS

Because of the common occurrence of unethical behavior and the negative impact it has on others, almost all societies have developed rules, codes, and laws to specify what is and is not acceptable behavior, and the punishments that will be meted out when violations occur.

Through the centuries, such codes and laws have been expanded, modified, and refined so that most forms of serious antisocial behavior are addressed and consequences for violations specified. These codes exist from a local to a global level. Several examples are given below.

- Most countries purport to abide by the Geneva Conventions, which govern certain types of conduct on an international scale.
- Most countries have national laws concerning murder, rape, theft, etc.
- In the United States, it is illegal to purchase alcohol unless you are 21 years of age. In England, the legal age is 18.
- In North and South Dakota, you can obtain a driver's license at age $14\frac{1}{2}$. In most other states, the legal age is 16.
- It is illegal to say "Oh boy!" in Jonesboro, Georgia.
- Nearly all states in America (41 of them) have ordinances prohibiting text messaging while driving.

These laws and guidelines apply to everyone within a population. However, even without a set of laws, people have a sense of right and wrong that guides their actions (even if we occasionally choose to ignore that sense). One person may feel very strongly that lying is absolutely wrong, while others may lie routinely. If a person finds a bag full of cash, he or she may take it to a local police station to see if someone has reported one lost, while someone else may simply take it – after all, 'finders keepers, losers weepers.'

morals
ethics

Many of these examples represent **morals** rather than **ethics**. Moral decisions are those that apply to an individual. Ethical decisions are those based on a right vs. wrong as defined for a population. For example, we would generally agree that stealing is wrong. However, some people would argue that taking a few dollars from a parent isn't terribly wrong. Some would say that, although we know stealing is wrong, we would do it in this case. Decisions of this type are based on a person's character.

An engineer may receive an offer like the following: in exchange for using a lesser grade of steel in a bridge project, a company may be happy to send that engineer on an all-expenses paid trip to Hawaii. There is certainly a moral component: One engineer may consider this more dishonest than another. However, this specific issue is addressed in engineering Codes of Ethics (we'll see these shortly). Engineers are not to take bribes, and are to put the safety of the public beyond other considerations. These rules are to be followed by a population of people: in this case, a population of engineers.

2.4.2 CODES OF ETHICS

Code of Ethics

In general, ethical decisions in engineering have a narrow focus specific to the problems that arise when designing and producing products or services of a technical nature. By the nature of their profession, engineers and scientists have a body of specialized knowledge that is understood only vaguely, if at all, by most of the population. This knowledge can be used for tremendous good in society, but can also cause untold mischief when used by unscrupulous practitioners. Various engineering organizations have thus developed codes of conduct specific to the profession. Perhaps the most well known is the **Code of Ethics** for Engineers developed by the National Society of Professional Engineers (NSPE). The entire NSPE Code of Ethics is rather long, so we list only the Engineer's Creed and the Fundamental Canons of the Code here.

Engineer's Creed

As a Professional Engineer, I dedicate my professional knowledge and skill to the advancement and betterment of human welfare. I pledge:

- To give the utmost of performance
- To participate in none but honest enterprise
- To live and work according to the laws of man and the highest standards of professional conduct
- To place service before profit, the honor and standing of the profession before personal advantage, and the public welfare above all other considerations

In humility and with need for Divine Guidance, I make this pledge.

Fundamental Canons

Engineers, in the fulfillment of their professional duties, shall

- Hold paramount the safety, health, and welfare of the public
- Perform services only in areas of their competence
- Issue public statements only in an objective and truthful manner
- Act for each employer or client as faithful agents or trustees
- Avoid deceptive acts
- Conduct themselves honorably, responsibly, ethically, and lawfully so as to enhance the honor, reputation, and usefulness of the profession

The complete code can easily be found online at a variety of sites. Use an online search tool to find the URL for the Code of Ethics from the NSPE and for your desired discipline.

2.5 TEAMING

A team is a group of two or more people who work together to achieve a common goal. When you are part of a team, you have access to all the knowledge, experience, and abilities of your teammates. Together you can have more ideas, achieve more goals, and solve more problems.

Most engineering projects require teamwork. They are usually multidisciplinary, which means they involve people from many different areas within a company and with outside resources. For example, a design engineer may work on a team with other engineers who specialize in production, testing, and sales and service. The team is likely to include financial managers, legal advisors, marketing representatives, and, of course, a project manager. There may even be client representatives on the team.

But, no matter who is on the team and what the team is doing, the interactions between team members will be most productive if the members of the team:

- Listen to each other
- Respect each other's opinions
- Recognize each other's skills and abilities
- Share the work load
- Share responsibility

2.5.1 ENGINEERING TEAMS

To prepare for a career as a professional engineer, students can work in teams to complete comprehensive projects. This provides the opportunity to learn, practice, and demonstrate the ability to function as a team member. Specifically, team members can:

- *Apply the design process as a team participant.* Team members count on each other to understand how to apply the design process to the task at hand.
- *Assume different roles as a team member within the project.* Each person on the team must be prepared to step in as necessary to help others and move the project toward successful completion.
- *Maintain an engineering notebook for the project.* An accurate and complete notebook shows the progress of each team member and provides documentation that work is completed.
- *Develop and test a model for the project.* Each team member should be prepared to participate in these steps which are vital to a project's success.
- *Prepare and present the project using effective communications skills.* Team members should all be able to use written and verbal communications skills to explain and demonstrate the project. This includes using tools such as word processing, graphing, and presentation software programs.

2.5.2 LEADERSHIP

Even when all members of a team have an equal role in decision making and problem solving, it is important to have a leader. A leader is a type of manager who knows how to use available resources to help others achieve their goals.

Leaders exhibit positive qualities that other people respect, such as self-confidence, honesty, effective listening and speaking skills, decisiveness, respect for others, and open-mindedness. They are organized and supportive of the other team members.

Although you might have heard that someone is a "born leader," that's not usually the case. Becoming a leader takes time and patience. Leaders have to prove that they can make decisions, set goals, and solve problems. You can develop leadership characteristics by recognizing and modeling positive leadership qualities. One way to do this is by participating in student leadership and professional development opportunities. For example, you can join clubs and activities such as your school's robotics team, and volunteer for leadership roles. You might also join a career technical student organization (CTSO) such as Technology Student Association (TSA) or SkillsUSA. CTSOs offer members a range of individual and group programs, activities, and competitions designed to build professional skills as well as leadership qualities.

2.5.3 QUALITIES OF AN EFFECTIVE TEAM MEMBER

While a strong leader is important to the success of a team, team members must also be committed to the group's success. An effective team member helps teammates if they need help, does not blame teammates for problems or mistakes, and offers ideas and suggestions instead of criticism. Effective team members use critical-thinking skills and interpersonal skills to identify and solve problems and achieve their goals.

You are a good team member if you are:

- Open minded
- Willing to compromise
- Cooperative
- Friendly
- Trustworthy

2.6 COMMUNICATION

Engineers spend a great deal of time communicating. They must speak, read, and write, and develop documents for communication such as reports, presentations, and, of course, their Engineering Notebook.

Effective communication is a process that involves an exchange of information between a sender and a recipient. The sender transmits the message with a specific intent. The recipient interprets the message and responds.

Effective communication occurs when the receiver interprets the message the way the sender intended. Ineffective communication occurs when the receiver misinterprets the message. Sometimes the sender thinks the message is clear, but in fact it might confuse or mislead the receiver.

In order to ensure effective communication:

- The message must be clear.
- The sender must be able to deliver the message in a concise and accurate manner.

- The receiver must be able to hear and receive the message.
- The receiver must be able to interpret the message correctly.

2.6.1 USING THE COMMUNICATION PROCESS

You can communicate effectively by using a six-step process:

1. *Be clear.* The receiver is more likely to get your message if you deliver it in a way he or she can understand. Speak slowly. Consider who he or she is. You probably use different language when you talk to a teacher than when you talk to a friend. You might even use a completely different type of communication—face-to-face instead of a social networking site.

2. *Be personal.* Use the other person's name—then he'll know you are communicating with him. Use an "I" statement—a statement that starts with the word "I"—to frame the statement in terms of you and your goals. An "I" statement indicates that you are taking responsibility for your thoughts and feelings. It helps the receiver understand your point of view and respond to you, instead of focusing on herself and how your statement affects her.

3. *Be positive.* Phrase your message in positive terms. Say what you want, not what you don't want.

4. *Get to the point.* Follow the "I" statement with an explanation of the message you are sending. Explain how or why you feel a certain way, or how or why you think a certain thing.

5. *Actively listen to the response.* Pay attention and make sure you hear the response.

6. *Think before you respond.* Make sure you understand the message. Repeat it, if necessary, and ask questions for clarification. Use critical thinking to make sure you are not letting emotions and preconceived ideas get in the way.

Once you master the communication process, you will be able to use it effectively for:

- Verbal communication, which includes speaking or writing
- Nonverbal communication, which includes visual messages that help put words into context, such as a smile or the tone of your voice, as well as visual aids such as slides, charts, images, and graphs
- Active listening, which is when you give the message sender your full attention and consideration.

2.6.2 VERBAL COMMUNICATION

Verbal communication is the exchange of messages by using sounds or words. For most of us, verbal communication is the most common way we stay in touch with other people in our lives; for example, we talk face to face or on the phone. In contrast, we use a wide variety of non-verbal communication (in the following cases, written communication) when we send text messages, e-mails, and instant messages. Moreover, we write blogs, pass notes, and send cards.

While written communication can be a useful mode of communication, talking can be more effective. When you speak clearly and use language the receiver understands, he or she will interpret the message the way you intend it. This is due to the additional information verbal communication embeds in the message through inflection – the sound of your voice. In addition, verbal communication is usually paired with nonverbal cues such as our posture, hand gestures, eye contact, and so on.

Our natural nonverbal communication helps put words into context. This form of communication includes visual messages that the receiver can see, like a smile when you are talking. It also includes physical messages, like a pat on the back. During a conversation, the tone of your voice and the language you use combine to provide context for the words. When you write, you lose some of the context, which can make the communication less effective. Techniques can be used to regain a sense of context, such as using emoticons to add emotion or expressions (smiles, laughter, sadness, etc.), or capitalizing each letter to signify shouting. However, there is still room for misinterpretation.

Exchanging written messages does not take place face to face; it might even be across great distances. When the receiver cannot hear your voice or see your nonverbal cues, he or she might misinterpret the message because, while speaking comes from the mouth, communication involves our entire bodies. With this in mind, be sure to craft any correspondence in the clearest manner possible. Being concise by not filling sentences with unneeded jargon or rhetoric will aid in conveying your messages without the aid of nonverbal cues.

2.6.3 ACTIVE LISTENING

Active listening is an important part of effective communication. When you are an active listener, you pay attention to the speaker, and make sure you hear and understand the message.

Use these skills to be an active listener:

- Show interest using eye contact and positive nonverbal messages.
- Let the other person finish speaking before you respond.
- Ignore distractions such as cell phones and other people.
- Set your predetermined opinions and emotions aside.
- Repeat the message that you hear out loud, to make sure you received it correctly.

Active listening is a sign of respect. It shows you are willing to communicate and that you care about the speaker and the message. When you listen actively, the other person is more likely to listen when you speak, too.

Figure 2-11 Active listening.

TECHNOLOGY IN THE REAL WORLD

Genetically Modified Foods

There is no question that genetic engineering lets us do amazing things. We can add vitamins to rice, producing better nutrition for millions of people. We can control the process of ripening in fruits—from tomatoes to coffee beans. We can even produce plants that "glow in the dark" when they need to be watered. All of these changes are produced by copying part of the genetic material from one organism to another.

Controlling a Sticky Pest

The tomato horn worm, shown to the right, or the bollworm, can cause serious damage to crops such as cotton, corn, and potatoes. For decades, organic farmers have been treating their crops with a naturally occurring pesticide produced by the bacteria (small, single-celled organisms) *Bacillus thuringiensis,* or Bt for short. Bt is deadly to certain plant-eating insects, including the bollworm. Even better, it is harmless to most other insects.

Using genetic engineering, scientists made copies of the Bt gene that is harmful to the bollworm. They inserted this gene into common crop plants. The genetically modified crops remained pest free.

A horn worm devours a tomato.

Some Unwanted Results

End of story? Not quite. The pests that were initially controlled by the Bt gene are starting to show signs of resistance to it. A greater percentage of each new generation of bollworms can tolerate the Bt gene. In addition, the Bt gene appears to be toxic to some beneficial insects, such as the monarch butterfly.

Scientists are quickly learning that genetic engineering can produce unanticipated and often unwanted results. Does this mean that we should stop using genetic engineering? Should we limit its use to certain types of organisms, such as plants and bacteria? Or should scientists be able to explore all the possibilities of genetic engineering, controlling and creating new life-forms at will? What do you think?

Genetic engineering is used widely to control the process of ripening in fruits, such as tomatoes.

CONNECTING TO STEM

How Safe Is Safe?

The next time you step into an elevator, consider how safe it is. Look around for a safety certificate that indicates the maximum number of passengers and the maximum weight the elevator can carry—for example, the posted number of passengers (also called the posted load) is 10 and the weight limit is 1,500 pounds. These loads assume that the average weight of each person is 150 pounds. What if, on that particular day, the 10 people are all football players who, together, weigh more than the posted limit? Or, what happens if 12 people crowd into the elevator? How safe would that be?

Putting a Number on Safety

The safety of any kind of structure is indicated by something called a safety factor. A **safety factor** is the ratio between the load that will cause the device to fail, and the posted load. For example, if the load that will cause the elevator to fail is 15,000 pounds and the posted load is 1,500 pounds, the safety factor is 10 (15,000 ÷ 1,500 = 10). The greater the safety factor, the less likely the device is to fail. Fast passenger elevators have safety factors of 11.9. The cables that support these elevators are designed to hold a weight at least 11.9 times greater than the posted load. Other safety factors are listed in the table.

Bridges and buildings are also assigned safety factors. They are essential to ensuring safety when the loads are extremely heavy. During rush hour, for example, many heavy trucks may be stuck on a bridge at the same time.

Is Safer Always Better?

If a greater safety factor is better, why don't engineers build structures with higher safety factors—say 100 or more? Because cost is a key factor as well. If a structure is made from stronger materials, it costs more to build and maintain. A product that is overly strong (and expensive) for its purpose will not be purchased by customers, so it will not survive in the marketplace.

Type of Structure	Safety Factor
Cable of a fast passenger elevator	11.9
Cable of a slow passenger elevator	7.6
Cable of a crane	6
Wooden building	6
Leg bone of a galloping horse	4.8
Steel building or bridge	2

Chapter 2 ASSESSMENT QUESTIONS

1. Please address the following questions in approximately one page. You may write, type, draw, sketch your answers . . . whatever form you would like to use.

(a) Where are you from?
(b) What type of engineering are you interested in? Why?
(c) Are there any engineers in the circle of your friends and family?
(d) How would you classify yourself as a student (freshman, sophomore, or junior)?
(e) What are your activities and hobbies?
(f) What are you most proud of?
(g) What are you fantastic at?
(h) What are you passionate about?
(i) What has been the most difficult aspect of high school so far?
(j) What do you expect to be your biggest challenge this year?
(k) Do you have any concerns about this class?
(l) Anything else you would like to add?

2. In 2008, the University of Memphis made national headlines when the Memphis Tigers played in the NCAA basketball national championship game. When interviewed by a local newspaper about how he helped his team "own" the tournament, Coach Calipari revealed that before each game, he had his star player write an essay about how the game would play out. This particular player was prone to nervousness, so to help him focus, the coach told him to mentally envision the type of plays and how he himself would react to them.

In this assignment, write a short essay on how you will prepare for and take the final exam. Consider the following:

(a) What will you do to study? What materials will you gather, and how will you use them? Where will you study? Will it be quiet? Will you play music?
(b) What kinds of things could go wrong on the day of the exam, and how would you avoid them? (List at least three.)
(c) What will the exam look like, and how will you work through it?

Thank you to Dr. Lisa Benson for contributing this assignment.

3. An article in *Science News* addressed the topics of nature and technology. In our electronic world, we are in constant contact with others, through our cell phones, iPods, Facebook, e-mail, etc. Researchers at the University of Washington have determined that this effect may create long-term problems in our stress levels and in our creativity.

Scenes of Nature Trump Technology in Reducing Low-Level Stress

Technology can send a man to the moon, help unlock the secrets of DNA, and let people around the world easily communicate through the Internet. But can it substitute for nature?

"Technology is good and it can help our lives, but let's not be fooled into thinking we can live without nature," said Peter Kahn, a University of Washington associate professor of psychology who led the research team.

We are losing direct experiences with nature. Instead, more and more we're experiencing nature represented technologically through television and other media. Children grow up watching Discovery Channel and Animal Planet. That's probably better than nothing. But as a species we need interaction with actual nature for our physical and psychological well-being.

Part of this loss comes from what the researchers call environmental generational amnesia. This is the idea that across generations the amount of environmental degradation increases, but each generation views conditions it grew up with as largely non-degraded and normal. Children growing up today in the cities with the worst air pollution often, for example, don't believe that their communities are particularly polluted.

"This problem of environmental generational amnesia is particularly important for children coming of age with current technologies," said Rachel Severson, a co-author of the study and a University of Washington psychology doctoral student. "Children may not realize they are not getting the benefits of actual nature when interacting with what we're calling technological nature."

[University of Washington (2008, June 16). "Scenes of Nature Trump Technology in Reducing Low-level Stress." Science Daily*]*

Go someplace quiet and spend at least 10 minutes clearing your head. Only after these 10 minutes, get out a piece of paper and sketch something you see; are thinking of; want to create or invent or imagine, etc. On the same sheet of paper, write a poem about your sketch; something you are thinking of; or quiet, wonder of the universe, lack of technology, etc.

Rules:

- Must be done all by hand.
- Must be original work (no copied poems or artwork).
- Ability does not count—draw like you are 5 years old!
- Draw in any medium you want: use pencil, pen, colored pencils, markers, crayons, watercolors.
- Poetic form does not matter: use rhyme, no rhyme, haiku, whatever you want it to be.
- It does not need to be elaborate; simple is fine.

4. Find a current job advertisement for an engineer on the Internet.
  (a) Identify the company for whom the engineer would work.
  (b) Give the Web address (the URL) for the advertisement.
  (c) Write 1-2 paragraphs describing the main responsibilities of this position. You may have to make some assumptions based on the description in the ad and the company placing the ad, but keep the assumptions realistic. Be concise!
  (d) Describe what the employer might look for on an applicant's resume. Be as specific as possible. The job ad should give valuable hints. Be clear and concise.
  (e) Describe what someone applying for this job should wear to the interview. Include a picture from the Internet.
  (f) Describe what this person would typically wear to work, and why that would be appropriate. Include a picture from the Internet.

5. Briefly describe the role an electrical engineer and a mechanical engineer would play in the design of the following items and systems. Be sure your answer is clear and effectively distinguishes electrical from mechanical:
  (a) Texas Giant roller coaster
  (b) Generator for drinkable water on the International Space Station
  (c) The driverless automobile
  (d) Small drones or quadcopters

6. Briefly describe the role an engineer, a scientist, and a technologist would play in the design, manufacture, maintenance. and operation of the following items and systems. Be sure your answer is clear and effectively distinguishes engineering from technology and science:
  (a) Texas Giant roller coaster
  (b) Small drones or quadcopters

7. For each of the following situations, indicate whether you think the action is ethical, unethical, or you are unsure. Do not read ahead; do not go back and change your answers.

Situation	Ethical	Unethical	Unsure
Speeding 5 miles per hour over the limit			
Killing a cockroach			
Speeding 15 miles per hour over the limit			
Killing a mouse			
Driving 90 miles per hour			
Driving 90 miles per hour taking an injured child to the hospital			
Killing a person			
Accepting a pen from a company trying to sell a new computer system to your company			
Using time at work to plan your friend's surprise party			
Accepting a wedge of cheese from a company trying to sell a new computer system to your company			
Taking a company pen home from work			
Borrowing company tools			
Accepting a Hawaiian vacation from a company trying to sell a new computer system to your company			
Letting a friend who has been sick copy your homework			
Cheating on a test			
Shortcutting by walking across the grass behind a house			
Helping people with their homework			
Letting somebody copy your homework			
Shortcutting by walking through a house			

8. For each of the following situations, indicate how great you feel the need is in the world to solve the problem listed. Do not read ahead; do not go back and change your answers.

Situation	Urgent	Great	Somewhat	Little	None
1. Teaching those who cannot read or write					
2. Helping starving children in poor nations					
3. Helping people locked in prisons					
4. Helping to slow population growth					
5. Helping to reduce dependence on foreign oil					
6. Helping to reduce greenhouse gas emissions					
7. Helping people persecuted for sexual orientation					
8. Helping to reduce gun ownership					
9. Helping those who are mentally disabled					
10. Helping to supply laptops to poor children					
11. Helping prevent prosecution of "victimless" crimes					
12. Helping to end bigotry					
13. Helping to prevent development of WMD (weapons of mass destruction)					
14. Helping prosecute "hate" crimes					
15. Helping to eliminate violence in movies					
16. Helping homeless people in your community					
17. Helping people with AIDS					
18. Helping people in warring countries					
19. Helping endangered species					

9. Find the IEEE Code of Ethics and the NSPE Code of Ethics online. Compare the IEEE Code of Ethics to the Fundamental Canons described in the NSPE Code of Ethics. What differences do you see? What similarities?

10. Discuss the possible actions, if any, that you would take in each of the following situations. In each case, use the four-step analysis procedure presented in Section 2.1 to help determine an appropriate answer.
    (a) You witness several students eating lunch on a bench outside school. When they finish, they leave their trash on the ground.
    (b) You see a student carving his initials in a large beech tree on the school's property.
    (c) You see a student writing graffiti on a trash dumpster.
    (d) There is a squirrel in the road ahead of a car you are driving. You know that a squirrel's instinct is to dart back and forth rather than run in a straight line away from a predator (in this case a vehicle) making it quite likely it will dart back into the road at the last instant.
    (e) You find a wallet containing twenty-three $100 bills. The owner's contact information is quite clear. Does your answer change if the wallet contained three $1 bills?

11. Choose an ordinary device that you use in your home. Make a timeline of innovations and inventions for this product. What future changes, or innovations, might be made?

12. Look online or in your newspaper for an example of a local, national, or global problem that uses technology as part of its solution. Explain how technological literacy can help you discuss the problem intelligently.

13. Describe a hybrid electric vehicle, shown below. Why is it considered environmentally friendly?

14. Explain why it is unethical to infect or damage other people's computers with viruses or to "hack" into other computer systems to gather or change information.

15. Imagine a world with little technical innovation. Identify two things in your life that would be dramatically different without technological advancement. Prepare a 90-second speech on these two innovations. The only visual aids you may use, if you wish, are the two devices themselves.

16. **(a)** Find the Internet address for your state Environmental agency (for example, in Texas, this is the Texas Commission on Environmental Quality) and the federal Environmental Protection Agency.
    (b) Investigate a material you may have at home or in the classroom (for example, unused paint or a specific cleaning fluid). Identify how you should safely dispose of this material according to each agency.

17. Identify three key procedures you are to follow for safety in your classroom. You may consult your school's Web site, other students, or your teacher for help with this question.

18. Identify 5 key differences between a group of people and a team.

19. Consider a time you were part of a team that was highly functional.

(a) How did you specifically contribute as a member of the team to achieve a positive outcome?

(b) How did you specifically work with others, or collaborate, to achieve a positive outcome?

20. Describe a time when you engaged as an active listener. What was the situation? What specifically made you an active listener?

21. Describe three things you are currently doing or could do this year to strengthen your leadership ability. For each, describe why this activity would help your leadership ability.

22. Select a discipline – either electrical, mechanical, chemical, or civil. Compare and contrast the career of someone working in one of these disciplines as an engineer, technologist, or scientist. What would each professional do that is similar among engineering, technology, and science? What would be some key differences?

23. Interview with a partner: You are interested in a career opening for an Environmental Engineer.

(a) Write your own ad for this Environmental Engineering position. You may use information from online job advertisements and may craft some characteristics on your own. The ad should be no more than ½ page.

Now, it's time to apply for this position. Swap job ads – you will apply for the position designed by your partner.

(b) Write a 1-page resume, targeting this specific ad as much as possible, and give this resume to your partner.

(c) Prepare for a mock interview: figure out a few key things you want to convey to the interviewer and a few questions you can ask your interviewer.

Pre-interview feedback:

(d) Critique the resume of this applicant: Do you see anything on your partner's resume that could be improved? Mark it on the resume.

The interview:

Each partner should interview the other.

(e) The interviewer: Be professional. Ask at least the following questions:

- **(i)** Why are you interested in this position?
- **(ii)** What qualifies you specifically for this position?
- **(iii)** Do you have any questions for me?

(f) The interviewee: Be professional. Either dress for a professional interview or describe what you should wear as part of the reflection. Answer each question. Engage in active listening. After the interview, write or email a thank-you to your interviewer.

Reflection:

(g) What did you do well during the process?

(h) What could you have improved upon?

Be sure to describe your professionalism: Include your dress (what you wore or what you should wear), your initial greeting, and your conduct and communication during the interview.

CHAPTER 3
ENGINEERING SYSTEMS AND DESIGN

Everyone solves problems in different ways. There are, however, some procedures and techniques that can help in developing a generally successful problem-solving approach. Although it is not possible to write down a specific recipe that will always work, some broad approaches will help.

3.1 PROBLEM SOLVING

SOLVEM

One problem-solving approach has been given the acronym **SOLVEM:**

Sketch
Observations or **O**bjectives
List
Variables and
Equations
Manipulation

Note that this approach is equally useful for problems involving estimation and more precise calculations. Each step is described below.

Sketch

Figure 3-1 illustrates how a drawing can help you visualize a problem. In sketching a problem, you are subconsciously thinking about it. Be sure to draw the diagram large enough so that everything is clear, and label the things that you know about the problem in the diagram. For some problems, a before-and-after set of diagrams may be helpful. In very complex problems, you can use intermediate diagrams or subdiagrams as well.

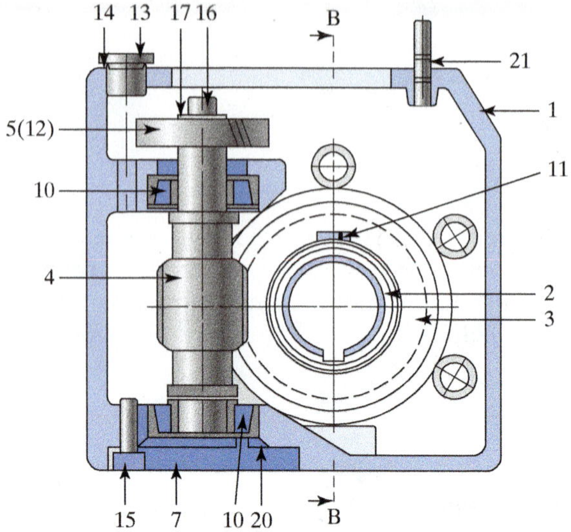

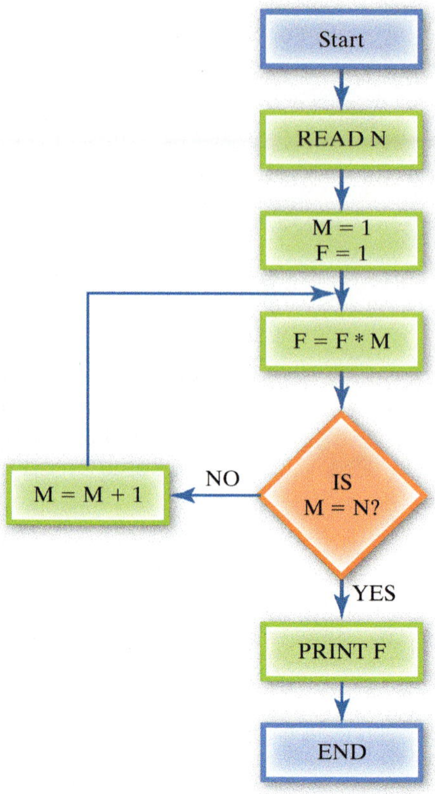

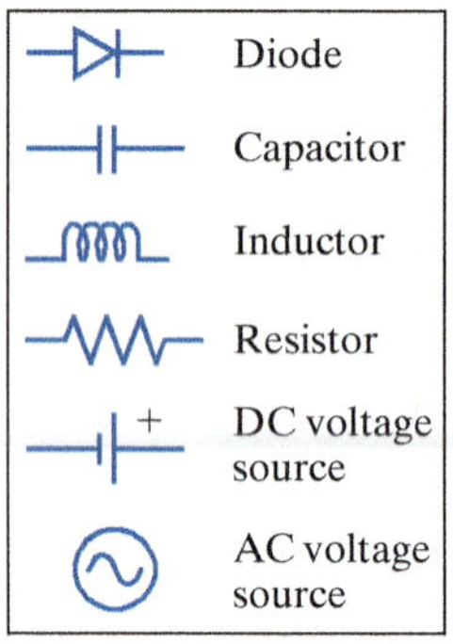

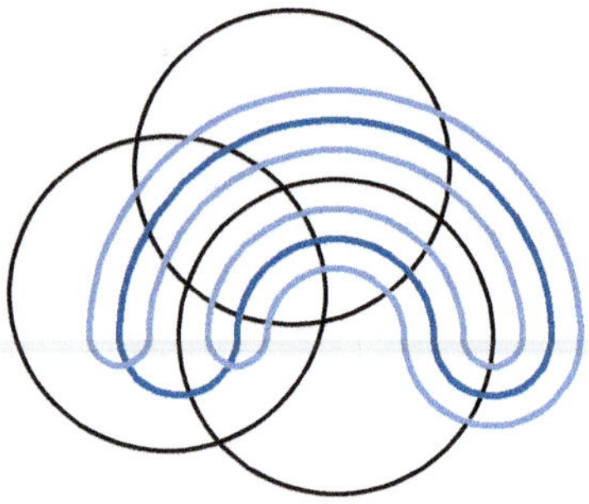

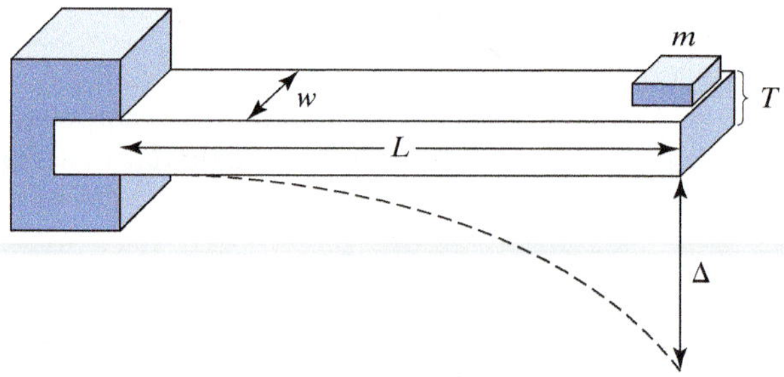

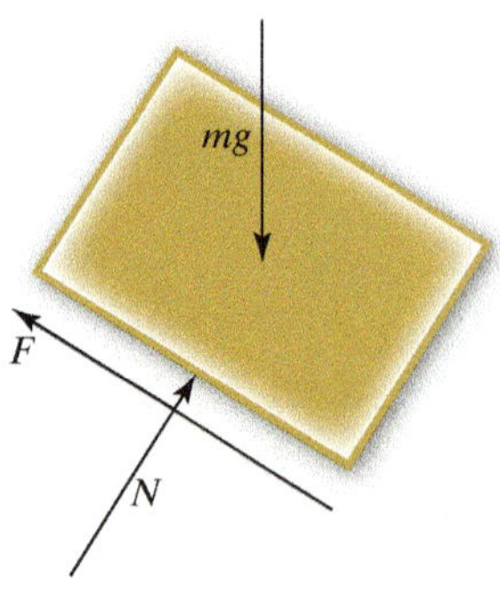

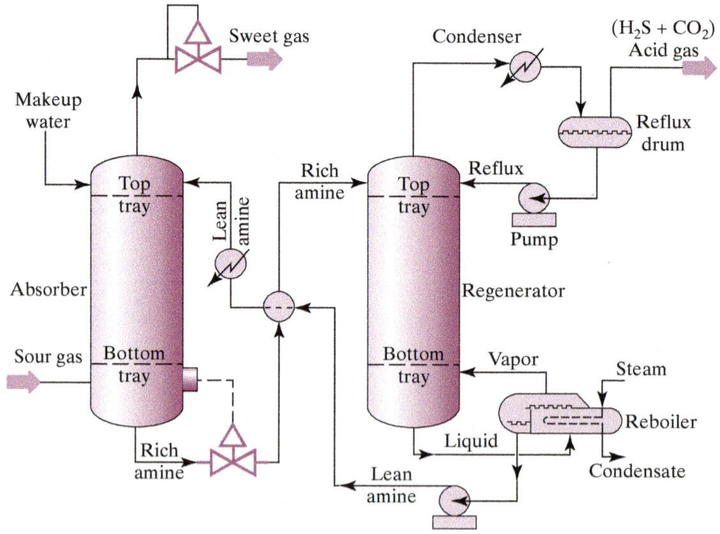

Figure 3-1 Sketches for seven problems.

Observations, Objectives

These can be in the form of simple statements, questions, or anything else that might acquaint you with the problem at hand. It often helps to divide your observations and objectives into several categories. Some of the easiest to remember are:

- Objective to be achieved
- Observations about the problem geometry (size, shape, etc.)
- Observations about materials and material properties (density, hardness, etc.)
- Observations about parameters not easily sketched (temperature, velocity, etc.)
- Other miscellaneous observations that might be pertinent

You will almost always find after writing down some observations that you actually know more about the problem than you originally thought.

Here are some typical examples:

Objectives

- Find the velocity, force, flow rate, time, pressure, etc., for a given situation.
- Profitably market the device for less than $25.
- Fit the device into a 12-cubic-inch box.

Observations

PROBLEM GEOMETRY

- The liquid has a free surface.
- The submerged plate is rectangular.
- The support is vertical.
- The tank is cylindrical.
- The cross-sectional area is octagonal.
- The orbit is elliptical.

MATERIALS AND MATERIAL PROPERTIES

- The gate is steel.
- The coefficient of static friction is 0.6.
- The specific gravity is 0.65.
- Ice will float in water.
- The alloy superconducts at 97 kelvin.
- The alloy melts at 543 degrees Fahrenheit.

OTHER PARAMETERS

- If depth increases, pressure increases.
- If temperature increases, resistance increases.
- The flow is steady.
- The fluid is a gas and is compressible.
- The pulley is frictionless.
- The magnetic field is decreasing.
- Temperature may not fall below 34 degrees Fahrenheit.

MISCELLANEOUS

- The force will act to the right.
- Gravity causes the ball to accelerate.
- The sphere is buoyant.
- Drag increases as the speed increases.

Remember to include those quantities whose value is zero! Often such quantities are hidden with terms such as:

- Constant (implies derivative $= 0$)
- Initially (at time $= 0$)
- At rest (no motion)
- Dropped (no initial velocity)
- At the origin (at zero position)
- Melts or Evaporates (changes phase, temperature is constant)

Finally, one of the most important reasons to make *many* observations is that you often will observe the "wrong" thing. For example, write down things as you read this:

> A bus contains 13 passengers.
> At the first stop, four get off and two get on.
> At the next stop, six get off and one gets on.
> At the next stop, nobody gets off and five get on.
> At the next stop, eight get off and three get on.
> At the next stop, one gets off.
> At the last stop, four get off and four get on.

After putting your pencil down and without looking again at the list, answer the following question:

QUESTION: How many stops did the bus make?

The lesson here is that often we may be observing the wrong thing.

List of Variables and Constants

Go over the observations previously determined and list the variables that are important. It may help to divide the list into several broad categories—those related to the geometry of the problem, those related to the materials, and a properties category—although for some types of problems those categories may not be appropriate. Include in your list the written name of the variable, the symbol used to represent the quantity, and, if the value of the variable is known, list the numeric value, including units. If a value is a constant you had to look up, record where you found the information.

Initial and Final Conditions

■ Initial temperature	(T_0)	60 [°F]
■ Initial radius	(r_i)	5 [cm]
■ Mass of the object	(m)	23 [kg]

Constants

■ Acceleration of gravity	(g)	32.2 [ft/s^2]
■ Ideal gas constant	(R)	8,314 [(Pa L)/(mol K)]

Geometry

■ Length of beam	(L)	[m]
■ Cross-sectional area of a pipe	(A)	[cm^2]
■ Volume of a reactor vessel	(V)	[gal]

Materials

- Steel
- Polyvinyl chloride (PVC)
- Plasma
- Gallium arsenide
- Medium-density balsa wood

PROPERTIES

■ Dynamic viscosity of honey	(μ)	2,500 [cP]
■ Density of PVC	(ρ)	1,380 [kg/m^3]
■ Spring constant	(k)	0.05 [N/m]
■ Specific gravity	(SG)	1.34

Equations

Only after completing the steps above (S-O-L-V) should you begin to think about the equations that might govern the problem. It is useful to make a list of the pertinent equations in a broad sense before listing specific expressions. For example:

- Conservation of energy
- Conservation of mass
- Conservation of momentum
- Frequency equations
- Ideal gas law
- Newton's laws of motion
- Stress–strain relations
- Surface areas of geometric solids
- Volumes of geometric solids
- Work, energy relations

You may need "subequations" such as:

- Distance = (velocity) (time)
- Energy = (power) (time)
- Force = (pressure) (area)
- Mass = (density) (volume)
- Voltage = (current) (resistance)
- Weight = (mass) (gravity)

For an equation, list the broad category of the equation (Hooke's law) and then the actual expression ($F = kx$) to help with problem recognition.

Do not substitute numerical values of the parameters into the equation right away. Instead, manipulate the equation algebraically to the desired form.

Manipulation

Most of the time, you need to manipulate pertinent equations before you can obtain a final solution. ***Do not substitute numerical values of the parameters into the equation right away.*** Instead, manipulate the equation algebraically to the desired form. Often you will discover terms that will cancel, giving you a simpler expression to deal with. By doing this, you will:

- Obtain general expressions useful for solving other problems of this type
- Be less likely to make math errors
- Be able to judge whether your final equation is dimensionally consistent
- Better understand the final result

The SOLVEM acronym does not contain a word or step for "numerical solution." In fact, this process helps you analyze the problem and obtain an expression or procedure so that you can find a numerical answer. The thought here is that engineers need training to be able to analyze and solve problems. If you can do everything except "substitute numbers" you are essentially finished—as an engineer, you will "be paid the big bucks" for analysis, not for punching a calculator.

When you have completed all the steps to SOLVEM, plug in values for the variables and constants and solve for a final answer. Be sure to use reasonableness. The final answer should include both a numeric value and its unit. In addition, it is often useful to write a sentence describing how the answer meets the objectives. Box your final answer for easy identification.

Repeated use of SOLVEM can help you develop a better "gut-level understanding" about the analysis of problems by forcing you to talk and think about the generalities of the problem before jumping in and searching for an equation into which you can immediately substitute numbers.

3.1.1 AVOIDING COMMON MISTAKES

Erroneous or argumentative thinking can lead to problem-solving errors. For example,

- **I can probably find a good equation in the next few pages.** Perhaps you read a problem and rifle through the chapter to find the proper equation so that you can start substituting numbers. You find one that looks good. You do not worry about whether the equation is the right one or whether the assumptions you made in committing to the equation apply to the present problem. You whip out your calculator and produce an answer. *Don't do this!*
- **I hate algebra, or I cannot do algebra, or I have got the numbers, so let us substitute the values right in.** Many problems become much simpler if you are willing to do a little algebra before trying to find a numerical solution. Also, by doing some manipulation first, you often obtain a general expression that is easy to apply to another problem when a variable is given a new value. By doing a little algebra, you can also often circumvent problems with different sets of units. *Do some algebra!*
- **It is a simple problem, so why do I need a sketch?** Even if you have a photographic memory, you will need to communicate with people who do not. It is usually much simpler to sort out the various parts of a problem if a picture is staring you right in the face. *Draw pictures!*
- **I do not have time to think about the problem, I need to get this stuff finished.** Well, most often, if you take a deep breath and jot down several important aspects of the problem, you will find the problem much easier to solve and will solve it correctly. *Take your time!*

EXAMPLE 3-1

Estimate how many miles of wire stock are needed to make 1 million standard paper clips.

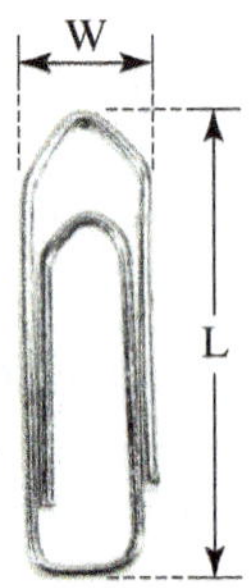

Sketch:

See the adjacent diagram.

Objective:

Determine the amount of wire needed to manufacture a million paper clips.

Observations:

- *Paper clips come in a variety of sizes.*
- *There are four straight segments and three semicircular sections in one clip.*
- *The three semicircular sections have slightly different diameters.*
- *The four straight sections have slightly different lengths.*

List of Variables and Constants:

- L — *Overall length of clip*
- W — *Overall width of clip*
- L_1, L_2, L_3, L_4 — *Lengths of four straight sections*

- D_1, D_2, D_3 — *Diameters of three semicircular sections*
- P_1, P_2, P_3 — *Lengths of three semicircular sections*
- A — *Total amount (length) of wire per clip*

Estimations and Assumptions:

- *Length of clip:* L = 1.5 in
- *Width of clip:* W = 3/8 in
- *Diameters from largest to smallest*
  - $D_1 = W = 3/8$ in
  - $D_2 = 5/6$ in
  - $D_3 = 1/4$ in
- *Lengths from left to right in sketch*
  - L_1 = To be calculated
  - $L_2 = 0.8$ in
  - $L_3 \approx L4 = 1$ in

Equations:

- *Perimeter of semicircle:* $P = \pi D/2$ *(half of circumference of circle)*
- $L_1 = L - D_1/2 - D_2/2$
- *Total length of wire in clip:* $A = L_1 + L_2 + L_3 + L_4 + P_1 + P_2 + P_3$

Manipulation:

In this case, none of the equations need to be manipulated into another form.

Length of longest straight side:	$L_1 = 1.5 - 3/16 - 5/32 \approx 1.2$ in
Lengths of semicircular sections:	$P_1 = \pi\,3/16 \approx 0.6$ in $P_2 = \pi\,5/32 \approx 0.5$ in $P_3 = \pi\,1/8 \approx 0.4$ in
Overall length for one clip:	$A = 1.2 + 0.8 + 1 + 1 + 0.6 + 0.5 + 0.4 = 5.5$ in/clip
Length of wire for 1 million clips:	$(5.5 \text{ in/clip})\ (1 \times 10^6 \text{clips}) = 5.5 \times 10^6$ in
Convert from inches to miles:	$(5.5 \times 10^6 \text{ in})\ (1 \text{ ft}/12 \text{ in})\ (1 \text{ mile}/5{,}280 \text{ ft}) \approx 86.8$ miles

---

One million, 1.5-inch paper clips require about 87 miles of wire stock.

IMPORTANT

Recall that you should manipulate the equations *before* inserting known values. Note that the final expression for 'total length of wire in clip' is given in terms of individual values L_1, L_2, L_3, L_4, P_1, P_2 and P_3. So you now have a general equation that can be solved for any values of these parameters. If you had begun substituting numbers into equations at the beginning and then wanted to obtain the same result for different starting values, you would have to resolve the entire problem.

EXAMPLE 3-2

Each of these items should be addressed by a team. List as many things about the specified items as you can determine by observation. One way to do this is to let each team member make one observation, write it down, and iteratively canvass the team until nobody can think of any more additions (it is fine to pass). Remember that you have five senses. Also note that observations are things you can actually detect during the experiment, *not* things you already know or deduce that cannot be observed.

To help, Examples A and B are given here before you do one activity (or more) on your own. Remember that not all observations will be important for a particular problem, but write them down anyway—they may trigger an observation that is important.

Example A: A loudspeaker reproducing music

- Electrical signal is sent to speaker.
- Speaker vibrates from electrical signal.
- Speaker is gray.
- A magnet is involved.
- Speaker is circular.
- Speaker is about 10 inches in diameter.
- Speaker diaphragm is made of paper; vibrating air creates sound.
- Gravity acts on speaker.
- Speaker is in a "box."

Example B: Drinking a soft drink in a can through a straw

- Liquid is cold.
- Straw is cylindrical.
- Can is cylindrical.
- Liquid assumes the shape of the container.
- Gravity opposes the rise of the liquid.
- Moving liquid has kinetic energy.
- As the liquid rises, it gains potential energy.
- Table supports the weight of the can, liquid, and straw.
- Can is opaque.
- Liquid is brown.
- Liquid is carbonated (carbonic acid).
- Liquid contains caffeine.
- Friction between straw and lips allows you to hold it.
- Plan view of the can is a circle.
- Silhouette of the can is the same from any direction.
- Silhouette of the can is a rectangle.
- Can is painted.
- Can is metal.
- Liquid surface is horizontal.
- Liquid surface is circular.

From the list below or others as selected by your teacher, list as many observations as you can about the following topic:

(a) An object provided by your teacher (placed on the desk)
(b) Candle placed on the desk and lit
(c) Ball dropped from several feet, allowed to bounce and come to rest

(d) A weight on a string pulled to one side and released; watched until it comes to rest
(e) Coin spinning on the desk
(f) Cup of hot coffee placed on the desk
(g) A glass of cold water placed on the desk
(h) Book pushed across the desk
(i) Large container guided smoothly up a ramp
(j) Ruler hanging over the desk
(k) Your computer (when turned off)
(l) Your chair
(m) Your classroom
(n) A weight tied on a string and twirled

Final Assignment: You have done several observation exercises. In these, you thought of observation as just a "stream of consciousness" with no regard to organization of your efforts. With your previous observation as a basis, generalize the search for observations into several (three to six or so) categories. The use of these categories should make the construction of a list of observations easier in the future.

3.2 ESTIMATION

Enrico Fermi was a Nobel laureate and one of many brilliant scientists and engineers involved in the Manhattan Project, which developed the first nuclear weapons during World War II. He wrote the following after witnessing the first test of an atomic bomb, called the Trinity Test (see Figure 3-2). Of particular note is the final paragraph.

On the morning of the 16th of July, I was stationed at the Base Camp at Trinity in a position about ten miles from the site of the explosion.

The explosion took place at about 5:30 A.M. I had my face protected by a large board in which a piece of dark welding glass had been inserted. My first impression of the explosion was the very intense flash of light, and a sensation of heat on the parts of my body that were exposed. Although I did not look directly towards the object, I had the impression that suddenly the countryside became brighter than in full daylight.

I subsequently looked in the direction of the explosion through the dark glass and could see something that looked like a conglomeration of flames that promptly started rising. After a few seconds the rising flames lost their brightness and appeared as a huge pillar of smoke with an expanded head like a gigantic mushroom that rose rapidly beyond the clouds probably to a height of the order of 30,000 feet. After reaching its full height, the smoke stayed stationary for a while before the wind started dispersing it.

About 40 seconds after the explosion the air blast reached me. I tried to estimate its strength by dropping from about six feet small pieces of paper before, during and after the passage of the blast wave. Since at the time there was no wind I could observe very distinctly and actually measure the displacement of the pieces of paper that were in the process of falling while the blast was passing. The shift was about 2½ meters, which, at the time, I estimated to correspond to the blast that would be produced by ten thousand tons of T.N.T.

**Citation: U.S. National Archives, Record Group 227, OSRD-S1 Committee, Box 82 folder 6, "Trinity."*

Figure 3-2 The first atomic bomb test, Alamogordo, New Mexico.

Before the test, no one knew what would happen. Speculation among the many people involved concerning the results of the test ranged from nothing (no explosion at all) to setting the planetary atmosphere on fire and destroying all life on Earth. When all the data from the blast were analyzed, the true strength of the blast was calculated to be 19 kilotons. By simply observing the behavior of falling bits of paper 10 miles from ground zero, Fermi's estimation of 10 kilotons was in error by less than a factor of 2.

Fermi problems

After the war, Fermi taught at the University of Chicago, where he was noted for giving his students problems in which so much information was missing that a solution seemed impossible. Such problems have been named **Fermi problems**, and in general they require the person considering them to determine an answer with far less information than would really be necessary to calculate an accurate value. Engineers are often faced with solving problems for which they do not have all the information. They must be adept at making initial estimates. This skill helps them identify critical information that is missing, develop their reasoning skills to solve problems, and recognize what a "reasonable" solution will look like.

Most practical engineering problems are better defined than Fermi problems and can be estimated more easily and accurately in general. The following are just a few examples of real engineering problems. See if you can estimate answers for these problems. These should be done without reference to the Web or any other source of information.

- How many cubic yards of concrete are needed to pave 1 mile of interstate highway (two lanes each direction)?
- How many feet of wire are needed to connect the lighting systems in an automobile?
- What is the average flow rate in gallons per minute of gasoline moving from the fuel tank to the fuel injectors in an automobile cruising at highway speed?

An accomplished engineer knows the answer to most problems before doing any calculations. This does not mean an answer to three significant figures, but a general idea of the range of values that would be reasonable. For example, if you throw a baseball as high as possible, how long will it take for the ball to hit the ground? Obviously, an answer of a few milliseconds is unreasonable, as is several months. Several seconds seems more realistic. When you actually do a calculation to determine this time, you should ask yourself, "Is my answer reasonable?"

Sample Fermi Problems

- Estimate the total number of hairs on your head.
- Estimate the number of drops of water in all of the Great Lakes.
- Estimate the number of piano tuners in New York City.

EXAMPLE 3-3

Every year, numerous people run out of fuel while driving their vehicle on the road. Determine how many gallons of gasoline are carried to vehicles with empty fuel tanks each year in the United States so that the vehicle can be driven to the nearest gas station. Do this without reference to any other material, such as the Internet or reference books.

Estimations

In almost all cases, the first step is to estimate unknown pieces of information that are not available. In general, there are numerous paths to a solution, and different people may arrive at different answers. Often, the answer arrived at is only accurate to within an order of magnitude (a factor of 10) or less. Nonetheless, such problems can provide valuable insight not only into the problem itself but also into the nature of problem solving in general.

Remember, these are estimates, not accurate values. Someone else making these estimates might make different assumptions.

- *Number of people in the United States: 500,000,000 persons.*
- *Fraction of people in the United States that drive: drivers per 10 persons.*
- *Average times a person runs out of gas per year: one "out of gas" per 4 years per driver.*
- *Fraction of "out of gas" incidents in which gas is brought to the car (rather than pushing or towing the car to a station): 24 "bring gas to car" per 25 "out of gas".*
- *Average amount of gas carried to car: 1.5 gallons per "bring gas to car".*

Calculation

This is where you combine your estimates to arrive at a solution. In the process, you may realize that you need further information to complete the computation.

$$\text{Drivers in United States} = (5 \times 10^8 \text{ people})\left(\frac{7 \text{ drivers}}{10 \text{ people}}\right) = 3.5 \times 10^8 \text{ drivers}$$

$$\text{Number out of gas per year} = (3.5 \times 10^8 \text{ drivers})\left(\frac{1 \text{ out of gas}}{(4 \text{ years})(1 \text{ driver})}\right)$$

$$= 8.75 \times 10^7 \frac{\text{out of gas}}{\text{year}}$$

$$\text{Number bring gas to car per year} = \left(\frac{8.75 \times 10^7 \text{out of gas}}{\text{year}}\right)\left(\frac{24 \text{ bring to car}}{25 \text{ out of gas}}\right)$$

$$= 8.4 \times 10^7 \frac{\text{bring gas to car}}{\text{year}}$$

$$\text{Amount of gas to cars} = \left(\frac{8.4 \times 10^7 \text{ bring gas to car}}{\text{year}}\right)\left(\frac{1.5 \text{ gallons of gas}}{\text{bring gas to car}}\right)$$

$$= 1.26 \times 10^8 \frac{\text{gallons}}{\text{year}}$$

Thus, we have estimated that about 125 million gallons of gas are taken to "out of fuel" vehicles each year in the United States. It would be perfectly valid to give the answer as "about 100 million" gallons, since we probably have only about one significant digit worth of confidence in our results, if that.

A few things to note about this solution:

- *"Units" were used on all numerical values, although some of these units were somewhat contrived (e.g., "out of gas") to meet the needs of the problem. Keeping track of the units is critical to obtaining correct answers and will be highly emphasized, not only in this text but also throughout your engineering education and career.*

- *The units combine and cancel according to the regular algebraic rules. For example, in the first computation, the unit "persons" appeared in both the numerator and the denominator, and thus canceled, leaving "drivers."*
- *Rather than the computation being combined into one huge string of computations, it was broken into smaller pieces, with the results from one step used to compute the next step. This is certainly not an immutable rule, but for long computations it reduces careless errors and makes it easier to understand the overall flow of the problem.*

General Hints for Estimation

As you gain more knowledge and experience, the types of problems you can estimate will become more complicated. Here we give you a few hints about making estimates.

- **Try to determine the accuracy required.**
  - Is order of magnitude enough? Is ±25%?
  - What level of accuracy is needed to calculate a satellite trajectory?
  - What level of accuracy is needed to determine the amount of paint needed to paint a specified classroom?

orders of magnitude

The term "**orders of magnitude**" is often used when comparing things of very different scales, such as a small rock and a planet. By far the most common usage refers to factors of 10; for example, three orders of magnitude refer to a difference in scale of $10^3 = 1{,}000$. If we wanted to consider the order of magnitude between 10,000,000 and 1,000, we would calculate the logarithm of each value ($\log(10{,}000{,}000) = 7$ and $\log(1{,}000) = 3$), thus there are $7 - 3 = 4$ orders of magnitude difference between 10,000,000 and 1,000.

- Remember that a **"ballpark" value is often good enough** for an input parameter.
  - What is the square footage of a typical house?
  - What is the maximum high temperature to expect in Dallas, Texas, in July?
  - What is the typical velocity of a car on the highway?
- Always ask yourself **if it is better to err on the high side or the low side.**
  - **Safety and practical considerations.** Will a higher or lower estimate result in a safer or more reliable result?
    - If estimating the weight a bridge can support, it is better to err on the low side, so that the actual load it can safely carry is greater than the estimate.
    - For the bridge mentioned above, if estimating the load a single beam needs to support, it is better to err on the high side, thus giving a stronger beam than necessary. Be sure you understand the difference between these two points.
  - **Estimate improvement.** Can the errors cancel each other?
    - If estimating the product of two numbers, if one of the terms is rounded low, the other should be rounded high to counteract for the lower term.
    - If estimating a quotient, if you round the numerator term on the low side, should the denominator term be rounded low or high?

- Do not get bogged down with **second-order or minor effects**.
  - If estimating the mass of air in the classroom, do you need to correct for the presence of furniture?
  - In most instances, can the effect of temperature on the density of water be neglected?

The best way to develop your ability to estimate is through experience. An experienced painter can more easily estimate how much paint is needed to repaint a room because experience will have taught the painter such things as how many coats of one paint color it will take to paint over another, how different paint brands differ in their coverage, and how to estimate surface area quickly. In *Outliers: The Story of Success*, Malcolm Gladwell provides examples from diverse career pathways that demonstrate 10,000 hours of practice are required to develop world-class expertise in any area. Fortunately for aspiring engineers, much of this expertise can be developed starting at a young age and outside of formal schooling. For example, how many hours have you spent observing the effects of gravity? Of course, some important engineering concepts stem from phenomena that are not so easily observed, and some lend themselves to misinterpretation. As a result, it helps to have a systematic approach to developing estimates—particularly where we have less experience to guide us. Estimating an approximate answer of a calculation including known quantities is a valuable skill—such as approximating the square root of 50 as about 7, approximating the value of pi as 3 for quick estimates, etc. These mathematical approximations, however, assume that you have all the numbers to begin with, and that you can use shortcuts to estimate the precisely calculated answer to save time or as a check against your more carefully calculated answer. Estimation is discussed here in a broader sense—estimating quantities that cannot be known, are complicated to measure, or are otherwise inconvenient to obtain. It is in these cases that the following strategies are recommended.

Estimation by Analogy

One useful strategy for estimating a quantity is by comparison to something else we have measured previously or otherwise know the dimension of. The best way to prepare for this approach to estimation is to learn a large number of comparison measures for each type of quantity you might wish to estimate. Each of these comparison measures becomes an anchor point on that scale of measurement. This book provides some scale anchors for various measurable quantities—particularly in the case of power and energy, concepts with which many people struggle.

Estimation by Aggregation

Another useful strategy is to estimate the quantity of something by adding up an estimate of its parts. This can involve multiplication in the case of a number of similarly sized parts, such as estimating the size of a tile by comparing it to your foot (estimation by analogy), counting the number of floor tiles across a room, and multiplying to estimate the total length of the room. In other cases, aggregation may involve adding together parts that are estimated by separate methods. For example, to estimate the volume of a two-scoop ice cream cone, you might estimate the volume of the cone and then separately estimate the volume of each scoop assuming they are each spheres.

● EXAMPLE 3-4

Estimate by aggregation the amount of money students at your school spend on pizza each year.

Ask students around you how often they purchase a pizza and how much it costs;
Convert this estimate into a cost per week;
Multiply your estimate by the number of weeks in an academic year;
Multiply that result by the number of students at your school.

Estimation by Upper and Lower Bounds

An important part of estimating is keeping track of whether your estimate is high or low. In the earlier example of estimating the volume of a two-scoop ice cream cone, we would have over-estimated, because one of the scoops of ice cream is pressed inside the cone. The effect of pressing the scoops together is not that important, because the same amount of ice cream is still there, but if the scoop is pressed into an ellipsoid, it may be difficult to estimate the original radius of the scoop.

Engineers frequently make "conservative" estimates, which consider the "worst-case" scenario. Depending on the situation, the worst case may be a lower limit (such as estimating the strength of a structure) or an upper limit (such as estimating how much material is needed for a project).

Estimation Using Modeling

In cases that are more complicated or where a more precise estimate is required, mathematical models and statistics might be used. Sometimes dimensionless quantities are useful for characterizing systems, sometimes modeling the relationship of a small number of variables is needed, and at other times, extrapolating even a single variable from available data is all that is needed to make an estimate.

3.2.1 SIGNIFICANT FIGURES

significant figures

decimal places

Significant figures or "sig figs" are the digits considered reliable as a result of measurement or calculation. This is not to be confused with the number of digits or decimal places. The number of **decimal places** is simply the number of digits to the right of the decimal point. Example 3-5 on the next page illustrates these two concepts.

The Meaning of "Significant"

All digits other than zero are automatically considered significant. Zero is significant when:

- It appears between two nonzero numbers
  - 306 has three significant figures
  - 5.006 has four significant figures
- It is a "terminal" zero in a number with a decimal point
  - 2.790 has four significant figures
  - 2000.0 has five significant figures

EXAMPLE 3-5

Number	Decimal Places	Significant Figures
376	0	3
376.0	1	4
376,908	0	6
3,760,000	0	3
3,760,000.	0	7
0.376	3	3
0.37600	5	5
0.0037600	7	5
376×10^{-6}	0*	3

** There is no universal agreement concerning whether numbers in scientific or engineering notation should be considered to have the number of decimal places indicated in the mantissa (as shown), or the number that would be present if the number were written out in standard decimal notation (6 in the last example above).*

For those who do not know the term mantissa, the two parts of a number expressed in either scientific or engineering notation are the mantissa and the exponent. The **mantissa** *is the part that gives the numerical values of the significant figures; the* **exponent** *specifies the location of the decimal point, thus the magnitude of the overall number. In the last row of the table above, the mantissa is 376, the exponent is −6. The mantissa can also contain a decimal point, for example, 3.76×10^{-4}: in this case the mantissa is 3.76 and the exponent is −4.*

mantissa
exponent

Zero is not significant when:

- It is used to fix a decimal place
  - 0.0456 has three significant figures
- It is used in integers without a decimal point that could be expressed in scientific notation without including that zero
  - 2000 has one significant figure (2×10^3)
  - 35,100 has three significant figures (3.51×10^4)

Calculation Rules

As an engineer, you will likely find that your job involves the design and creation of a product. It is imperative that your calculations lead to the design being "reasonable." It is also important that you remember that others will use much of your work, including people with no technical training.

Engineers must not imply more accuracy in their calculations than is reasonable. To assist in this, there are many rules that pertain to using the proper number of "significant figures" in computations. These rules, however, are cumbersome and tedious. In your daily life as an engineer, you might use these rules only occasionally. The rules given below provide a reference if you ever need them; in this text, however, you are simply expected to be *reasonable*. In general, asking yourself if the number of significant figures in your answer is reasonable is usually sufficient. However, it is a good idea to be familiar with the rules, or at least know how to find them and use them if the need ever arises.

Multiplication and Division

A quotient or product should contain the same number of significant figures as the number with the fewest significant figures. Exact conversions do not affect this rule.

EXAMPLE 3-6

$$(2.43)(17.675) = 42.95025 \cong 43.0$$

- *2.43 has three significant figures.*
- *17.675 has five significant figures.*
- *The answer (43.0) has three significant figures.*

Addition and Subtraction

The answer resulting from an addition or subtraction should show significant figures only as far to the right as the least precise number in the calculation. For addition and subtraction operations, the "least precise" number should be that containing the lowest number of decimal places.

EXAMPLE 3-7

$$1725.463 + 489.2 + 16.73 = 1931.393$$

- *489.2 is the least precise.*
- *The answer should contain one decimal place: 1931.4.*

Reasonableness

In the preceding discussion of estimation, the word *reasonable* was mentioned in several places. We consider two types of reasonableness in answers to problems in this section.

- **Physically reasonable.** Does the answer make sense in light of our understanding of the physical situation being explored or the estimates that we can make?
- **Reasonable precision.** Is the number of digits in the answer commensurate with the level of accuracy and precision available to us in the parameters of the problem?

When Is Something Physically Reasonable?

Here are a few hints to help you determine if a solution to a problem is physically reasonable.

- First, ask yourself if the answer makes sense in the physical world.
  - You determine that the wingspan of a new airplane to carry 200 passengers should be 4 feet. This is obvious rubbish.
  - You determine that a sewage treatment plant for a community of 10,000 people must handle 100,000 pound-mass of sewage effluent per day. Since a gallon of water weighs about 8 pound-mass, this is about 1.25 gallons per person per day, which is far too low to be reasonable.

- You are designing playground equipment, including a swing set. The top support (pivot point) for the swings is 10 feet above the ground. You calculate that a child using one of the swings will make one full swing (forward, then backward) in 3.6 seconds. This seems reasonable.

- If the final answer is in units for which you do not have an intuitive feel, convert to units for which you do have an intuitive feel.
  - You calculate the speed of a pitched baseball to be 2×10^{13} millimeters per year. Converting this to miles per hour gives over 1,400 miles per hour, obviously too fast for a pitched baseball.
  - You are told by a colleague that a ¾-inch pipe supplying water to a chemical process delivers 10 cubic meters of water per day. Converting to gallons per minute gives 1.8 gallons per minute, which seems completely reasonable.

3.2.2 PRECISION AND ACCURACY

First, we need to differentiate between the two terms: *accurate* and *precise*. Although laypersons tend to use these words interchangeably, an aspiring engineer should understand the difference in meaning as applied to measured or calculated values.

accuracy

Accuracy is a measure of how close a calculation or measurement (or an average of a group of measurements) is to the actual value. For measured data, if the average of all measurements of a specific parameter is close to the actual value, then the measurement is accurate, whether or not the individual measurements are close to each other. The difference between the measured value and the actual value is the error in the measurement. Errors come about due to lack of accuracy of measuring equipment, poor measurement techniques, misuse of equipment, and factors in the environment (e.g., temperature or vibration).

repeatability

Repeatability is a measure of how close together multiple measurements of the same parameter are, whether or not they are close to the actual value.

precision

Precision is a combination of accuracy and repeatability, and is reflected in the number of significant figures used to report a value. The more significant figures, the more precise the value is, assuming it is also accurate.

To illustrate these concepts, consider the distribution of hits on a standard "bulls-eye" target. Figure 3-3 shows all four combinations of accuracy and repeatable.

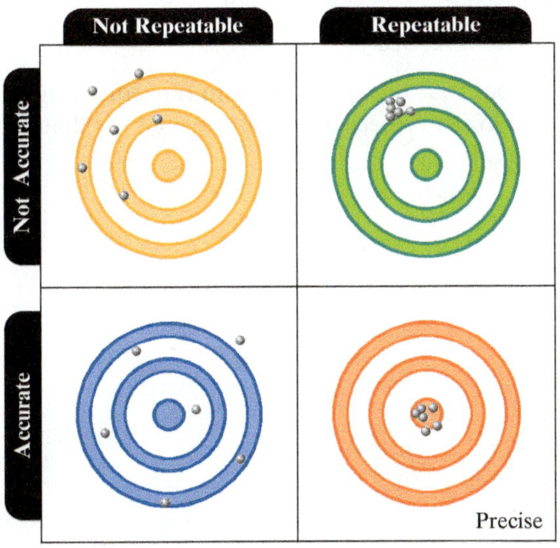

Figure 3-3 Precision and accuracy.

- Neither repeatable nor accurate.
- Repeatable, but not accurate.
- Accurate, but not repeatable.
- Both repeatable and accurate. This is called precise.

When considering if the precision of a numeric value is reasonable, always ask yourself the following questions:

- How many significant figures do I need in my design parameters? The more precision you specify in a design, the more it will cost and the less competitive it will be unless the extra precision is really needed.
  - You can buy a really nice 16-ounce hammer for about \$20. If you wanted a 16 ± 0.0001-ounce hammer, it would probably cost well over a hundred dollars, possibly thousands.

- What are the inherent limitations of my measuring equipment? How much is the measurement affected by environmental factors, user error, etc.?
  - The plastic ruler you buy at the discount store for considerably less than a dollar will measure lengths up to 12 inches with a precision of better than 0.1 inch, but not as good as 0.01 inch. On the other hand, you can spend a few hundred dollars for a high-quality micrometer and measure lengths up to perhaps 6 inches to a precision of 0.0001 inch.

You should report values in engineering calculations in a way that does not imply a higher level of accuracy than is known. Use the fewest number of decimal places without reducing the usefulness of the answer. Several examples, given below, illustrate this concept.

- We want to compute the area (A) of a circle. We measure the diameter (D) as 2.63 centimeters. We calculate

$$A = \frac{1}{4}\pi D^2 = \frac{1}{4}\pi(2.63)^2 = 5.432521 \text{ cm}^2$$

  The value of π is known to as many places as we desire, and ¼ is an exact number. It seems reasonable to give our answer as 5.4 or 5.43 square centimeters since the original data of diameter is given to two decimal places. Most of the time, reporting answers with two to four significant digits is acceptable and reasonable.

- We want to compute the area (A) of a circle. We measure the diameter (D) as 0.0024 centimeters. We calculate

$$A = \frac{1}{4}\pi D^2 = \frac{1}{4}\pi(0.0024)^2 = 0.0004446 \text{ cm}^2$$

  If we keep only two decimal places, our answer would be 0.00 square centimeters, which has no meaning. Consequently, when reporting numerical results, particularly those with a magnitude much smaller than 1, we use significant figures, not decimal places. It would be reasonable to report our answer as 0.00044 or 4.4×10^{-4} square centimeters.

- We want to determine a linear relationship for a set of data, using a standard software package such as Excel®. The program will automatically generate a linear relationship based on the data set. Suppose that the result of this exercise is

$$y = 0.50236x + 2.0378$$

  While we do not necessarily have proof that the coefficients in this equation are nice simple numbers or even integers, a look at the equation above suggests that the linear relationship should probably be taken as

$$y = 0.5x + 2$$

3.3 UNITS

As aspiring engineers you must learn to distinguish among many terms that laymen tend to use interchangeably. You must also understand the technical meaning of terms that are misunderstood by those untrained in science or engineering. One term that is often misunderstood is "dimension." To most people, a dimension refers to a straight line (length, one dimension), a flat surface (area, two dimensions), or a solid object (volume, three dimensions). Some slightly more educated folks might include time as a fourth dimension. The number of concepts classified as dimensions is far broader.

dimension
unit

A **dimension** is a measurable physical idea; it generally consists solely of a word description with no numbers. A **unit** allows us to quantify a dimension—to state a number describing how much of that dimension exists in a specific situation. Units are defined by convention and related to an accepted standard.

- Length is a dimension. There are many units for length, such as mile, foot, meter, light-year, and fathom.
- Time is a dimension. There are many units for time, such as second, minute, hour, day, fortnight, year, and century.
- Temperature is a dimension. There are many units for temperature, such as Celsius, Fahrenheit, and kelvin.

The Metric System

The SI system (Le Système International d'Unités), commonly known as the metric system, is the standard set of units for most of the world. Originally developed by French scientists under King Louis XVI, the SI system was finalized by the international scientific community as the standard unit system in 1971. This system defines seven base units, from which all others are derived. The seven base units and their corresponding fundamental dimensions are shown in Table 3-1.

Table 3-1 Fundamental dimensions and base units

Dimension	Symbol	Unit	Symbol
Length	L	meter	m
Mass	M	kilogram	kg
Time	T	second	s
Temperature	Θ	kelvin	K
Amount of substance	N	mole	mol
Light intensity	J	candela	cd
Electric current	I	ampere	A

SI (Système international, or International System) Prefixes

SI prefix

The SI system, or the metric system, is based upon multiples of 10. By using an **SI prefix** when reporting numbers, we avoid scientific notation or long strings of zeros. For example, instead of saying, "The distance to Atlanta is 198,000 meters," we would say, "The distance to Atlanta is 198 kilometers."

For a list of SI prefixes, refer to Table 3-2. Note that the abbreviations for all SI prefixes from kilo- down to yocto- are lowercase, whereas from Mega- up to Yotta- are uppercase.

Determining the appropriate SI prefix to use becomes simple when the number is placed in engineering notation: just examine the exponent.

Table 3-2 SI prefixes (example: 1 millimeter [mm] = 1 × 10^{-3} meters [m])

Numbers Less than One			Numbers Greater than One		
Power of 10	Prefix	Abbreviation	Power of 10	Prefix	Abbreviation
10^{-1}	deci-	d	10^{1}	deca-	da
10^{-2}	centi-	c	10^{2}	hecto-	h
10^{-3}	milli-	m	10^{3}	kilo-	k
10^{-6}	micro-	μ	10^{6}	Mega-	M
10^{-9}	nano-	n	10^{9}	Giga-	G
10^{-12}	pico-	p	10^{12}	Tera-	T
10^{-15}	femto-	f	10^{15}	Peta-	P
10^{-18}	atto-	a	10^{18}	Exa-	E
10^{-21}	zepto-	z	10^{21}	Zetta-	Z
10^{-24}	yocto-	y	10^{24}	Yotta-	Y

EXAMPLE 3-8

Express the following values using scientific notation, and using the correct SI prefix.

Standard	Scientific	With Prefix
(a) 43,480,000 m	4.348×10^{7} m	43.48 Mm
(b) 0.0000003060 V	3.060×10^{-7} V	306.0 nV
(c) 9,860,000,000 J	9.86×10^{9} J	9.86 GJ
(d) 0.0351 s	3.51×10^{-2} s	35.1 ms

Other Unit Systems

Prior to the adoption of the SI unit system by the scientific community, several other systems of units were used and are still used today, particularly in the United States. The other countries that use non-SI units are Liberia and Myanmar. Great Britain officially converted to metric in 1965, but it is still common there to see nonmetric units used in communications for the general public.

It is important to know how to convert between all unit systems. Table 3-3 compares several systems. The system listed as AES (American Engineering System) is in common use by the general public in the United States. The USCS (United States Customary System) is commonly called "English" units.

Table 3-3 Comparison of unit system, with corresponding abbreviations

Dimension	SI (MKS)	AES	USCS
Length {L}	meter [m]	foot [ft]	foot [ft]
Mass {M}	kilogram [kg]	pound-mass [lb_m]	slug
Time {T}	second [s]	second [s]	second [s]
Relative temperature {Θ}	Celsius [°C]	Fahrenheit [°F]	Fahrenheit [°F]
Absolute temperature {Θ}	kelvin [K]	Rankine [°R]	Rankine [°R]

3.3.1 CONVERSION PROCEDURE FOR UNITS

We use conversion factors to translate from one set of units to another. This must be done correctly and consistently to obtain the right answers.

LENGTH

1 m = 3.28 ft

1 km = 0.621 mi

1 ft = 12 in

1 in = 2.54 cm

1 mi = 5,280 ft

1 yd = 3 ft

Let us examine the conversions found for the dimension of length, as shown in the box, beginning with the conversion: 1 meter [m] = 3.28 feet [ft]. By dividing both sides of this equation by 3.28 feet, we obtain

$$\frac{1\text{ m}}{3.28\text{ ft}} = 1$$

or in other words, "There is 1 meter per 3.28 feet." If we divide both sides of the original expression by 1 meter, we obtain

$$1 = \frac{3.28\text{ ft}}{1\text{ m}}$$

or in other words, "In every 3.28 feet there is 1 meter."

The number 1 is dimensionless, a pure number. *We can multiply any expression by 1 without changing the expression.* We do this so as to change the units to the standard we desire.

For example, on a trip we note that the distance to Atlanta is 123 miles [mi]. How many kilometers [km] is it to Atlanta? From the conversion table, we can find that 1 kilometer [km] = 0.621 miles [mi], or

$$1 = \frac{1 \text{ km}}{0.621 \text{ mi}}$$

By multiplying the original quantity of 123 miles by 1, we can say

$$(123 \text{ mi})(1) = (123 \text{ mi})\left(\frac{1 \text{ km}}{0.621 \text{ mi}}\right) = 198 \text{ km}$$

Note that we could have multiplied by the following relationship:

$$1 = \frac{0.621 \text{ mi}}{1 \text{ km}}$$

We would still have multiplied the original answer by 1, but the units would not cancel and we would be left with an awkward, meaningless answer.

$$(123 \text{ mi})(1) = (123 \text{ mi})\left(\frac{0.621 \text{ mi}}{1 \text{ km}}\right) = 76 \frac{\text{mi}^2}{\text{km}}$$

As a second example, we are designing a reactor system using 2-inch [in] diameter plastic pipe. The design office in Germany would like the pipe specifications in units of centimeters [cm]. From the conversion table, we find that 1 inch [in] = 2.54 centimeters [cm], or

$$1 = \frac{1 \text{ in}}{2.54 \text{ cm}}$$

By multiplying the original quantity of 2 inches by 1, we can say

$$(2 \text{ in})(1) = (2 \text{ in})\left(\frac{2.54 \text{ cm}}{1 \text{ in}}\right) = 5 \text{ cm}$$

In a final example, suppose a car travels at 40 miles per hour (abbreviated mph). Stated in words, "a car traveling at a rate of 40 mph will take 1 hour to travel 40 miles if the velocity remains constant." By simple arithmetic this means that the car will travel 80 miles in 2 hours or 120 miles in 3 hours. In general,

$$\text{Distance} = (\text{velocity})(\text{time elapsed at that velocity})$$

Suppose the car is traveling at 40 mph for 6 minutes. How far does the car travel? Simple calculation shows

$$\text{Distance} = (40)(6) = 240$$

Without considering units, the preceding example implies that if we drive our car at 40 mph, we can cover the distance from Charlotte, North Carolina, to Atlanta, Georgia, in 6 minutes! What is wrong? Note that the velocity is given in miles per hour, and the time is given in minutes. We need to apply the conversion factor that 1 hour = 60 minutes. If the equation is written with consistent units attached, we get

$$\text{Distance} = \left(\frac{40 \text{ mi}}{\text{h}}\right)\left(\frac{6 \text{ min}}{} \middle| \frac{1 \text{ h}}{60 \text{ min}}\right) = 4 \text{ mi}$$

It seems more reasonable to say "traveling at a rate of 40 miles per hour for a time period of 6 minutes will allow us to go 4 miles."

Unit Conversion Procedure

1. Write the value and unit to be converted.
2. Write the conversion formula between the given unit and the desired unit.
3. Make a fraction, equal to 1, of the conversion formula in Step 2, such that the original unit in Step 1 is located either in the denominator or in the numerator, depending on where it must reside so that the original unit will cancel.
4. Multiply the term from Step 1 by the fraction developed in Step 3.
5. Cancel units, perform mathematical calculations, and express the answer in "reasonable" terms (i.e., not too many decimal places).

Conversions Involving Multiple Steps

Sometimes, more than one conversion factor is needed. We can multiply by several conversion factors, each one of which is the same as multiplying by 1, as many times as needed to reach the desired result. For example, suppose we determined that the distance to Atlanta is 123 miles [mi]. How many yards [yd] is it to Atlanta? From the conversion table, we do not have a direct conversion between miles and yards, but we see that both can be related to feet. We can find that 1 mile [mi] = 5,280 feet [ft], or

$$1 = \frac{5{,}280 \text{ ft}}{1 \text{ mi}}$$

We can also find that 1 yard [yd] = 3 feet [ft], or

$$1 = \frac{1 \text{ yd}}{3 \text{ ft}}$$

By multiplying the original quantity of 123 miles by 1 using the first set of conversion factors, we can say:

$$(123 \text{ mi})(1) = (123 \text{ mi})\left(\frac{5{,}280 \text{ ft}}{1 \text{ mi}}\right) = 649{,}440 \text{ ft}$$

If we multiply by 1 again, using the second set of conversion factors and applying reasonableness:

$$(649{,}440 \text{ ft})(1) = (649{,}440 \text{ ft})\left(\frac{1 \text{ yd}}{3 \text{ ft}}\right) = 216{,}000 \text{ yd}$$

This is usually shown as a single step:

$$(123 \text{ mi})\left(\frac{5{,}280 \text{ ft}}{1 \text{ mi}}\right)\left(\frac{1 \text{ yd}}{3 \text{ ft}}\right) = 216{,}000 \text{ yd}$$

LESSONS OF THE MARS CLIMATE ORBITER

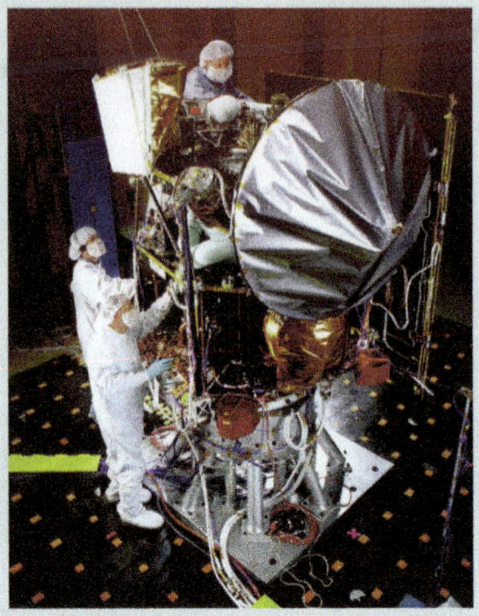

Some of you may have heard that the loss of the Mars Climate Orbiter (MCO) spacecraft in 1999 was due to a unit conversion error. The complete story is rather more complicated and illustrates a valuable lesson in engineering design. Most engineering failures are not due to a single mistake, since built-in redundancies and anticipation of failure modes make this unlikely. Three primary factors (plus bad luck) conspired to send the MCO off course.

First, the spacecraft was asymmetrical, with the body of the spacecraft on one side and a large solar panel on the other. You might think shape is not an issue in the vacuum of space, but in fact it is, and the NASA engineers were well aware of the potential problems. The panel acted like a sail, causing the craft to slowly change its orientation and requiring the MCO to make occasional small corrections by firing thrusters onboard the craft. This was a perfectly manageable "problem."

Second, the software on the spacecraft expected thruster data in SI units, requiring the force expressed in newtons. On the Earth, a separate system calculated and sent instructions to the MCO concerning when and how long to fire its thrusters. The Earth-based system relied on software from an earlier Mars mission, and the thruster equations had to be modified to correct for the thrusters used on the new spacecraft. The original software had been written correctly, with the conversion factor from pound-force to newtons included. However, this conversion was neither documented nor obvious from the code, being buried in the equations. When the equations were rewritten, the programmers were unaware of the conversion factor and it was left out of the new code. This sent incorrect thruster-firing data to the MCO, specifically being too small by a factor of 4.45. This problem alone was manageable by comparing the calculated trajectory with tracking data.

Finally, after the third trajectory correction, the MCO entered "safe mode" while adjusting the solar panel, indicating a fault on the craft. At about the same time, the preliminary indications that the spacecraft trajectory was flawed began to come in. Unfortunately, the engineers spent the next several weeks trying to determine what caused the craft to enter safe mode, falsely assuming the preliminary trajectory data was in error and waiting for longer-term tracking to give a better estimate. In the end, the spacecraft arrived at Mars about 100 kilometers off course.

Here is where the bad luck comes in. Other configurations of the craft or trajectory might have caused the 100 kilometer error to be away from Mars or parallel to the surface, in which case the trajectory could have been corrected later. Unfortunately, the trajectory was 100 kilometers lower than expected, and the MCO was probably destroyed by heating and stresses as it plunged through the Martian atmosphere. Cost: well over \$100 million.

EXAMPLE 3-9

Convert 55 miles per hour [mph] to units of meters per second [m/s].

Note that we have two units to convert here, miles to meters, and hours to seconds.

Method	Steps
(1) Term to be converted	55 mph
(2) Conversion formula	1 km = 0.621 mi; 1 km = 1,000 m; 1 h = 60 min; 1 min = 60 s
(3) Make fractions (equal to one) (4) Multiply	$\frac{55\ \text{mi}}{\text{h}} \left\| \frac{1\ \text{km}}{0.621\ \text{mi}} \right\| \frac{1{,}000\ \text{m}}{1\ \text{km}} \left\| \frac{1\ \text{h}}{60\ \text{min}} \right\| \frac{1\ \text{min}}{60\ \text{s}}$
(5) Cancel, calculate, be reasonable	24.6 m/s

In the past, many units were derived from common physical objects. The "inch" was the width of a man's thumb, and the "foot" was the heel-to-toe length of a king's shoe. Obviously, when one king died or was deposed and another took over, the unit of "foot" changed, too. Over time, these units were standardized and have become common terminology.

New units are added as technology evolves; for example, in 1999 the unit of katal was added as an SI derived unit of catalytic activity used in biochemistry. As you proceed in your engineering field, you will be introduced to many "new" units. The procedures discussed here apply to *any* unit in *any* engineering field.

Table 3-4 Common units in the SI system

Dimension	SI Unit	Base SI Units	Derived from
Force (F)	newton [N]	$1\ \text{N} = 1\frac{\text{kg m}}{\text{s}^2}$	$F = ma$ Force = mass times acceleration
Energy (E)	joule [J]	$1\ \text{J} = 1\ \text{N m} = 1\frac{\text{kg m}^2}{\text{s}^2}$	$E = Fd$ Energy = force times distance
Power (P)	watt [W]	$1\ \text{W} = 1\frac{\text{J}}{\text{s}} = 1\frac{\text{kg m}^2}{\text{s}^3}$	$P = E/t$ Power = energy per time
Pressure (P)	pascal [Pa]	$1\ \text{Pa} = 1\frac{\text{N}}{\text{m}^2} = 1\frac{\text{kg}}{\text{m s}^2}$	$P = F/A$ Pressure = force per area
Voltage (V)	volt [V]	$1\ \text{V} = 1\frac{\text{W}}{\text{A}} = 1\frac{\text{kg m}^2}{\text{s}^3\ \text{A}}$	$V = P/I$ Voltage = power per current

- A note of caution: One letter can represent several quantities in various engineering disciplines. For example, the letter "P" can indicate pressure, power, or vertical load on a beam. It is important to examine and determine the nomenclature in terms of the context of the problem presented.
- Always remember to include the units in calculations.

Special Unit: Radian

radian

The derived unit of **radian** is defined as the angle at the center of a circle formed by an arc (S) equal in length to the radius (r) of that circle. In a complete circle there are 2π radians. Since by definition a radian is a length (S) divided by a length (r), it is a dimensionless ratio.

$$1 \text{ radian } [\text{rad}] = S/r$$

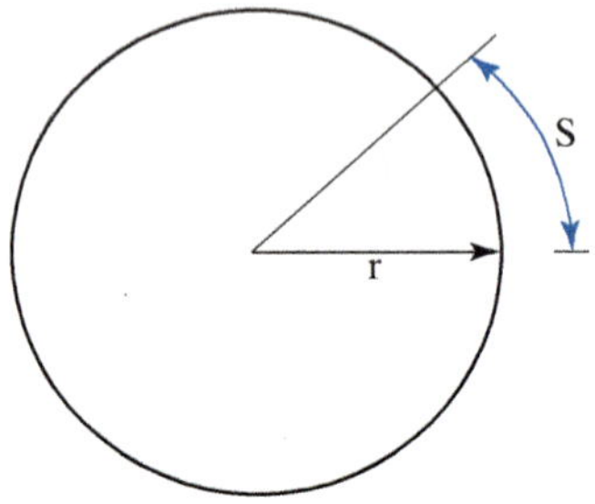

Figure 3-4

Thus, an angle has units, but is dimensionless! A dimensionless quantity is formed by a grouping of parameters in such a way that all the dimensions (such as mass, length, time, temperature) "cancel" out, through division or multiplication.

For example, a weight (which is a force) divided by a frictional force is dimensionless. A common dimensionless quantity is π (ratio of length of circumference to length of diameter of a circle). In addition to radians, another common unit used for angle is the degree [°]. There are 360° in a complete circle

$$\mathbf{360°2\pi \text{ radians}}$$

IMPORTANT CONCEPT: **UNIT LAW**

Every term in an equation must have the same units so that the arithmetic operations of addition and subtraction can be carried out.

COROLLARY TO UNIT LAW

A dimensionally consistent and unit consistent equation is not necessarily a valid equation in terms of physical meaning.

3.3.2 CONVERSION INVOLVING EQUATIONS

Engineering problems are rarely as simple as converting from one set of units to another. Normally, an equation is involved in the problem solution. To minimize the likelihood of mistakes, we adopt the following procedure for all problems. While this procedure may seem to overanalyze simple problems, it is relatively foolproof and will become more and more useful as the material progresses in difficulty.

Equation Procedure

1. Given a problem, first convert all parameters into base SI units, combinations of these units, or accepted non-SI units. Use the five-step conversion procedure previously described.
2. Perform all necessary calculations, as follows:
   (a) Determine the appropriate equation.
   (b) Insert the known quantities and units. Be sure to carry the units through until the end!
   (c) Calculate the desired quantity.
   This gives the answer in SI units.
3. Convert the final answer to the required units and express the answer in "reasonable" terms.

EXAMPLE 3-10

On a trip from Alphaville to Betaville, you can take two main routes. Route 1, which goes through Gammatown, is 50 kilometers [km] long; however, you can only drive an average speed of 36 miles per hour [mph]. Route 2 travels along the freeway, at an average speed of 50 mph, but it is 65 km long. How long does it take to complete each route? State the time for each route in minutes [min].

NOTE

$$\frac{\text{km}}{\frac{\text{km}}{\text{h}}} = \frac{\text{km h}}{\text{km}} = \text{h}$$

Step One: Convert to Base SI Units		
Method	**Route 1**	**Route 2**
(1) Term to be converted	36 mph	50 mph
(2) Conversion formula (3) Make a fraction (equal to one) (4) Multiply	$\frac{36 \text{ mi}}{\text{h}} \Big\vert \frac{1 \text{ km}}{0.621 \text{ mi}}$	$\frac{50 \text{ mi}}{\text{h}} \Big\vert \frac{1 \text{ km}}{0.621 \text{ mi}}$
(5) Cancel, calculate	58 km/h	81 km/h
Step Two: Calculate		
Method	**Route 1**	**Route 2**
(1) Determine appropriate equation	Distance = (velocity) (time) which can be rewritten as . . . Time = distance/velocity	
(2) Insert known quantities	$\text{Time} = \frac{50 \text{ km}}{58 \frac{\text{km}}{\text{h}}}$	$\text{Time} = \frac{65 \text{ km}}{81 \frac{\text{km}}{\text{h}}}$
(3) Calculate	Time = 0.86 h	Time = 0.8 h
Step Three: Convert from Base SI Units to Desired Units		
Method	**Route 1**	**Route 2**
(1) Term to be converted	0.86 h	0.8 h
(2) Conversion formula (3) Make a fraction (equal to one) (4) Multiply	$\frac{0.86 \text{ h}}{} \Big\vert \frac{60 \text{ min}}{1 \text{ h}}$	$\frac{0.8 \text{ h}}{} \Big\vert \frac{60 \text{ min}}{1 \text{ h}}$
(5) Cancel, calculate, be reasonable	52 min	48 min

3.4 SYSTEMS

Every field of study includes a number of key ideas that set that field apart from others. In mathematics, key ideas relate to algebra, geometry, and measuring and analyzing data. In science, key ideas relate to the living environment and the physical setting. In technology, key ideas involve systems, resources, requirements, trade-offs and optimization, processes, and controls.

system

A **system** is a group of interrelated parts that work together to produce a desired result. A system can be any imaginable size, but its size does not affect its complexity. One example of a very small system is a computer chip, which includes millions of electronic parts (Figure 3-5). A system can also be as large as the global network that connects telephones all over the world.

Figure 3-5 A computer chip is actually a small system.

Models and Modeling

A model is a representation of a system, process, product, structure, or environment. Models can be *graphic* (such as a drawing), *physical* (such as a scale model), virtual (such as a computer simulation), or *mathematical* (such as a graph, chart, or equation). Modeling alternative designs allows engineers and designers to think about advantages and disadvantages of possible solutions before they invest time and money in producing their designs. In engineering and technology, using models and modeling techniques are important in helping to represent, analyze, predict, and explain the performance of a device or system.

Very often, physical models are built before the final product is produced. For example, a model is often created by architects and structural engineers before a building is built.

There are a variety of model types. Physical models, such as the architectural model in Figure 3-6a, model cars, and model rockets are one type. Drawings and diagrams, such as the model of the toilet tank flushing mechanism in Figure 3-6b, are also models.

A computer simulation that shows how a system works is another kind of model. Even a mathematical equation is a model. Mathematical models are used often by scientists and engineers to predict the behavior of a system. For example, a formula (known as Ohm's Law) that relates current flow (I) to the amount of voltage (E) and electrical resistance (R) in a circuit ($I = E/R$) allows a designer to predict how much current will flow through a light bulb when it is plugged into an electrical outlet that provides 120 volts.

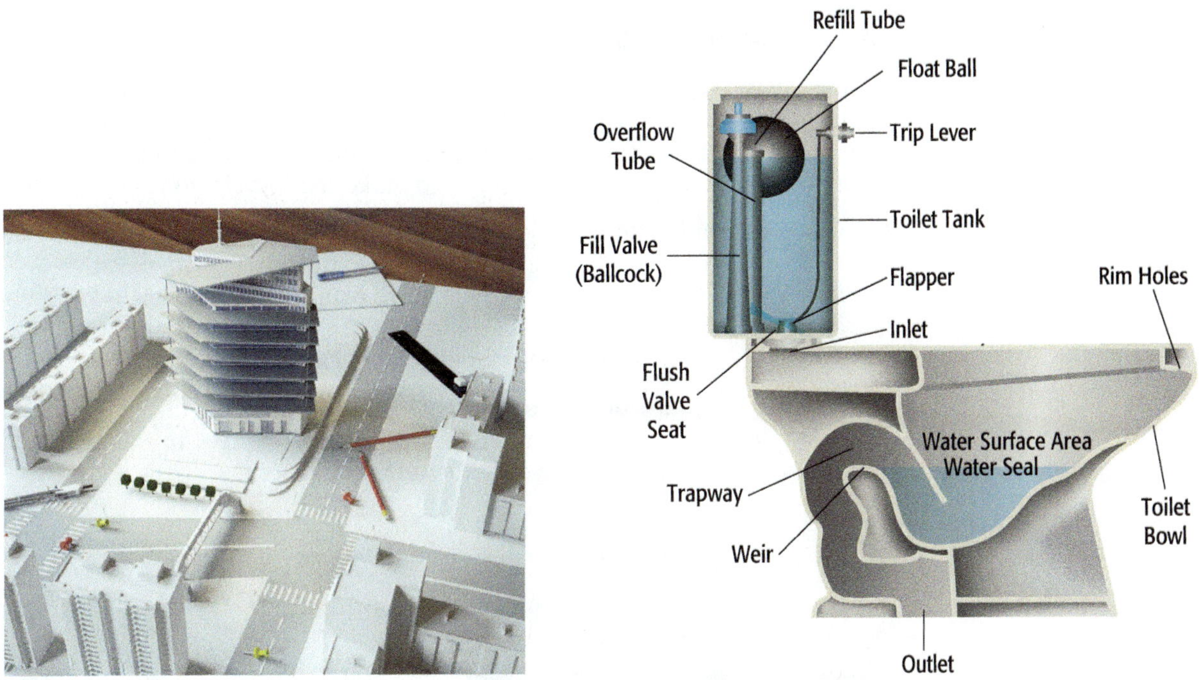

Figure 3-6 (a) Architectural model of a planned building and (b) a toilet tank flushing mechanism.

As engineers, we understand processes, interactions within machinery, circuitry, and every day occurrences with the same framework. Each scenario has the following notion in common: Something enters the process or "machine" on one side and something else emerges on the other end. Our concern in this chapter involves the middle step, the "machine" or process yielding the output – a *system*. Whether or not we realize it, the result of engineering is almost always a system which provides a desired output. Consider the example in which we view a bicycle as a system.

3.4.1 THE UNIVERSAL SYSTEM MODEL

system

A **system** is any group of interrelated parts designed collectively to achieve a desired goal. The desired goal can be anything from driving an automobile to making toast to calling a friend. Different parts of a system work together to produce a specific result.

A technological system produces results through the use of technology. A car, for example, is a technological system used for traveling from one place to another. A telephone is an electronic or cellular system used for speaking to someone in a different location. A system may be huge, such as the space shuttle, or it may be small, such as a pocket calculator. All systems include an input, a process, and an output. The universal system model describes any technological system. It shows how the parts of a system work together.

A Bicycle as a System

A bicycle is a type of system that allows us to travel from one place to another more quickly than on foot. It can be designed and made for different purposes—dirt riding, street and ramp riding, long-distance racing, or all-around use. It has numerous components, many of which are used to make the bicycle go faster or slower.

- Gears determine how fast you need to pedal to achieve a certain speed.
- Brakes let you slow down gradually or stop quickly.
- Shock absorbers can make the ride more comfortable, especially on bumpy roads or trails.
- Wider tires provide more support, but they can also create a slower ride.

And you—the rider—are also part of the system. You control how fast the bicycle goes, where it goes, and how well it is maintained. In this chapter, you will learn how a bicycle and many other systems work.

Figure 3-7

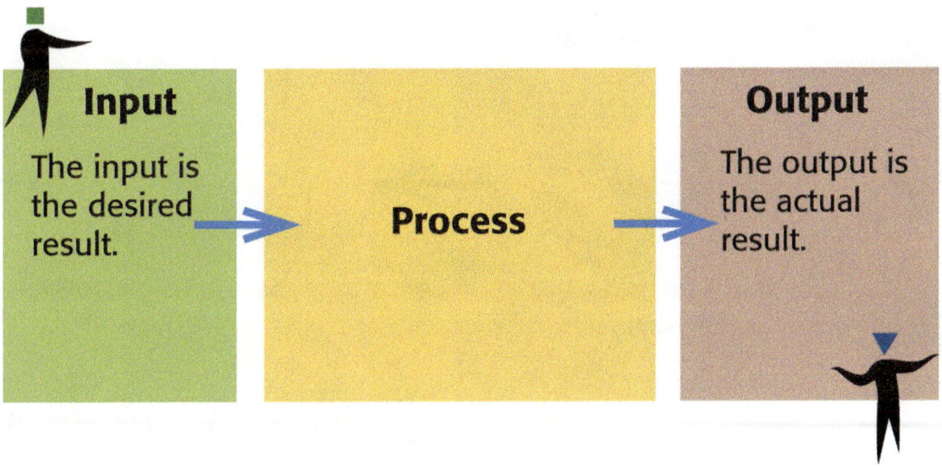

Figure 3-8 All systems include an input, a process, and an output. Each part of a system affects the other.

Input

input

An **input** is the command given to a system. It is also the desired result. When we turn on a television set, for example, we are giving it a command: "Give us a picture and sound." With a car, we provide the input by driving. A car moves when we "tell it to" by stepping on the gas. The input command (or desired result) might be: "Go 30 miles an hour." Other inputs to the system include resources such as time, energy, and materials.

Process

process

The **process** is the action part of a system. It combines resources and produces a result, which is the output. The process is a response to the input command. During the process, the seven technological resources are used.

In the driving process, both the car (machine) and the driver (people) are essential. The driver gives the input command ("Go 30 miles an hour") and then steps on the accelerator pedal. Gas is converted to energy to make the engine run and move the car forward. The car and the gasoline cost money, or capital. The driver uses information to make the car produce the desired result.

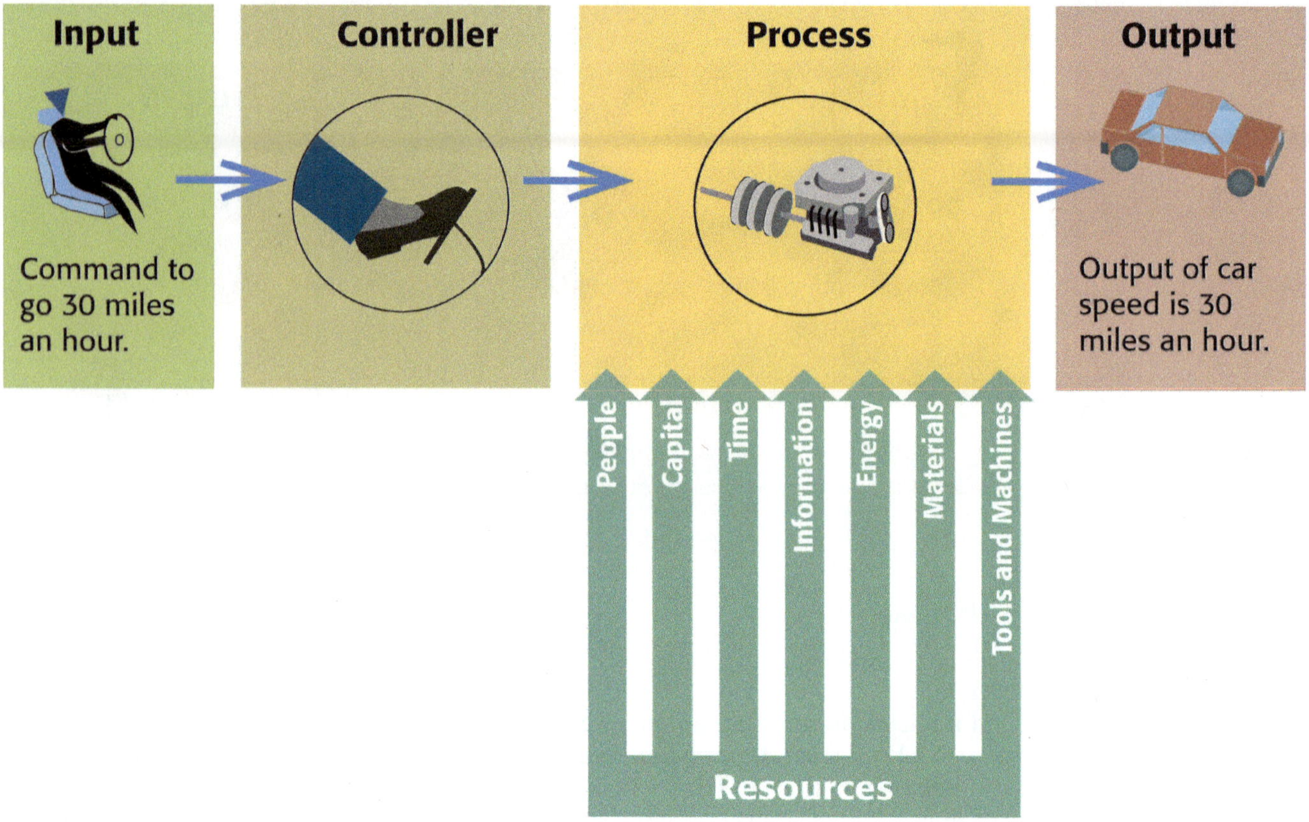

Figure 3-9 In response to the input command, a car uses resources to produce an output.

Output

output

The **output** is the product of a system. It is the actual result. If a system works correctly, the output matches the input, or desired result. With the car system example, we wanted the car to go 30 miles an hour. A properly functioning car will produce this output in response to the input command.

Most systems have more than one output. Outputs can be desirable and expected (for example, the car moves), or they can be undesirable and unexpected (the car produces exhaust and heat). Let's take a look at the outputs of a power plant that produces electricity (Figure 3-10).

1. The expected, desirable outputs are electricity and heat.
2. The expected, undesirable outputs are noise and smoke.
3. At one power plant, an unexpected, desirable output is produced by heat that is discharged into the river. Tropical fish flourish, creating an attraction for employees and nearby residents.
4. An unexpected, undesirable output is acid rain. Acid rain can damage the environment.

When designing systems, engineers and designers must consider the unexpected outputs as well as the expected outputs. They must understand how the different parts of a system can produce unexpected results. Sometimes, designs are changed to modify these outputs even if some of the desired output is lost when the unwanted outputs are reduced. Every system involves this trade-off between desired and undesired outputs.

Figure 3-10 Outputs from a system can include expected or unexpected, desirable or undesirable results.

Subsystems

subsystem

Systems are often made up of many smaller systems. Each of these smaller systems is called a **subsystem**. To understand a large system, it might be helpful to break it into subsystems. Then, you can study each one separately.

Suppose that you want to look at a transportation system that carries goods by truck from Los Angeles to New York City. You could break down the large system into smaller subsystems, such as the vehicle system, the management system, and the communication system. On a smaller scale, a home entertainment center is another example of a complex system (Figure 3-11). Combining different subsystems produces more

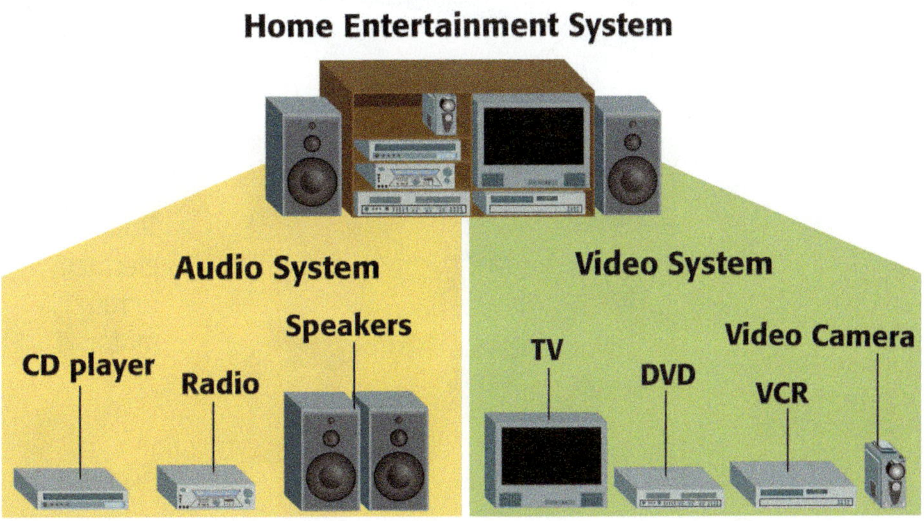

Figure 3-11 Even a home entertainment system is a complex system that includes many different subsystems.

powerful and complex systems that can perform tasks that independently operating subsystems could not accomplish.

3.4.2 OPEN-LOOP SYSTEMS

If you have ever sat in a car waiting for the traffic light to turn green, even when there are no other cars around, you have used an open-loop system.

open-loop system

A system that has no way of monitoring or adjusting itself is called an **open-loop system**. An open-loop system cannot alter its output based on changing conditions. A simple traffic light is set to stay red for a specific period of time, regardless of whether or not there are cars present.

A model railroad with an on/off switch is another example of an open-loop system (Figure 3-12). The train runs at a fixed speed on a level track. Traveling uphill, however, the train slows down. Downhill, it speeds up. The speed is not maintained at a constant level. In such an open-loop system, the output (speed) changes when conditions change. The changing conditions are the different slopes of the track. Another example of a technological open-loop system is a home heating system that has no thermostat. In such a system, the heater is turned on and continues to run. The house gets warmer and warmer, but nothing tells it to turn off. Without a thermostat to detect the increase in room temperature, the system has no way to respond to changes that have taken place.

Open-loop systems can be controlled to some degree. Timers and other kinds of programmed controls can modify an operation based on expected conditions. For example, an oven can be set to turn off after a certain period of time, or the lights in your house can be set on timers. Although these systems are controlled by timers, they do not change when conditions change. The oven will turn off at the set time even if the food is not fully cooked.

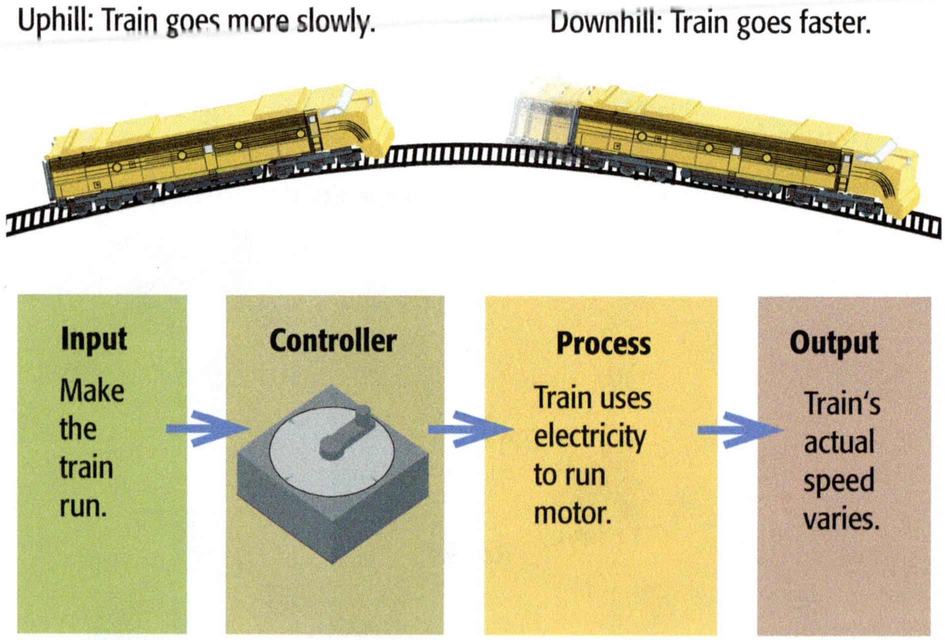

Figure 3-12 An open-loop system, such as this train, has no way to regulate its output.

3.4.3 CLOSED-LOOP SYSTEMS

closed-loop system
feedback

Some systems may require the use of feedback and control subsystems. A system that can respond to changes and adjust its inputs or outputs is called a **closed-loop system**. Closed-loop systems operate using feedback. **Feedback** is information about a system's output that is used to adjust the system. It "closes the loop" from output to input (Figure 3-13).

Comparing Open-Loop and Closed-Loop Systems		
System	**Open-Loop**	**Closed-Loop**
Bicycle steering	Rider is blindfolded.	Rider watches road and adjusts pedaling or braking.
Car speed	Driver does not look at speedometer.	Driver looks at speedometer regularly and adjusts speed as needed.
Model train	Train is either on or off; there is no speed control.	Operator adjusts speed of train to maintain constant speed.

Figure 3-13 Open-loop systems can be changed to closed-loop systems by adding feedback.

For example, a traffic light that uses feedback can adjust its output based on the amount of traffic.

Feedback Loops

Let's return to our car system example. How does the driver know when the car is traveling 30 miles an hour? He or she checks the speedometer. The speedometer, which shows how fast the vehicle is traveling, provides the driver with feedback—information about the car's speed. Based on this feedback, the driver monitors the speed and controls it manually by pressing the accelerator or the brake pedal. The combination of the speedometer, the driver's eye, and the driver's brain forms the feedback loop (Figure 3-14).

One purpose of feedback is to improve the operation of a system; therefore, some type of feedback can be added to most systems. A model train can be equipped with a human-operated speed control, which allows the operator to control the speed when conditions change (as when the train goes up an incline). The human operator provides the feedback when he or she monitors and adjusts the speed of the train in order to keep it on track.

Similarly, a person or a thermostat can make an open-loop heating system closed. They provide data or information that adjusts the system. Feedback from the person or the thermostat turns the heater off or on, keeping the room at a constant temperature.

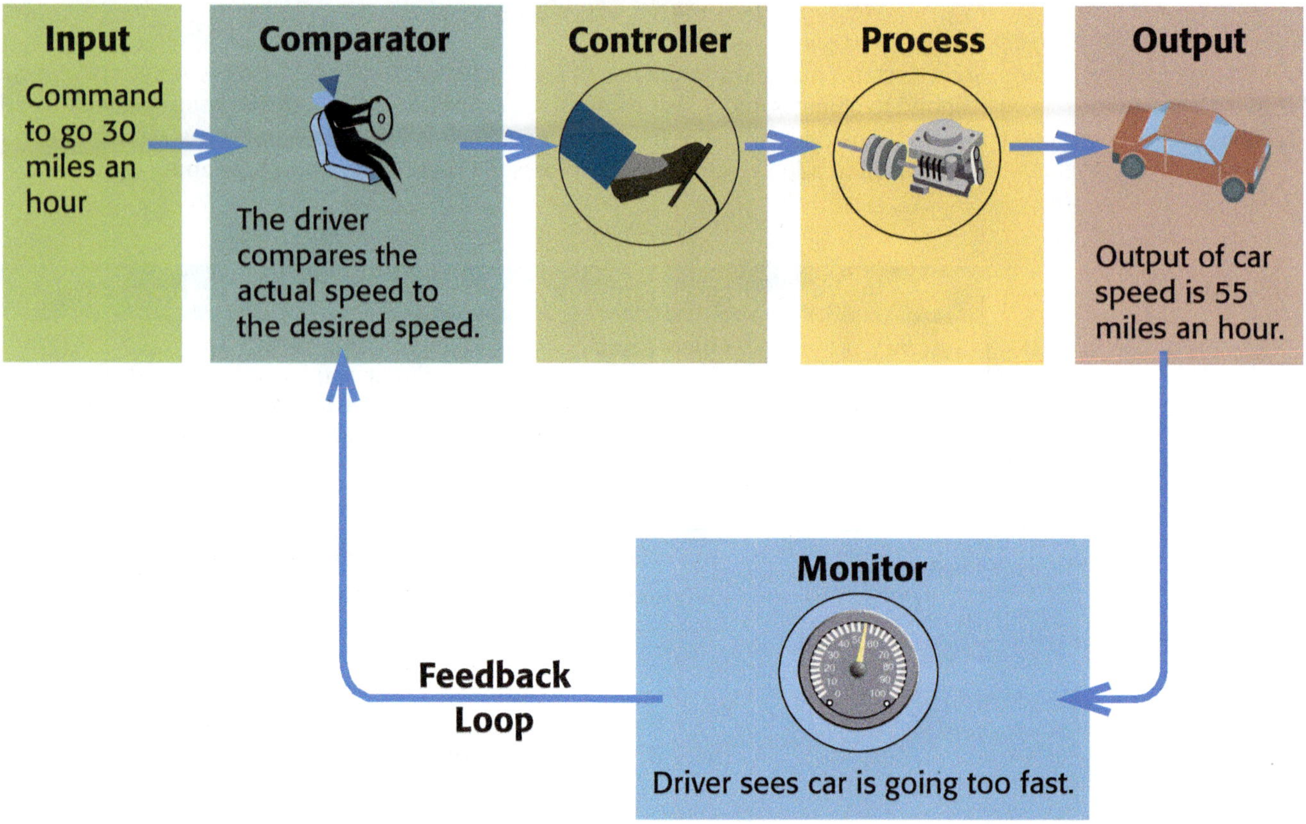

Figure 3-14 A closed-loop system has a feedback loop that helps regulate the system's output.

Types of Feedback

Feedback comes in many forms. It can come from other people, as when your parents ask you to turn down the volume on the television or radio. Feedback can come from touch or sight. It can also come from a mechanical device, such as a cash machine. Whenever you use your fingers to pick up or hold something, you are receiving tactile feedback. Feedback is also used in school. Your teacher uses homework and tests to monitor how well you are learning. The grades you receive on your work are your feedback, indicating how you are doing so that, if needed, you can change your study methods and work to improve your test-taking skills.

TECHNOLOGY IN THE REAL WORLD

Human Feedback: Driving and Reaction Times

Imagine that you are driving and you suddenly see an animal in the road ahead. Your eyes act as a sensor. They send a signal to your brain (the controller), which sends a signal to your leg and foot muscles. You take your foot off the accelerator pedal and step on the brake. The car slows down.

Calculating Reaction Time

This process takes time before braking occurs, usually about 3/4 of a second. How far does the vehicle travel during this time? Calculate the distance for speeds of 20 miles per hour (mph), 40 mph, and 60 mph. First, convert miles per hour (miles/hour) to feet per second (feet/second). Let's do this for 1 mph. There are 5,280 feet in 1 mile and 3,600 seconds in 1 hour. (Notice that the units for miles and hours cancel.)

$$1 \text{ mile/hour} = \frac{5{,}280 \text{ feet/mile}}{3{,}600 \text{ seconds/hour}} = \frac{5{,}280 \text{ feet}}{3{,}600 \text{ seconds}} = 1.467 \text{ feet/second}$$

At 20 mph, the car travels at twenty times this speed, or

$$(20)\,(1.467 \text{ ft/sec}) = 29.3 \text{ ft/sec}$$

In 3/4, or 0.75, of a second—the time it takes us to start braking—the distance traveled is

$$(0.75 \text{ sec})\,(29.3 \text{ ft/sec}) = 22 \text{ feet}$$

When we are driving at 20 mph, then, the car travels 22 feet before we begin braking.

Making a Graph

By performing similar calculations for 40 mph and 60 mph, we can create the graph shown. The graph describes how our driving speed affects the distance (number of feet) we travel before braking.

Effect of Driving Speed on Reaction Time

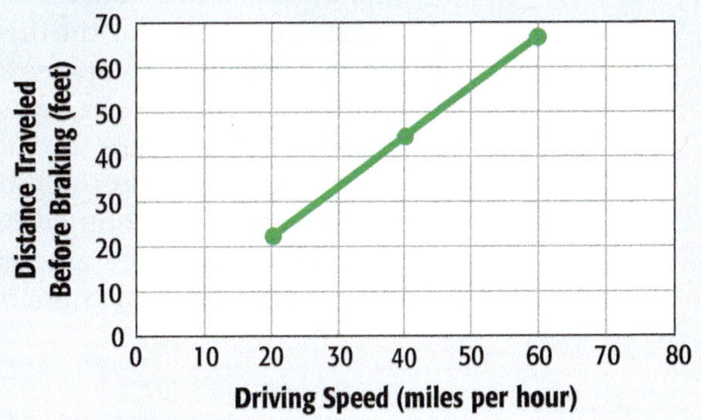

3.4.4 THREE ELEMENTS OF A FEEDBACK LOOP

You have learned that some systems use feedback to adjust their output based on changing conditions. Systems that use feedback are closed-loop systems. In these systems, feedback loops have three essential parts: sensors, comparators, and controllers. The three parts are sometimes combined into one mechanism, but all three must be present for the loop to work. Let's look at each one in more detail.

Sensor

Figure 3-15 Pressure gauges are used for diving, high-pressure gas storage, and taking blood pressures.

sensor

A **sensor** is a system component that monitors a system's output. (A sensor can also be called a monitor.) Sensors with which you may be familiar include your eyes, a thermometer, a compass, and a pressure gauge (Figure 3-15). Many systems use several kinds of sensors at the same time. Our bodies, for example, sense the temperature, humidity, light, and noise around us. Computer systems can also monitor these conditions using electronic sensors.

Some sensors send back information directly to the system. Others change it into another form first. Information is most often changed into electrical signals, which can be sent easily and quickly.

A device that changes information from one form to another is called a transducer. Pressure sensors, for example, can change mechanical pressure readings into electrical voltage. In a stereo system, the laser of a CD player is a transducer that translates the information on the surface of a CD into music. Once translated, the information is used as part of a system's feedback loop.

Comparator

comparator

A **comparator** compares the actual output of a system with the desired output or result. It makes this comparison using information from the sensor.

If a system's actual and desired outputs are not the same, the comparator indicates that there is a difference. For example, if a car has an automatic speed control system, such as cruise control, the driver does not need to monitor the speed. The cruise control device constantly compares the actual speed with the desired speed set by the driver.

Some comparators measure the difference between the actual and the desired results. For example, a thermostat in a home heating system decides whether the room temperature matches the setting. Another example is a computer in a car engine that monitors whether enough gas is going into the engine. It can decide how much to add in order to reach the amount needed. Comparators may be simple, or they may be complex devices that contain computers.

Controller

controller

A controller changes a system's process to make the actual output closer to the desired output. A **controller** is a device that turns a process on or off or that changes it in some way.

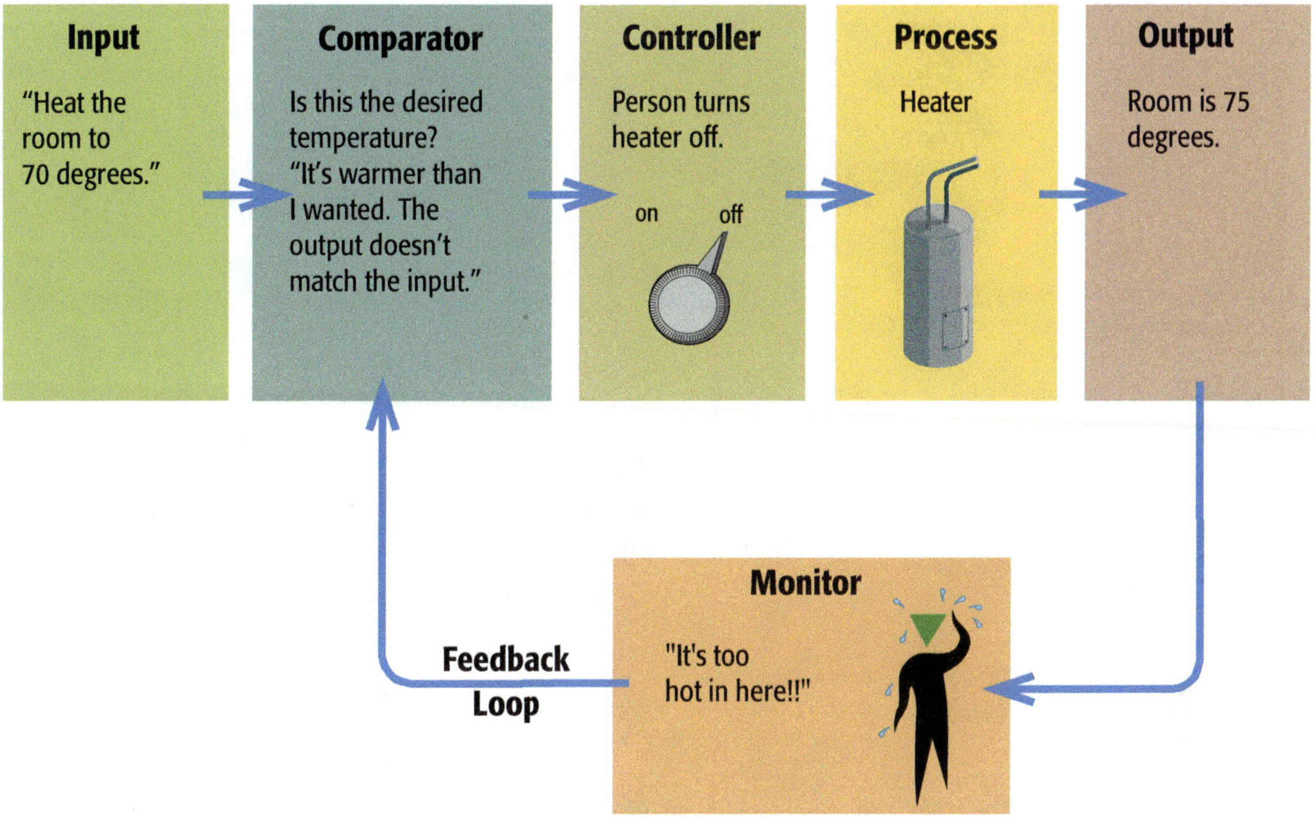

Figure 3-16 With manual control of a heating system, a person acts as sensor, comparator, and controller.

Closed-loop systems can be controlled either manually or automatically. Consider a home heating system. If the system is controlled manually, a person acts as the sensor, comparator, and controller (Figure 3-16). Once the heater is turned on, the person may determine that the room is warmer than the desired temperature (the input command). The person must constantly monitor the temperature of the room to adjust the process (turn the heater on or off).

In an automatic home heating system, a thermostat determines whether the room temperature matches the temperature setting (Figure 3-17). The metal sensor, a bimetallic strip in the thermostat, changes when the temperature changes.

A bimetallic strip is made of two different types of metals. These metals expand or contract when heated or cooled, but no two metals expand or contract at the same rate. As a result, the bimetallic strip bends as one metal expands more than the other. The degree of the bend indicates the temperature. Using this simple mechanism, a thermostat continuously monitors a room's temperature and connects or disconnects with the furnace to turn it on or off.

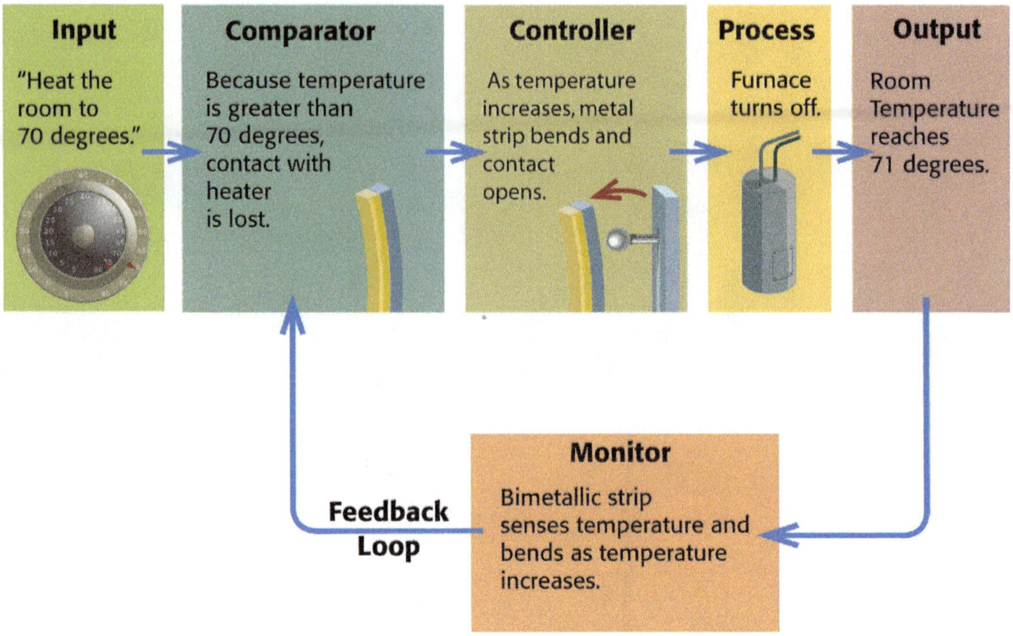

Figure 3-17 With automatic control of a heating system, a thermostat regulates the system's process and output.

3.4.5 CONTROL OF NONTECHNOLOGICAL SYSTEMS

You have been learning how technological systems are controlled. Nontechnological systems must be controlled as well. Very often, information used to make decisions is processed using computers. However, some systems are so complex that the information is incomplete. Experience or even educated guesses can help people make decisions.

An economic system is an example of a system that relies on people's experience to control it. In a free-market system, people are free to decide what products to produce, whom to sell them to, and at what price to sell them. A nation's economic system is also controlled in part by the amount of money in circulation. A government may decide to print and distribute more money to help the economy grow. In that case, the sensor is a set of economic indicators that tell how the country's economy is doing. Economic indicators are statistics such as unemployment rates that are used to monitor and measure how well the economy is doing. The monitor is the group of people

Figure 3-18 Prices rise and fall based on the strength of an economic system. Some states enact laws stating which gas drivers must use.

who study the indicators and decide whether to print more money. The controller makes the decision to print more money. All of these people—and the information they use—make up a complex system.

In another type of system, such as a large company, managers are the decision makers, or controllers. They study company data, such as profits, debt, market forecasts, and many other factors. They make long- and short-term decisions. Should the company hire new people? Should it spend more money on research? The number of people hired, the amount spent on research, and the types of new products developed are some of the controlling factors in a company.

Laws and regulations also provide controls. Laws are made as a result of feedback. For example, when tests showed that our air was becoming polluted by the exhaust from trucks, cars, and buses, new laws were passed to require changes in fuel and design to reduce pollutants. The laws were used as a controller. Other laws govern toxic waste disposal, nuclear power plants, and genetic research.

TECHNOLOGY IN THE REAL WORLD

The Hubble Space Telescope: Controlling a System from Miles Away

One amazing example of how a technological system uses controls is the Hubble Space Telescope. The Hubble orbits 600 kilometers above Earth, using a precise control system and state-of-the-art instruments to provide stunning views of the universe. These images cannot be made using ground-based telescopes or other satellites.

How Does It Work?

The telescope uses sophisticated instruments that serve as electronic eyes. These instruments include cameras and spectrographs. The Hubble's cameras do not use film. They collect light using electronic detectors similar to those in home video cameras. The spectrographs separate starlight into a rainbow of colors, just as a prism splits sunlight. By carefully studying the colors, astronomers can decode a star's temperature, motion, composition, and age.

The Hubble Space Telescope in orbit above Earth.

An Incredibly Accurate System

The Hubble must be pointed at the object to be studied. It must hold very steady in position, or the image will blur. The system is so accurate that if the telescope were located in Washington, D.C., it could steadily focus a laser beam on a dime atop the Empire State Building in New York City. Once the Hubble locks onto an object, its sensors check for movement 40 times a second. Commands sent from Earth keep the telescope pointed at its target using a feedback control system. If movement occurs, constantly spinning wheels change speeds to smoothly bring the telescope back into position.

Hubble's photo of a galaxy that lies 150 million light years from Earth.

3.5 THE ENGINEERING DESIGN PROCESS

The Need for Problem Solving

Every day, we are faced with situations that require thoughtful solutions. Some of these situations, such as designing video games, are fairly short-term and involve a specific goal. Others are more general and long-term. Some problems, such as the following, involve society and the environment:

- Disposing of wastes without harming the environment.
- Producing enough energy to meet our increasing needs.
- Ensuring a continuing supply of clean, safe water.

Other problems have to do with people's needs or wants, as follows:

- Providing more independence for people with vision or hearing impairment or other disabilities (Figure 3-19).
- Improving the sound quality of recorded music.
- Allowing automobile drivers to communicate with each other to prevent accidents.

Each of these problems can be solved using science, technology, and design.

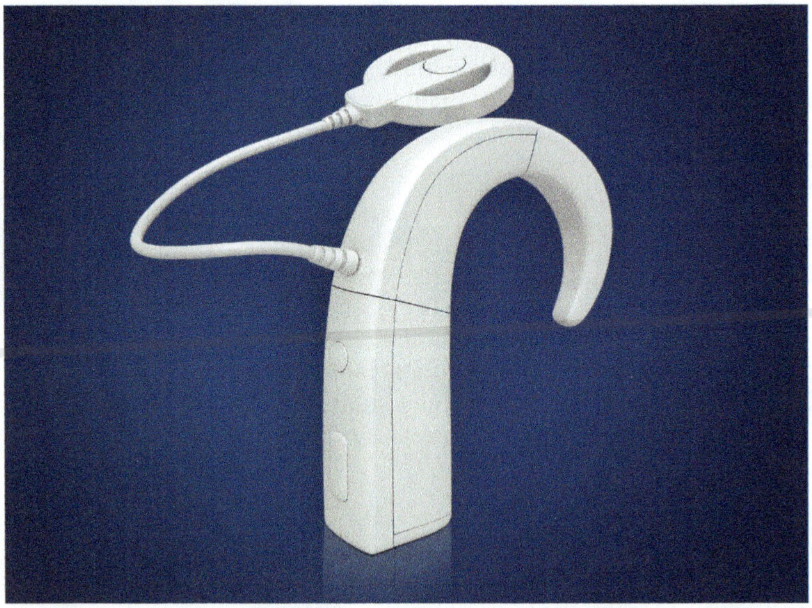

Figure 3-19 Cochlear implants can restore hearing in many cases.

Good Design in Problem Solving

Solving a problem is rarely quick and easy. It often takes time, thought, and creativity. We must understand exactly what is needed to solve the problem. More money, extra time, or other people's help may be needed. Solutions or products should be as inexpensive and easy to use as possible.

Problems in technology are seldom well-defined when we first encounter them. For example, if the problem is to design a house for a family, many questions need to be considered. How big will the house be? What styles and materials will be used? How will it be heated and cooled? How much money and time are available? The architect,

Figure 3-20 Before construction begins, many details and questions need to be discussed.

builder, and customer must work together to clarify each problem and arrive at the best possible solution (Figure 3-20).

To solve any kind of problem, decisions and compromises need to be made. This is done through a process called informed design.

Engineering Design

Engineering design is a systematic, creative, and iterative process for addressing challenges. It is a process for meeting human needs and wants and leads to useful products and systems. Engineering design is an **informed design process.** It is a method of making choices based on knowledge about the design problem, and then coming back to refine or revise those choices at any point in the process as often as necessary. Using the steps to make choices and decisions, more detail is added, resulting in a better defined problem, which leads to an acceptable solution.

informed design process

An informed design process is used by engineers, architects, craftspeople, and artists. All of these people are designers. The choices that they make to create their designs often involve compromises. Cheaper materials may be used, or a simpler, less sophisticated design may be required to finish the project on time. Compromises are also known as trade-offs. Designers often make trade-offs when selecting the best design from all the choices.

Clarify the Design Problem

In the first step of the informed engineering design process, shown in Figure 3-21, the designer describes the problem clearly and fully. Design requirements indicate criteria for success, specifications, constraints (or limits) on the solution. There are often trade-offs between competing requirements.

specifications
design challenge

The **specifications** for a design problem are the performance requirements that the solution must fulfill. The **design challenge** states the problem and lists the specifications

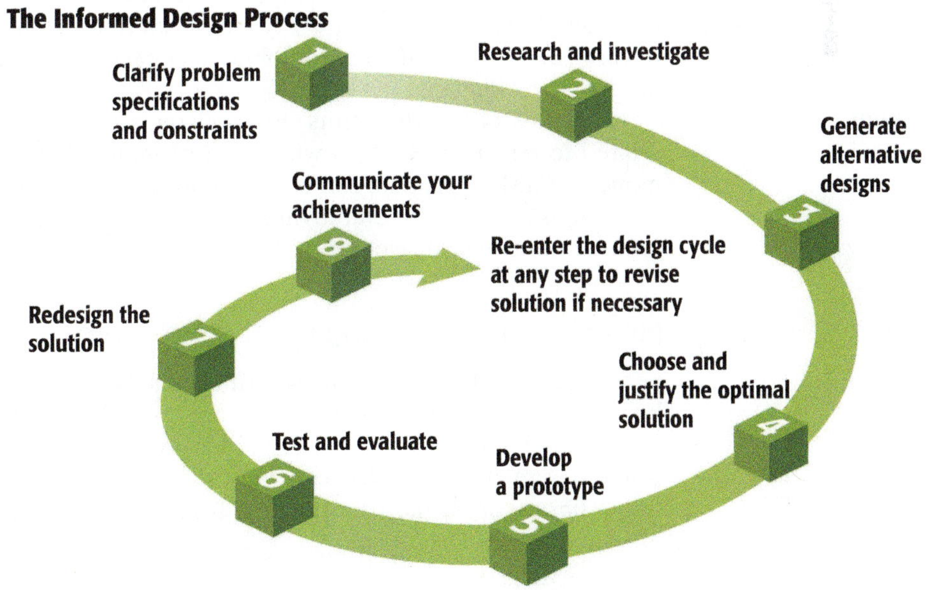

Figure 3-21 Design process steps are usually completed in order, but sometimes certain phases need to be repeated.

Design Brief

Design Problem: Design a durable, comfortable, and moderately priced backpack to carry materials for school and other activities.

Requirements
- Backpack should be able to hold at least 3 textbooks, a CD player, a thin jacket or fleece, a palm pilot, and a cell phone.
- Backpack should have several smaller pockets for money, keys, pens, and pencils.
- Material should be brightly colored, waterproof, and durable.

Constraints
- Backpack cannot cost more than $35 retail.
- Backpack must be adjustable so one size can be used by 5th–8th graders.

Figure 3-22 This figure describes the specifications for a new backpack.

for a design. Specifications are different for each type of design project. The design specifications for toothpaste might be that it must clean plaque from teeth, taste good, and be squeezed easily out of a tube. A design specification for a certain type of bicycle might be that it is lightweight and can be used on paved roads as well as rocky trails.

Let's look at a specific example. You have decided that you need a new backpack to carry your things to and from school. The backpack needs to comfortably hold your school supplies. You also want separate compartments for your calculator and your phone.

Design specifications may also include safety considerations. A passenger elevator, for example, might require a safety factor ten times greater than the load it is expected to carry, or the front of an automobile may need to withstand a collision at 5 miles per hour without damage.

constraints

Constraints are the limits imposed on a design solution. Constraints are often related to resources, such as what kind of materials the designer can use, how much money a finished product can cost, or how much time is available to produce it. Other limitations may involve employing only workers with special skills or limiting negative effects on the environment.

Research and Investigate the Problem

Search for and discuss existing solutions to solve the problem at hand or similar problems. Through this process, you will learn a lot about good and bad solutions.

Through your research, you can identify problems, issues, and questions that relate to the design challenge. You can gather information from other people, from research in the library or on the Internet, and from visits to stores that sell products similar to the one you are designing.

You will also need to take measurements and collect data about different needs the solution must fulfill. As you research, always keep the design criteria in mind. They will help you identify the important questions that you need to answer.

For your backpack, you will need to research several options. How large should the backpack be? How many pockets should it have? What types of materials should be used? Should it have zippers, Velcro®, or some other type of fastener? You may want to collect data about people your age and how much weight they can carry. You may also want to look at existing products to get new ideas.

Generate Alternative Designs

Don't stop at the first solution that might work. Continue to approach the challenge in new ways. As you continue your research, you should develop one or more potential solutions to your problem (Figure 3-23). The ideas can be totally different, or they may be improvements on your first idea. These different ideas are called alternatives—two or more possibilities from which choices can be made. The process of generating different designs is not the same as solving a mathematics problem in your textbook. Usually, mathematics problems are well-defined, with only one correct answer.

brainstorming

Because there are many possible design solutions, good designers are rarely satisfied with the first idea that pops into their minds. **Brainstorming** is a method by which a person or group of people can come up with alternative solutions. During brainstorming, each person in a group suggests ideas. One person writes down all the ideas,

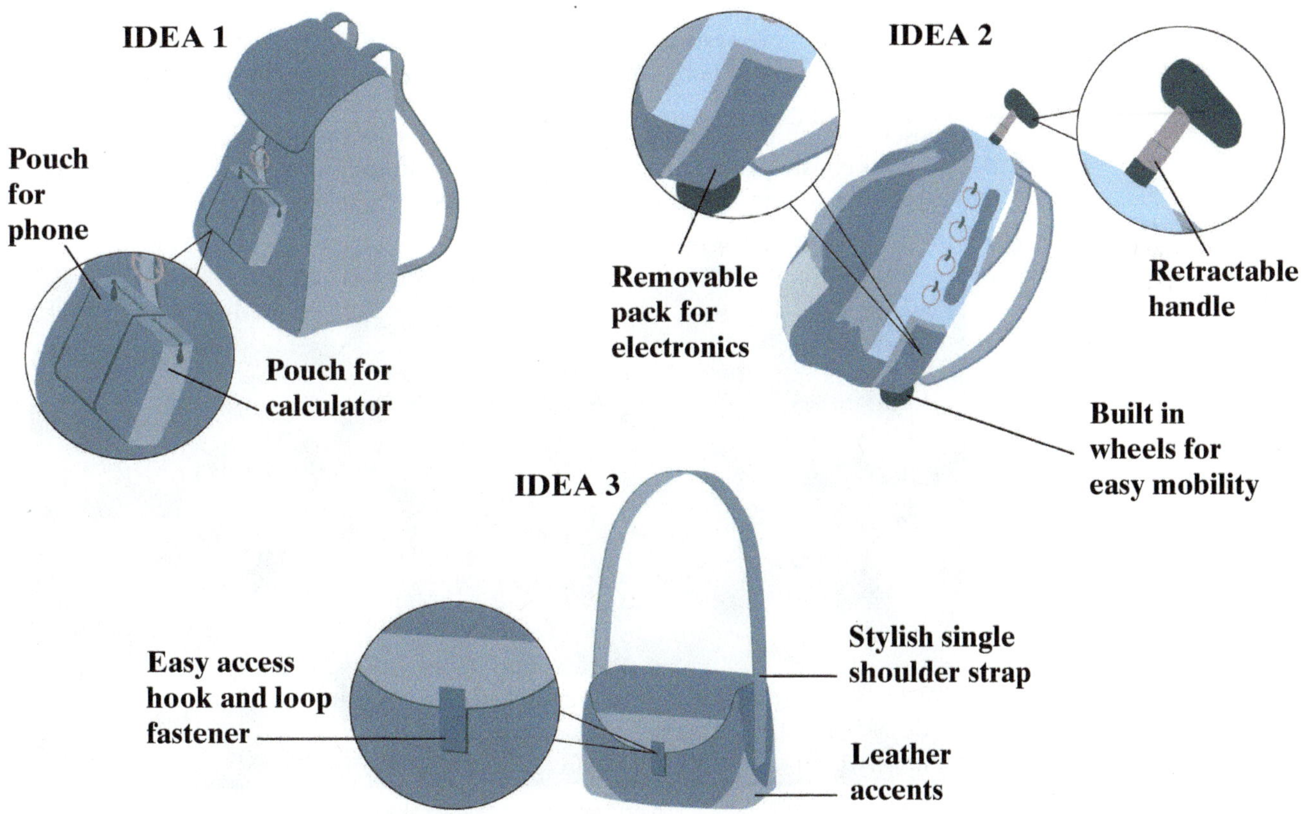

Figure 3-23 Each of these sketches represents a possible design solution.

and no one is allowed to laugh at or criticize them—no matter how foolish or unusual they might seem. After many ideas have been proposed, the group reviews them all. The best ideas are developed further.

Develop a Prototype

scale model

prototype

Make either a model or a prototype of the design solution. A **scale model,** which may be larger or smaller than the final product, is a model of an object with all parts in correct proportion to those of the object. For example, on a scale model, 1 inch may represent 1 foot. A **prototype** is a full-scale, fully operational version of the solution (Figure 3-24).

Models are important for several reasons. The solution or product may be very large and costly. Alternatively, the individual product might be inexpensive, but plans to make several thousand would involve a large expense. Sometimes, the proposed solution might need evaluation of its possible risks to people or the environment. For example, a model of a nuclear power plant would be built and debugged before construction of an actual power plant. In all of these instances, a scale model or prototype makes it easier for a designer to modify and refine the design.

mathematical model

Another kind of model is a mathematical model. In a **mathematical model,** equations describe how a product will function or perform. For an electrical device, a mathematical model might show how current in a circuit is directly related to the amount of voltage and inversely related to the amount of resistance.

Test and Evaluate the Design Solution

Develop one or more tests to assess the performance of the design solution. Test the design solution, collect performance data, and analyze the data to see how well the design satisfies the original constraints and specifications.

variables

Observing, or monitoring, the results of tests may suggest improvements to a design or its construction. You should note each different factor that affects the performance of your design. These are called **variables.**

Figure 3-24 Skilled craftspeople and technicians often make prototypes before full-scale production is started.

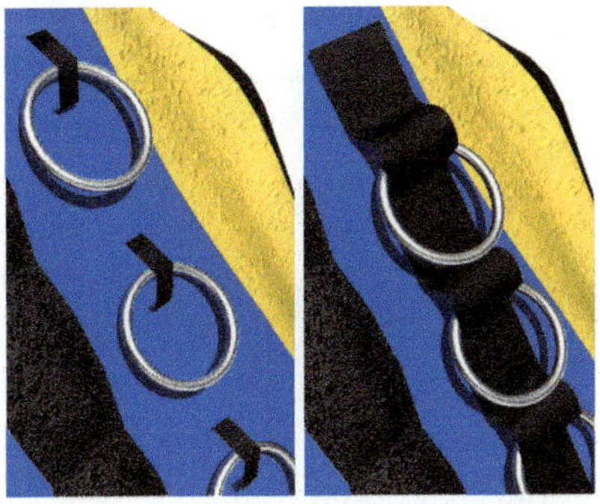

Figure 3-25 Slight modifications can improve a design and facilitate its production.

Sometimes, tests are run on a computer using computer simulations, in which a computer tests different models to predict how a product will perform. Computer simulations are most useful when a large number of calculations must be carried out.

Redesign the Solution with Modifications

Examine your design critically. Note how other students' designs perform to see where improvements might be made to yours. Even during this phase, you should be willing to consider new ideas.

During the redesign phase, you will need to identify and change any variables that affect the performance of your design. Think about any science or math concepts that underlie these variables.

In the backpack design, for example, you may have to modify a feature in order to reduce the cost or make the backpack lighter. In Figure 3-25, for example, the rings now have stronger attachments.

Communicate Your Achievements

Present your final design, along with a summary of how it meets each of your specifications. This is your opportunity to "show off" your final design. Keep in mind that your presentation is as important as the design itself. Describe the final product and how it will be used.

3.5.1 PRINCIPLES OF ENGINEERING DESIGN

Functionality

functionality

Functionality refers to the capability of a product, system, or process to fulfill its intended purpose over the course of its desired life span. For example, a light bulb designed to give 1,000 hours of service is not expected to burn out before that amount of time, and a software program should complete all the tasks it was designed to perform.

The form of an object is often determined by its function. Dining tables hold plates of food, so they are made to be flat. Because drinking glasses are usually held in one hand, they are designed to be higher than they are wide. It would make no sense to design dining tables with slanted tops or drinking glasses too wide to grasp.

Quality

quality

The product, system, or process must be designed to meet certain minimum standards of quality. **Quality** is the degree of excellence with which a product is made. For example, it is essential that medicines use the same formulation in every batch produced. Products made in different locations all need to meet the same standard for quality.

Ergonomics

ergonomics

Ergonomics is the science of adapting the work environment to people. Also called human factors engineering, it deals with designing products so they can be used easily and comfortably. Ergonomics combines an understanding of the human body with the techniques of design and engineering. A good design fits the user's size and capabilities. A desk chair, for example, must be designed for human comfort. The design solution would differ depending on the specifications—whether it's for a teenager, a man, a

PEOPLE IN TECHNOLOGY: PETER LUCAS AND MAYA DESIGN, INC.

CREATING USER-FRIENDLY DESIGNS

Peter Lucas is the co-founder of a company called the MAYA Design in Pittsburgh, Pennsylvania. MAYA Design, Inc. creates user-friendly designs for very complex products—items such as DVD players, remote controls, medical instruments, and computer software. According to Peter, design that is based on a product's function or technology is "hot, and will remain hot. . . . No one wants a remote control with a zillion buttons. It just won't sell."

The Importance of Design

MAYA stands for **M**ost **A**dvanced **Y**et **A**cceptable design. It is a phrase coined by a famous designer named Raymond Loewy. He used it to describe an ideally designed product—one that is advanced enough to take advantage of new technology but is still user-friendly.

One of Peter's central beliefs is that design must play an important role in the development of a new product. Rather than letting a product's design evolve slowly, Peter believes that companies need to treat design more seriously. Because the impact of design on a final product is so great, companies need to consider it as an important part of the way industry does business.

Figure 3-26 Pueblo Bonito ruins and a kiva

Creativity through Brainstorming

At MAYA Design, Inc., designers work in areas as diverse as consumer electronics, multimedia, and product design. They are encouraged to be as creative as possible.

To brainstorm new ideas, they gather in a circular room with white, washable walls to write on. The room is called a kiva. It gets its name from the circular underground rooms that the Pueblo Indians built for their religious ceremonies. "You can work your way around the erasable white board. When you get back to where you started, you are done with your idea!" jokes Peter. Or, more likely, they start erasing and keep going with even better ideas.

woman, a child, or all four. Another example of ergonomic design is an automobile's dashboard. All the information is displayed so that the driver can read it in a glance.

anthropometry

Anthropometry is the science of measuring people. It provides information about the average size and shape of people's bodies. Most often, designers accommodate the middle range of body sizes. They ignore the bottom 5 percent and the top 5 percent of dimensions. For this reason, a pocket calculator may have buttons too small for people with very large fingers.

Safety

When failure occurs, a pump should not explode and an electric motor should not burst into flame. Ensuring people's safety is a critical part of a designer's job.

Aesthetics

aesthetics

Designers know that if a product is going to sell well, people must like the way it looks. **Aesthetics** is the study of the way something looks and how that affects people's feelings about it. For some products, aesthetics is used as the overriding principle in its design. In other cases, greater quality or functionality also makes a product more aesthetically pleasing.

Maintainability

Equipment and structures need to be maintained. A machine part needs periodic lubrication. Measurements (such as electrical resistance) need to be taken. A design has to allow for access to certain parts for maintenance or repair and to others—the filter on a clothes dryer, for example—for regular cleaning.

3.5.2 APPLYING THE DESIGN PROCESS TO SOLVE REAL-WORLD PROBLEMS

Social and Environmental Concerns

Figure 3-27 Design can affect society and the environment.

Most problems that can be solved in your technology class do not involve issues of the real world. Often, they are a bit simpler, such as designing a backpack or making a product more ergonomic. As you learn to use problem solving, you will see that solutions to real-world problems often involve more limitations. Engineers and designers must keep in mind how their solutions will affect society and the environment. If an airport is being planned, it cannot be built too near a residential area because the noise from arriving and departing planes would be annoying (Figure 3-27).

In real-world problem solving, the needs of society or the community must be considered. When a technological solution has potentially negative effects, people must decide whether the solution's benefits outweigh its risks. A good technological solution meets the needs of people and has little adverse effect on the environment and creates a win-win situation for the economy and the environment.

Figure 3-28 Industrial pollution is often a political issue.

Politics

Often, solutions to real-world problems are affected by politics (Figure 3-28). Groups of people join together to promote a special interest. For example, some groups oppose the building of nuclear power plants. They believe that if an accident were to happen, radioactive material could poison the environment. They point out that, for now, there is no safe way to get rid of radioactive waste.

Other groups favor nuclear power plants. They believe that nuclear power is an important source of energy. They think nuclear power will make us more self-sufficient and less dependent on other countries for oil to fuel our power plants. They are convinced that the risk of an accident is small, and that the benefits of a reliable source of energy are worth the risk.

Values

values

Our **values,** or what we think is important, influence all of our decisions. They are a factor in our decisions for or against something. For example, if we look at cars only as transportation, we might decide to buy a basic vehicle that gets good gas mileage. If we think that cars are neat and that driving is fun, we might decide to buy a sports model.

Risk/Benefit Trade-off

risk/benefit trade-off

A common trade-off made in solving complex problems is a **risk/benefit trade-off.** To obtain desired benefits, designers must accept some risk, but they try to keep the risk as low as possible. They may not implement a solution if the risks are too high.

For example, when we travel by car, we accept a very low risk of being hurt in an accident. In exchange for this risk, we receive the benefit of being able to travel quickly and comfortably. It is never possible for products and projects to be 100 percent risk free, but a well-designed product presents the minimum possible risk to people as well as to the environment.

Determining risk requires a combination of human knowledge and mathematics. The goal is to determine the probability of some negative event (the risk).

Continued Monitoring

Solutions to real-world problems must be monitored and maintained to make sure there are no undesirable outcomes. A material may not be as durable as expected, or a product may be used in ways that were not anticipated. Information from monitoring should be obtained and evaluated over the life of the product (or solution) to ensure that it continues to meet the design criteria.

This information may indicate that changes need to be made to a product's design. When this occurs, the product is recalled. For example, the windshield wiper motors on a car might have to be redesigned because information from customers indicates that the original motor presents a safety hazard.

Other types of recall are more serious. Certain medicines, for example, have been found to produce serious and often harmful side effects. Proper testing and continuous monitoring reduce the likelihood of negative effects.

Chapter 3 ASSESSMENT QUESTIONS

1. Create a list of variables and constants for the following problem:

Calculate the mass in kilograms of gravel stored in a rectangular bin 18.5 feet by 25.0 feet. The depth of the gravel bin is 15 feet, and the density of the gravel is 97 pound-mass per cubic foot.

2. Find a partner and read the following together:

I am standing on the upper deck of the football stadium. I have an egg in my hand. I am going to drop it and you are going to try to catch it. You are standing on the ground. Explain to your partner why you would not want to stand directly underneath me, where you should stand instead, and why. If you can run at 20 feet per second and I am at a height of 100 feet, how far away can you stand and still catch the egg if you start running when I let go?

3. Water drips from a faucet at the rate of three drops per second. What distance separates one drop from the following drop 0.65 seconds after the leading drop leaves the faucet? How much time elapses between impacts of the two drops if they fall onto a surface that is 6 feet below the lip of the faucet?

Your sketch should include the faucet, the two water drops of interest, and the impact surface. Annotate the sketch, labelling each item shown and denote the relevant distances in symbolic form. For example, you might use d_1 to represent the distance from the faucet to the first drop.

4. If you are to estimate how many gallons of paint are needed to paint the room you are in, what assumptions will you need to make? Where will you need to make assumptions to ensure that you have enough paint without running out?

In estimating the wall area, you should round up the length and height.

Noting that paint (for large jobs) is sold in 5-gallon pails, you will want to round your final estimate to the next whole 5-gallon pail.

Close estimates allow for subtracting 21 square feet per doorway (if the doors are not being painted the same color). In making a rough estimate, if there are not a lot of doorways, it would be conservative to leave in the door area.

5. Determine the number of significant figures and decimal places for each value.

(a) 0.0050
(b) 3.00
(c) 447×10^9
(d) 75×10^{-3}
(e) 7,790,200
(f) 20.000

6. Express the answer to the following, using the correct number of significant digits.

(a) $102.345 + 7.8 - 169.05 =$
(b) $20. * 3.567 + 175.6 =$
(c) $(9.78 - 4.352)/2.20 =$
(d) $(783 + 8.98)/(2{,}980 - 1{,}387.2) =$

7. **(a)** Estimate by aggregation how many gallons of gasoline are used by cars each year in the United States.

(b) Estimate by aggregation the volume of a person. A rough approximation of the volume of a person would be a cylinder approximately 1.75 m tall with a radius of 0.25 m. How different was your estimate from the cylindrical approximation?

8. Use the following materials and procedures for this problem:

Materials
Ruler
Tape measure
Calipers

Procedure
The following measurements and estimations are to be completed individually, not using one set of measurements per team. This activity is designed to help you learn the size of common items to help you with future estimates. Be sure to give both the value and the unit for all measurements.

Measure the following:

(a) Your height in meters
(b) Your arm-span (left fingertip to right fingertip) in meters
(c) The length of your index finger, in centimeters
(d) The width of your thumb, in centimeters
(e) The width of the palm of your hand, in centimeters
(f) The length of your shoe, in feet
(g) The length of your "pace," in feet. A pace is considered as the distance between the toe of the rear shoe and the toe of the lead shoe during a normal step. Think about how to make this measurement before doing it.

Determine the following relationships:

(h) How does your arm span measurement compare to your height?
(i) How does your knee height compare to your overall height?
(j) How does the length of your index finger compare to the width of your thumb?

Determine through estimation:

(k) The height of a door in units of feet
(l) The length of a car in units of yards
(m) The area of the floor of the classroom in units of square meters
(n) The volume of your body in units of gallons

9. In each of the cases below, either display a value to the requested number of significant figures or display a calculated value to the appropriate number of significant figures.

(a) Round to three significant figures: 0.70973 kilograms
(b) Area of a box measuring 1.15 centimeters long by 1.62 centimeters wide. Area is determined by length times width.
(c) You measure something using a ruler that has markings every 1/16 inches. You measure an object to 3 and 11/16 inches, commonly written as 3-11/16". Report its length in decimal format in inches.
(d) (536,000 meters) × (6,576 meters)

10. The highest mountain in the world is Mount Everest in Nepal. The peak of Mount Everest is 29,029 feet above sea level. Convert the height from feet [ft] to miles [mi].

11. To be considered a full time employee, companies in the United States require you to work more than 30 hours in a week. Convert the time 30 hours [h] into units of minutes [min].

12. Officially, a hurricane is a tropical storm with sustained winds of at least 74 miles per hour. Convert this speed into units of kilometers per minute.

13. Many toilets in commercial establishments have a value printed on them stating the amount of water consumed per flush. For example, a label of 2 Lpf indicates the consumption of 2 liters per flush. If a toilet is rated at 3 Lpf, how many flushes are required to consume 20 gallons of water?

14. A hogshead is a unit of volume describing a large barrel of liquid. Convert 10 hogsheads into units of cubic feet. Conversion factor: 1 hogshead = 63 gallons.

15. A category F5 tornado can have wind speeds of 300 miles per hour [mph]. What is this velocity in units of meters per second?

16. A new hybrid automobile with regenerative braking has a fuel economy of 55 miles per gallon [mpg] in city driving. What is this fuel economy expressed in units of feet per milliliter?

17. The AbioCor™ artificial heart pumps at a rate of 10 liters per minute. Express this rate in units of gallons per second.

18. Which of the following is the longest distance?
   (a) 26.4 miles
   (b) 40 kilometers
   (c) 2,500 yards
   (d) 100,000 feet

19. Which of the following is the largest volume?
   (a) 50 gallons
   (b) 100 liters
   (c) 1.5 cubic meters
   (d) 2.5 cubic feet

20. In 2001, the first iPod™ by Apple had a rated battery life of 10 hours to run audio files. The 6th model, introduced in 2009, had a rated battery life of 36 hours to run audio files. If the average song is 3.5 minutes, how many more songs can you listen to using the 6th model iPod rather than the original iPod on a single battery charge?

21. One of the National Academy of Engineering Grand Challenges for Engineering is **Provide Access to Clean Water**. Only 5% of water is used for households—the majority is used for agriculture and industry.
   (a) It takes 240 gallons of water to produce one pound of rice. How many liters of water are needed to produce one kilogram of rice?
   (b) It takes 1,680 gallons of water to produce one pound of grain-fed beef. How many cubic feet of water are needed to produce onc kilogram of beef?
   (b) It is estimated that 528 gallons of water are required to produce food for one person for one day. How many liters per year are required to feed one person?

22. What are the dimensions of the constant coefficient (k) in the following equations?
   (a) Energy = k * (mass) * (temperature)
   (b) Force = k * (pressure)
   (c) Pressure = k * (temperature)/(volume)

23. The largest hailstone is the United States was 44.5 centimeters in circumference in Coffeyville, Kansas. What is the diameter of the hailstone in units of inches?

24. What is a system input? What is a system output?

25. Think of a system to squeeze orange juice. Identify the input, process, and output.

26. What is feedback? Explain feedback in the example of driving a car.

27. What systems can you think of that are open-loop systems? Would they be improved if feedback was added to the loop? Why or why not?

28. Compare and contrast a technological system and a nontechnological system.

29. Does monitoring a system always allow control over it? Why or why not?

30. Give an example of a controllable system, and explain what makes it controllable.

31. You want to play a movie at home on a DVD player. Identify the input, process, and output in using the system you choose.
32. Choose a system that you are familiar with (e.g., a toaster, a home heating system, an automobile cruise control) and diagram it using the universal systems model.
33. What is the difference between specifications and constraints? Use examples to explain the difference.
34. What is a prototype? Why is it used in the design process?
35. What is probability? How is it related to design?
36. What are the eight steps of the informed design process? Illustrate these steps with the design of a piece of new playground equipment.
37. What is ergonomics? Give three examples of products that use ergonomic design.
38. What are the six main principles of design? Select a piece of sports equipment that you have used, such as a tennis racket, and describe each of the principles in terms of that product.
39. You are in charge of market research for a new backpack. Design a questionnaire you will use to gather information from your classmates.
40. Write a 1-page paper on past participation in, or participate in and assume a leadership role in, a design or science competition (e.g., A Technology Student Association, Lego League, First Robotics, Science Olympiad, or Science Fair competition). How did you demonstrate leadership?

CHAPTER 4
ENGINEERING NOTEBOOKS: DESIGNING A ROBOTIC ARM

Designing new creations is one of the highlights of an engineering career. The opportunity to imagine, transform, and create something that addresses a need is one reason people choose to become engineers. There are times when, given a pile of components, engineers (and nearly everyone else) cannot wait to dive right in and start putting things together. As we have seen with the design process, it is to our advantage to understand our constraints, understand the criteria that will define a successful design, plan and assess potential solutions, and - once something is built/designed/prototyped - test the design to ensure it works and meets the criteria.

4.1 ENGINEERING DOCUMENTATION: A HISTORICAL PERSPECTIVE

Leonardo da Vinci (1452–1519) was widely known as an artist, inventor, mathematician, and engineer. Many of da Vinci's designs were eventually realized: for example, he designed a flying machine that was similar to a modern helicopter. While he did not live to see a flying machine, he studied and designed models that would plausibly allow humans to fly. Da Vinci documented these designs famously in a series of journals and loose notes (Figure 4-1). Imagine if he had not documented these ideas: Someone widely regarded as one of the most intelligent people to have lived may have been forgotten, or turned into some kind of folk legend!

Figure 4-1 Entry from Leonardo da Vinci's design notebook.

Another one of the most famous inventers and engineers, Thomas Edison (1847–1931), had many inventions that framed life as we know it. Imagine life without light bulbs or the recording of sound. Edison held more than 1,000 patents, documenting his inventions and protecting his intellectual property. A patent on the light bulb means that someone cannot steal the idea and claim that they invented it; the honor belongs to Edison alone. Although Edison began to invent and engineer in the late 1860s, he began to systematically keep written documentation in 1871. One of his journals from 1870 contains the following note: "All new inventions I will here after keep a full record." Eventually, Edison and his team produced over 3,000 notebooks with designs and notes to document the design process.

Figure 4-2 Edison's phonograph: one invention that changed society.

Both of these famous engineers/inventors documented their work to prove that they were indeed working toward a solution or an invention. This does not mean that everything worked or that the idea being documented is going to be an important or influential invention by any means. In fact, these notebooks contain all work – including some errors! Fortunately, there are procedures to deal with errors, changes in plans, and even success!

4.2 THE ENGINEERING NOTEBOOK

The engineering notebook provides you the opportunity to document the entire thought process, development, investigation into criteria and constraints faced, mistakes, and successes. It is the place to record calculations, ideas, and sketches leading to the final solution. The purpose of the notebook is to outline a detailed, chronological record of the development of the design.

Figure 4-3 Record the entire design process in your engineering notebook.

Why would we care about a notebook of this type? Wouldn't it be easier to use a 3-ring binder or an electronic document? If the design becomes patentable (or it is an idea that can make you rich and famous), this notebook serves as a record that the idea was yours. Since it is chronological, it shows the complete thought process behind the design. Further, if you unexpectedly leave to join a prestigious university or to begin a high-salary career, the notebook will provide a record to pass to the person who will take over the project. Surely the design you are working on is valuable and deserves to be preserved and continued. Without your expertise, this notebook will serve as the tool the new team needs to complete your design.

Documentation — material providing official information or evidence — is incredibly important in the design process. We can think of engineering documentation a few different ways, one of which is as a permanent record as we just discussed. In the same way we can peruse the drawings of da Vinci and Edison and spot mistakes or poor ideas along the beaten path of the design process, your notebook should not only catalogue your success, but your struggles as well. Saving the poor designs could provide inspiration for a redesign, like a phoenix re-emerging from the ashes.

The notebook should be bound: While it would be easier to use a binder, a binder also allows someone to remove and insert pages. As a result, the record isn't chronological anymore. A quad-ruled notebook would be ideal, but a composition book – or even plain paper – will work. The style of the notebook is the preference of the designer.

What are some key features of the engineering notebook? Your teacher may have specific requirements, or the type of project or the team you work with might dictate some specific things in the notebook, but a general set of guidelines is as follows:

- The first page or two should be reserved for a Table of Contents (TOC). When you start a new section, enter it into the TOC.
- Each page should be numbered consecutively, beginning with the first page after the TOC. If you use one side of a piece of paper, don't write on the back of any of the other pages – you should be consistent. If both sides are used, both sides should be numbered.
- Your work should be clear and neat. Remember that one purpose of the notebook is to allow you to share the work or pass it to someone else, who could then duplicate

your work. On the other hand, this isn't meant to be a perfect record suitable for a book: It is meant to capture the ideas and work of the author. Treat the notebook as though your work will lead to a new invention, and the notebook will need to be the documentation when you go for the patent.

- You should only use ink – not pencil. Pencil can be erased, which gives the appearance of tampering with the record.
- Each page should have the date and each page or each major section should have the name of the author.
- Entries should be written in chronological order. Don't try to add sections between existing pages. Use the pages in order. When a new idea starts, enter it in the TOC.
- Start on a new page each day. If there is room at the bottom of a page at the end of the day, cross out the empty section (see below).
- Try not to leave large, blank sections. If the section of work is ended and you are ready to start on a new page, use a single line to cross out the blank area. It's often good to sign the line to show that it was intentionally drawn there.
- In the case of an error, cross out the incorrect work with a single line. Don't try to destroy the work, try to erase it, or use correction fluid. Documenting the mistakes along the way help to establish that the notebook is a record of your progress.
- The notebook should include all calculations, sketches, and work. For all calculations and measurements, use consistent units.

Other notebook guidelines can be found online; your teacher may also have additional ideas. Example pages are shown in Figure 4-4.

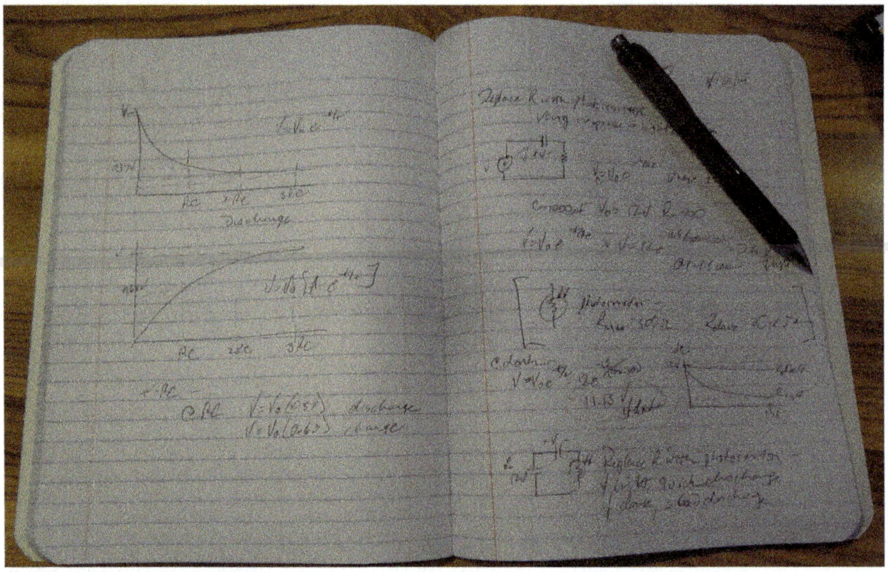

Figure 4-4 Sample engineering notebook.

4.3 THE DIGITAL PORTFOLIO

In many ways, the digital portfolio is similar to the engineering notebook: It contains sketches and can contain chronological notes, ideas, even mistakes. However, it is different in many ways as well.

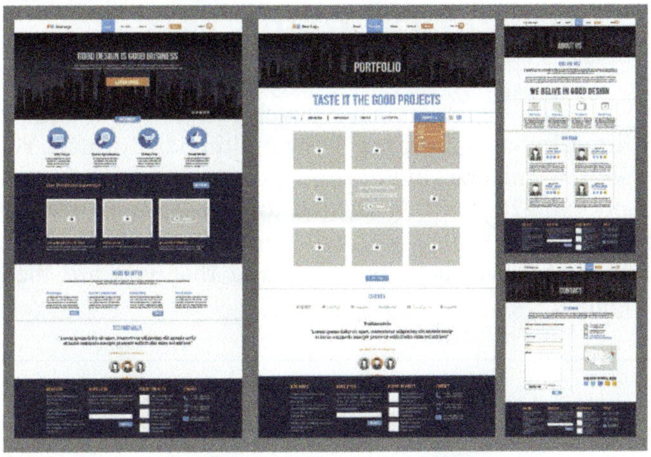

Figure 4-5 The front page of your digital portfolio should give a good overview.

A digital portfolio is typically a Web site or a set of shared documents detailing a design. It may be specific to one design, or it may be a record of multiple designs for a single person or team. In some cases, a person or team may choose to build a PowerPoint file and call this the digital portfolio in its original PowerPoint form. It could also be saved as a read-only document such as a .pdf file.

Because it is digital, the portfolio is expected to be effective as a visual presentation. Therefore, aesthetic ideas like layout, background, and organization can be important as you try to get your message across. In addition, a digital portfolio can be edited and rearranged, so it is not considered a chronological presentation of your design process. Instead, it is more useful as a tool to share your design. The reader doesn't need to see the details to see significant accomplishments or final ideas.

Further, because it is digital, the portfolio can be far more interactive than an engineering notebook. Imagine da Vinci's notebooks if they could have links and animation!

Teams can create a digital notebook through the use of shared document space on the Cloud, such as Google docs. Some sites allow individuals to build their own Web sites; one of these could host your digital portfolio.

4.3.1 ARRANGING INFORMATION IN A DIGITAL PORTFOLIO

If you have browsed through pages on the Internet, you have probably seen some pages that look "right" and others that are simply unappealing or difficult to use. Trust your instincts when designing your Web-based digital portfolio. If you want an audience to discover, view, and share the resource, the layout should not detract from your message. Here are some tips that may be helpful:

- Create a visually appealing homepage with clear links to projects.
- Make your best products appear first to draw the attention of the viewer.

- Describe your solution.
- Consider your audience as you describe your solution.
- Be sure to link back to the original problem and/or to reflect on what the solution may mean to you or to society.

4.4 COMMUNICATION REVISITED

Most successful engineers have developed good communication skills, not only oral and written, but also those involving multimedia formats. You might have the best idea in the known universe, but at some point you are going to have to convince someone to supply the $200 million needed to develop it. You *must* be able to communicate effectively not only that you have this great idea for a practical antigravity device, but also that it will actually work and that you are the person to lead the team developing it.

WISE WORDS: HOW IMPORTANT ARE COMMUNICATION SKILLS AT YOUR JOB?

When working with clients, we deliver our approach, analysis structure, status, findings, and final deliverable by presentation.

M. Ciuca, ME

Absolutely essential. While I thankfully have a job that allows me to dig into the math and analysis, I still deal with a lot of people. Being able to communicate effectively, where two people (or more) really understand what each is saying is very important on a complex project, and is often more difficult than one would think; or at least it involves more active participation than many are inclined to put into it.

J. Dabling, ME

I have to write design documentation, edit customer specs, produce design review presentations, and sometimes present to our customers.

E. D'Avignon, CpE

Being able to connect with customers, and internal team members, is the means to develop relationships and win new business.

B. Dieringer, ME

Absolutely critical. I can think of people that I have worked with who do not have these skills, and they are not easy or fun to work with.

S. Forkner, ChE

Social skills are crucial to success when working at a company. Many times the person who is most successful is not necessarily the one who has the best or brightest ideas, but rather the individual who has solid ideas and is able to communicate in a manner that allows others to easily understand the vision and path to goal achievement.

L. Gascoigne, CE

Good social skills are very important because they improve productivity, teamwork, and goodwill. I have also found social skills can be a deciding factor in job advancement.

R. Holcomb, IE

4.4.1 BASIC PRESENTATION SKILLS

Most students consider giving an oral presentation a more daunting task than submitting written documents.

With this in mind, you need to focus on several factors when planning a presentation. Note that the first item is *who* the audience is, although the other factors mentioned are equally important.

Preplanning

- **Who is my audience?** Know the age group, demographics, prior knowledge about the topic, and what positions or opinions they may hold.
- **What is my purpose?** What do I hope to accomplish? What response do I expect? What will the audience get out of my speech?
- **Where is all the equipment I need?** Where will the talk be held?
- **When am I on the program agenda?** Will I be the first presenter (when audience is most alert) or the last one before lunch (when they are becoming restless) or after lunch (when they are sleepy)? What will I need to do to keep my listeners attentive?
- **Why am I giving this talk?** Why is the audience here?
- **How long should I talk?** Remember that only few people can focus for more than 20 minutes. Trim your talk so that people will ask for more information rather than thinking "When will he sit down?"

NOTE

5 Ws and 1 H

- Who
- What
- Where
- When
- Why
- How

Preparing the Verbal Elements

The preceding list focused primarily on logistics. In addition to these considerations, the structure of your presentation is vital. As a simple example, which of the following two sentences is easier to understand?

> Sentence A: While perambulating in the antithesis of the metropolis to evade the intemperate brouhaha thereof, my visual cortex perceived an ophidian.
>
> Sentence B: I saw a snake while taking a relaxing walk in the woods.

Although sentence A may be phrased in a more intriguing manner, it tends to obscure the underlying meaning. This is perhaps desirable in poetry or fancy fiction but generally detrimental to a professional engineering presentation.

To help you avoid such pitfalls, we offer the 4-S formula for structuring presentations.

- **Shortness:** Use short sentences, avoid too many details, and do not talk too long.
- **Simplicity:** Avoid wordy, lengthy phrases.
- **Strength:** Use active voice and action verbs, not passive voice and "to be" verbs.
- **Sincerity:** Convey empathy, understanding, and respect for the audience.

Keep in mind the purpose of discrete elements of a speech.

- **Introduction:** Purpose: to capture the interest of the audience. Your first task is to *hook your audience*. What is it about your subject that *they* (and not necessarily you) would find most interesting and relevant?
- **Body:** Purpose: to keep your audience interested. They will continue to pay attention if you keep the material interesting and relevant to them.
  - Divide the presentation into two or three main points.
  - Use one or more simple examples to illustrate each major point.
- **Conclusion:** Purpose: to pull it all together.
  - Summarize major points.
  - Show appreciation for your audience's attention.
  - Allow for a few questions, but be sensitive to your audience and the other speakers.

Preparing Visual Aids

Most of you have seen presentations that used slides with unreadable text, incomprehensible graphics, or annoying special effects. Well-designed graphics can greatly enhance your presentation, not only making it easier for the audience to understand, but also keeping their attention focused. Although our focus is on PowerPoint presentations, these suggestions apply to your digital portfolio as well.

Helpful Hints

- Keep each slide simple, with one concept per slide. As a rule, use no more than *six lines per slide*. Each slide should correspond to an average of *60 seconds* of speech.
- If possible, make slides in *landscape* format.
- Present data in simple graphs rather than in lists or tables. Avoid excessively complex graphs with extensive data. If you must present tables, divide them among several slides.
- Pictures, diagrams, and video simulations all may enhance your presentation. Be sure that all are large enough to be seen by the audience, and have color schemes that do not appear washed out when projected. Often, such items are designed for viewing on a small screen and do not project well. Be sure to test them prior to your presentation.
- Use bullet points with important phrases to convey ideas. Avoid complete sentences.
- Large size text is best. A font size of at least 18 points and preferably no less than 24 points should be used. This includes *all* objects, such as axis and legend captions, table headings, figure symbols, and subscripts.
- Use high-contrast colors. Avoid fancy fonts, such as cursive, or light colors, such as yellow or other pastels. Avoid using all capitals.
- Use a light background and dark print to keep the room brighter.
- Keep background styles simple and minimize animation to avoid distracting from the presentation. Keep all the slide backgrounds the same throughout a single presentation.

4.4.2 SAMPLE PRESENTATIONS

To illustrate the visual aids caveats, we critiqued two student presentations.

EXAMPLE 4-1 Sample Student Presentations and Digital Portfolio Pages

Original Presentation: Critique

- Slide 1 font is difficult to read, poor choice of abbreviation for approximately.
- Slide 2 dates and text appear disjointed due to text size and graphic; graphic is too large.
- Slide 3 graph is difficult to read.
- Slide 4 too many words; graphic is too large.

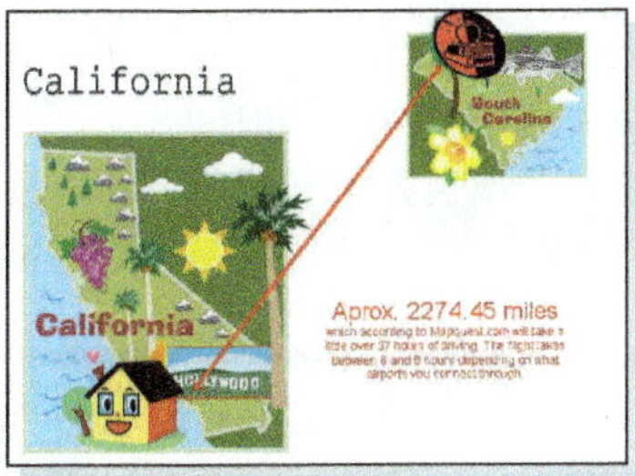

1

2

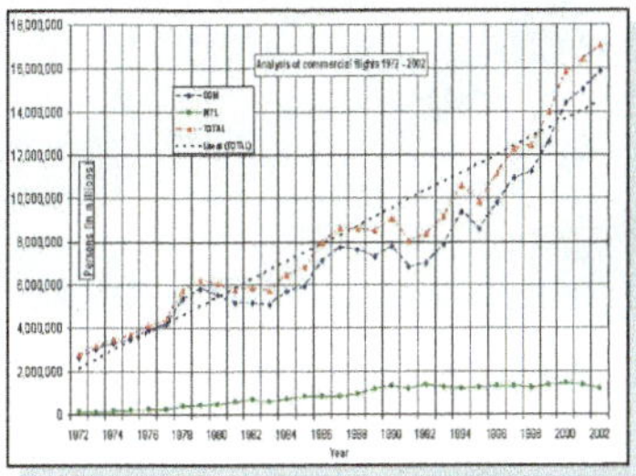

3

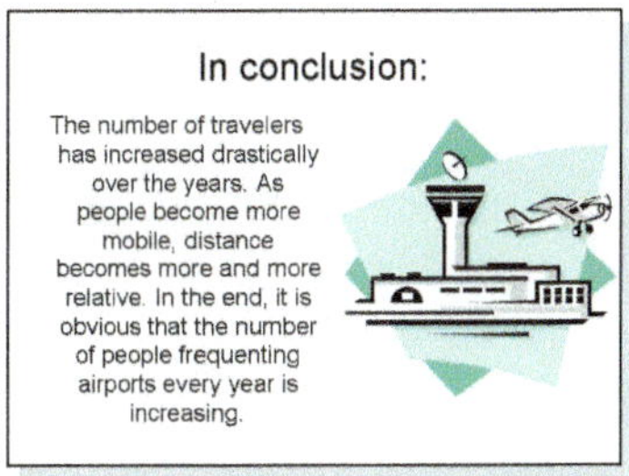

4

Improved Presentation:

1

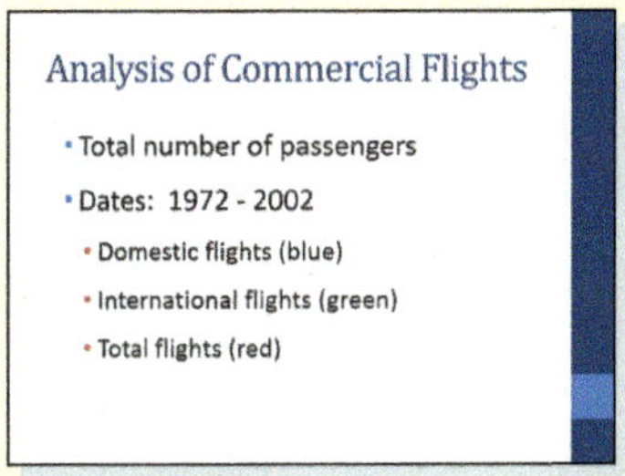

2

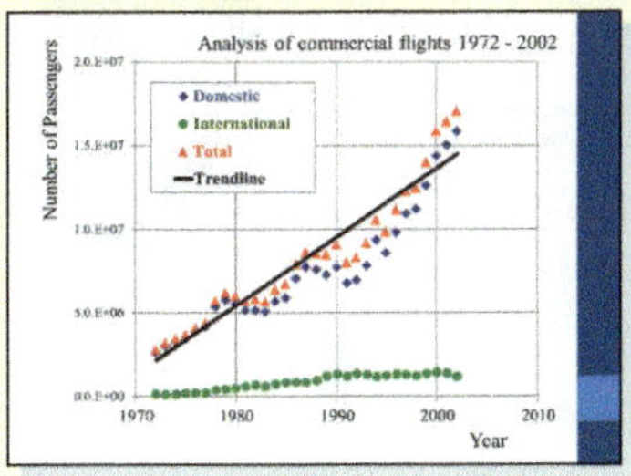

3

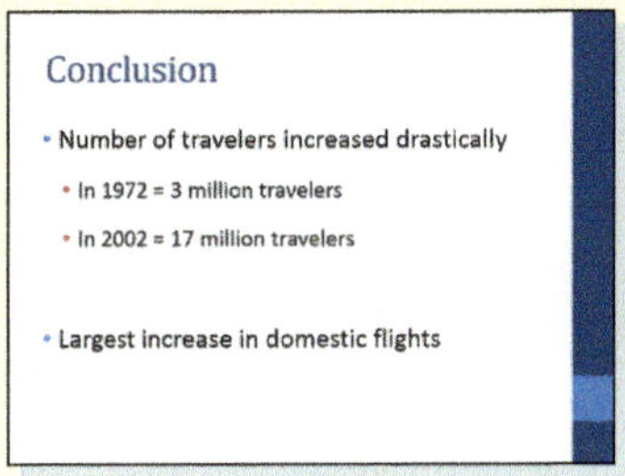

4

● EXAMPLE 4-2 Sample Student Presentations and Digital Portfolio Pages 2

Original Presentation: Critique

- Green backgrounds with white text do not project well; blue and red text is especially hard to read.
- Slides 2 and 4 have too many words; should use bullets, not sentences.
- No graphics; some pictures of acid rain damage would be helpful.

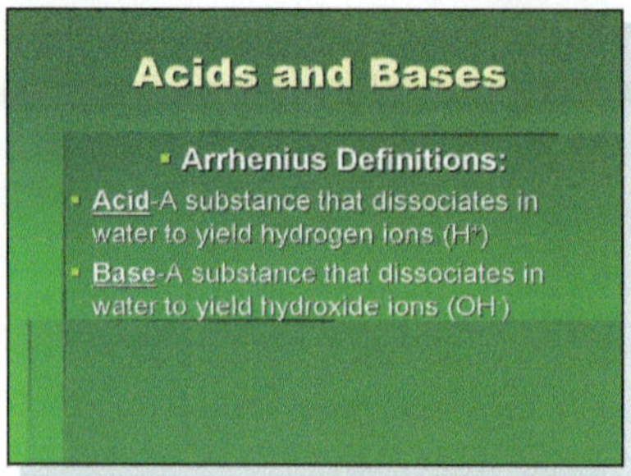

(1)

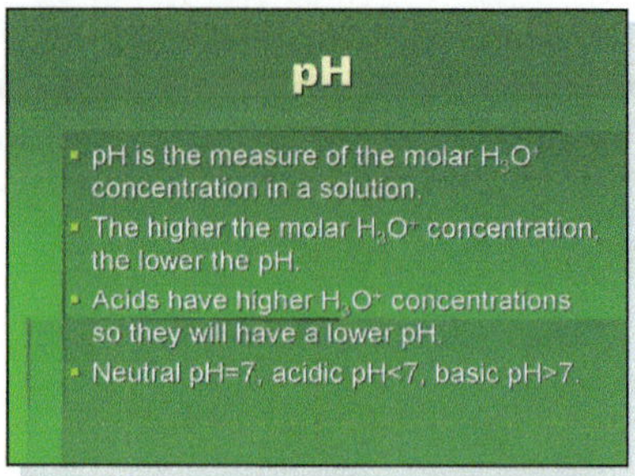

(2)

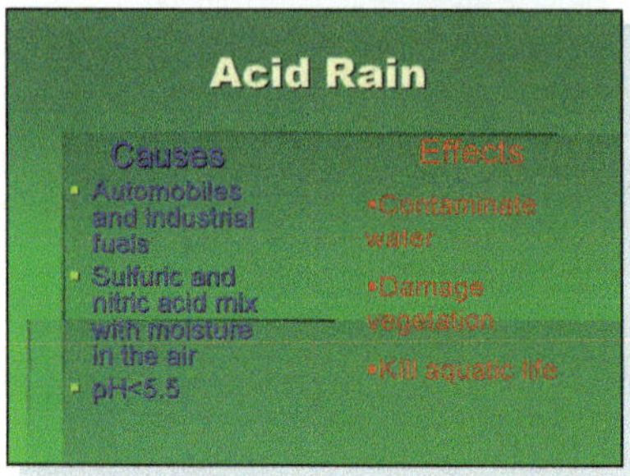

(3)

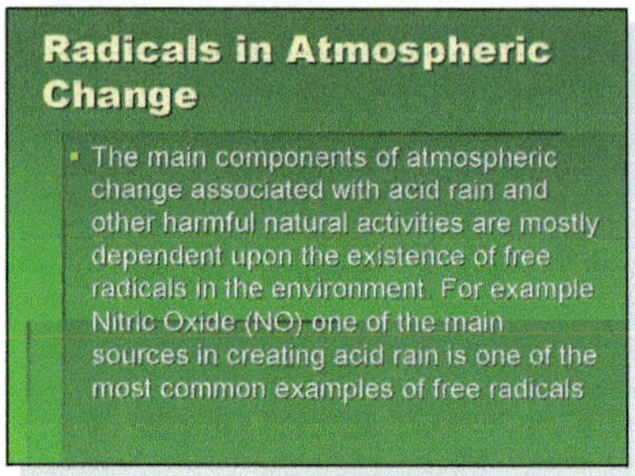

(4)

Improved Presentation:

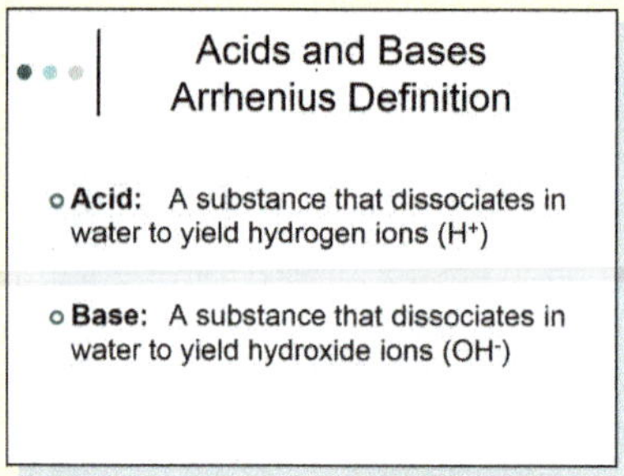

(1)

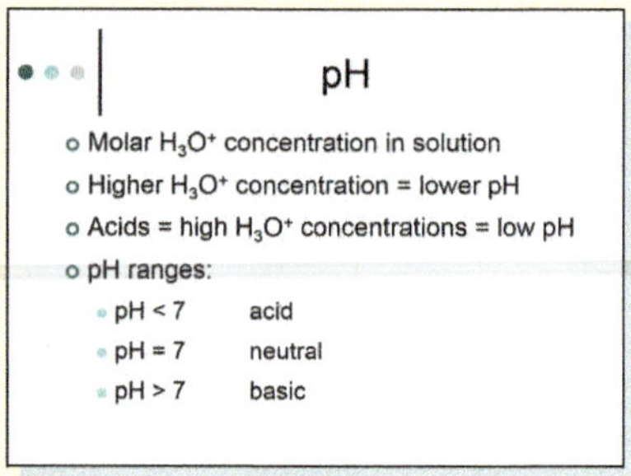

(2)

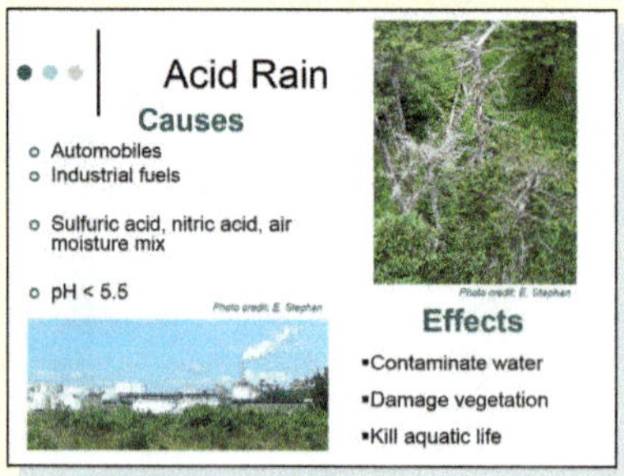

(3)

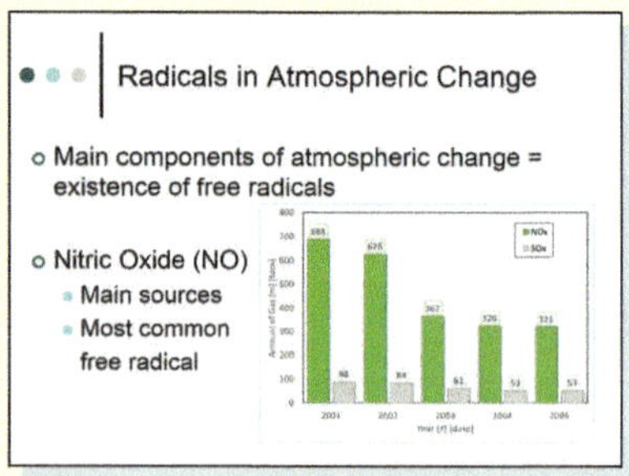

(4)

Making the Presentation

Oral presentations present several challenges for effective communication. How many of you have seen a presentation in which the presenter simply reads the contents of the slides with no embellishment? Do not inflict such things on your own audiences.

Presentation Dos and Don'ts

When delivering a presentation, do:

- Relax!
- Speak slowly and clearly, making good eye contact.
- When your hands are not busy, drop them to your sides.
- Rehearse your presentation *out loud* multiple times. If possible, have a friend critique.
- Arrive early enough to make sure that all technology is present and working, and resolve any problems you may discover.

When delivering a presentation, do not:

- Lean on your surroundings, turn your back to the audience, or cover your mouth while speaking.
- Read your presentation from a prepared text.
- Tell inappropriate jokes.
- Stammer, overuse the pronoun "I," or repeatedly say "um" or "uh." Do not be afraid of a little silence if you need to glance at notes or collect your thoughts.
- Chew gum, remove coins from pockets, crack your knuckles, etc.
- Shuffle your feet or slouch; move repetitively, for example, pace back and forth or sway.
- Play with your notes.

4.4.3 BASIC TECHNICAL WRITING SKILLS

Although most of you probably consider written communications much easier than oral presentations because the fear factor is largely absent, technical documents you produce will often be far more important to your company and your career than a live presentation of the same information.

Although we made some of these points earlier, effective technical writing requires its own set of guidelines.

- **Be clear**; use precise language. Keep wording efficient without losing meaning. Do not exaggerate.
- Ensure that the finished copy logically and smoothly **flows** toward a conclusion. Beware of "choppiness" or discontinuity. Avoid extremely long sentences because they may confuse the reader.
- If possible, use **10- or 12-point font size and 1 to 1.5 line spacing**.
- **Generally, prefer past tense verbs.** Keep verb tenses in agreement within a paragraph.
- **Define any terms** that might be unfamiliar to the reader, including acronyms and symbols within equations.
- **Present facts or inferences** rather than personal feelings.
- **Maintain a professional tone.** Do not be emotional or dishonest.
- **Number and caption all tables, figures, and appendices.** Refer to each from within the body of the text, numbering them in order of appearance within the text.
    - Tables are numbered and captioned *above* the table.
    - Figures are numbered and captioned *below* the figure.

Table 4-1 Example of a properly formatted table

Current (I) [A]	2	6	10	14	16
Energy of Inductor #1 (E1) [J]	0.002	0.016	0.050	0.095	0.125
Energy of Inductor #2 (E2) [J]	0.010	0.085	0.250	0.510	0.675
Energy of Inductor #3 (E3) [J]	0.005	0.045	0.125	0.250	0.310

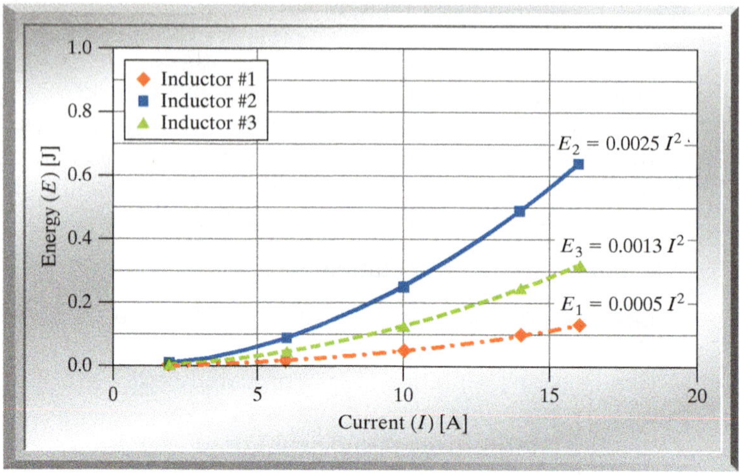

Figure 4-6 Example of a properly formatted figure.

- **Proofread and edit several times.**
  - **Remember to include** headings, figures, tables, captions, and references.
  - Do not assume that the spell check on the computer will catch everything! It will not distinguish between such words as *whether* and *weather*, or *was* and *as*.
- **Read it twice:** once for technical content and once for flow.
  - As you proofread, look for and *remove* the following: unnecessary words; sentences that do not add to the message.
  - Do one proofreading aloud. When you encounter commas, semicolons, colons, or periods, pause. Read a comma as a brief pause. Read a colon or semicolon as a longer pause. Read a period as a complete stop before the next sentence. Read what is actually written, not what you "think" it should say. If the text sounds stilted or blurred when read, you probably need to reconsider your use of these punctuation marks.
  - If possible, have someone not associated with the project (a friend or a mentor) read it, and ask that person for suggestions.
- **Spell out a number that starts a sentence.** If the number is large (e.g., a date), reword the sentence.
  - 23 points were outliers. (Unacceptable)
  - Twenty-three points were outliers. (OK)
- **Keep the leading zero with a decimal.**
  - The bridge cost .23 dollars per gram. (Unacceptable)
  - The bridge cost $0.23 per gram. (OK)
- **For long numbers, do not spell out.**
  - The average was one thousand, two hundred fifty-five grams. (Unacceptable)
  - The average was 1,255 grams. (OK)

Proper Use of References

Contributed by Ms. Jan Comfort—Engineering Reference Librarian, Clemson University Libraries

In today's wired age, most students immediately go to the Internet to find information. Although this can be an excellent source, particularly for preliminary research, there are definite risks associated with using online sources, since essentially anyone can put anything they want on the Web. For example, type "flat earth society" into your favorite search engine and check out the "truth" concerning our home planet, or explore how you can save the endangered Pacific Northwest Tree Octopus.

When making presentations or writing reports, it is important to verify the veracity of any sources you consult. These guidelines will help you avoid egregious errors in your own technical communications.

The **ABCs of evaluating information** offer a useful start.

- **Authority:** Is it clear who is responsible for the site? What are the author's credentials? Is the author an expert in the field? Is it a .com or .gov or .edu site?
- **Bias:** What is the purpose of the article? Is it free of obvious bias? Is the author presenting an objective view of the subject matter?
- **Currency:** When was the information created or last updated?

But there is more to evaluating resources than that. Good students take it to the next level. Here is how you can, too.

- **Use sources that have been reviewed by experts.** Instead of searching for hours trying to find Websites that meet stringent requirements, try using library sources to identify good quality sources that have already been through a review process.
- **Secure a peer review:** An expert in the appropriate field evaluates something proposed (as for research or publication).[1] Academic Search Premier and Expanded Academic ASAP are the names of two very good multisubject databases that contain scholarly (peer-reviewed) as well as popular articles. One or both of them should be available at your library.
- **Compare the information found in your article or Website with content from other Websites, or from reviewed sources.** Comparing sources can also alert you to controversial information or bias that will need further study. Are facts from one Website the same as those of another? How about depth of coverage? Maybe one site has better-quality information. Does the site have photos or other unique features that make it a good choice? Or perhaps a journal article from a library database is a better source. Until you compare several sources, you will not know what you are missing!
- **Corroborate the information.** Verify the facts from your source—regardless of where you found it—against one or more different sources. Do not take the word of one person or organization. A simple rule might be: "Do not use information unless you have corroborated it. Corroboration with varied and reviewed sources increases the probability of success."[2]

[1]Peer review. (2009). In *Merriam-Webster Online Dictionary*. Retrieved May 13, 2009, from http://www.merriam-webster.com/dictionary/ peer review

[2]Meola, Marc. (2004). "Chucking the checklist: A contextual approach to teaching undergraduates web-site evaluation." Portal: *Libraries and the Academy*, 4(3), 331–344.

- **Use the dollar symbol.**
  - The bridge cost four thousand dollars. (Unacceptable)
  - The bridge cost $4,000. (OK)
- **Watch for significant figures. Keep it reasonable!**
  - The bolt is 2.5029 inches long. (Unacceptable)
  - The bolt is 2.5 inches long. (OK)

Memo (One Page Limit)

The following template is for a one-page memo. Your teacher may ask you to adhere to this format, or may suggest a different one.

To:	Dr. Engineering
From:	Ima Tiger, Section 000 (IMT@school.edu)
Subject:	Memo Guidelines
Date:	May 21, 2015

Introduction: The first three or four sentences should explain the purpose behind the memo. You should attempt to explain what you were asked to do, what questions you are trying to solve, what process you are attempting to determine, etc.

Paste Special

Results: Place any experimental results, in tabular and/or graphical format, here. As space is limited, this normally only includes two items: two tables, two figures, or one table and one figure. Be sure that each is clear enough to stand alone, with one to two sentences of explanation. Be aware each table and figure should illustrate a *different* idea. Include a table caption at the top of each table, and a figure caption at the bottom of each figure. The caption should include a number and a word description. When a figure is used within a document, a title is not necessary on the graph and is replaced by the caption. The two items should be pasted side by side using the **Paste Special > JPEG** command or similar picture format command (PNG, Bitmap) and then sized appropriately.

Discussion: In this section, discuss how you obtained your data, the meaning of any trends observed, and significance of your results. Refer to the tables or figures shown in results by name (Table 1 or Figure 1). Explain any errors in your data (if possible) and how your data differs from theory. If you are deciding among several alternatives, in addition to justifying your final selection be sure to explain why you did NOT choose the other options.

References: List any sources you use here. You may use a new page for references if necessary. Any reference format is acceptable; Modern Language Association (MLA) citation style is preferred.

Remember: This document has a ONE PAGE limit!

NOTE

You should use a 10-point font such as Times New Roman or Verdana, with 1 to 1.5 line spacing.

Margins should be set to 1 inch all around.

Be sure to use correct spelling and grammar.

Include the headings given here, in bold.

Be sure to keep this memo to a ONE PAGE limit.

EXAMPLE 4-3 Sample of a poorly written student memo

To: Dr. Engineering
From: Ima Student
Subject: Memo
Date: April 1, 2016

Introduction: We are given the job to analyze the cost of upgrade a machine line, which produces widgets. We were given three companies to choose from, to figure out witch would be the best for the cost and its production. Just by graphing the variables would allow us to find our answer.

Results:

	Klein Teil	
Varible Cost	0.95	
Fixed Cost	5.00E+06	
Material Cost	0.75	
Energy Cost	0.15	
Labor Cost	0.05	
Selling Price	3	
apacity per da	6500	
antity Produc	ein Teil Total Co	Revenue
0.00E+00	5.00E+06	0.00E+00

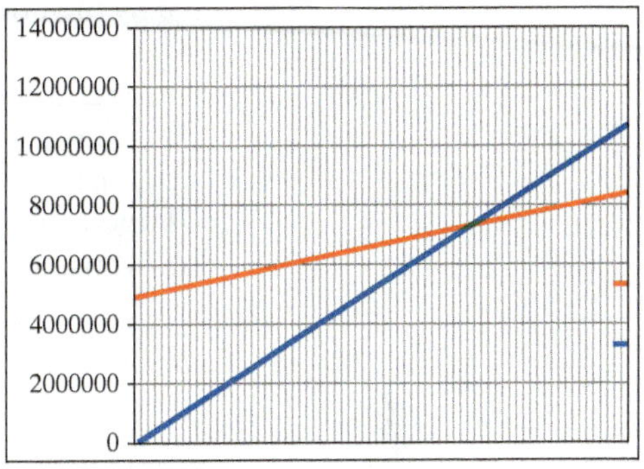

Discussion: We got the data by taking all variables from the information provided, then graphing the results together. This allowed us to see which machine line would provide the better outcome for the situation at hand. Considering the cost of the machine, material, labor, and energy into consideration with what would produce the quantity and quality product we're striving for. Figure 1 display all three solutions': total cost, revenue for us, and our breakeven point to ensure us of our choice. From observation of the graph we see that the Klein Teil machine is better. Its breakeven is at $2,400,000 and the profit is twice as much.

Summary: So to answer the question, Klein Teil would be our best option. The results yielded that the Klein Teil machine would give us the most quality for its price, a better production rate, and more money in return. From this research I hope you choose to take the Klein Teil machine.

Comments on this memo:
There are so many problems with this submission that we address only the major problems.

- *The subject line simply informs us that this is a memo.*
- *The introduction tells us very little about what the memo will address.*
- *The same data is presented in both the graph and the table. The formatting of the table is very poor; the formatting of the graph is worse.*
- *The discussion does not explain how the data was analyzed, and the justification of the final recommendation is essentially nonexistent.*
- *Similarly, the summary says almost nothing. Although not an appropriate summary, the three words "Buy Klein Teil" would probably have been more effective.*

How many more problems can you find in this sample memo?

Short Report (Two to Four Pages)

The following template is for a short report. Again, your teacher may ask you to follow these guidelines or provide a somewhat modified version.

Introduction Type the introduction here. This should be four or five sentences. What is the problem that will be addressed in this memo?

Procedure Type the procedure here. This should be at most ¾ page. It may be in bulleted format. You should generalize the procedure used to include the basic steps, but you do not need to include every detail. The reader should gain an understanding of how you collected your data and performed your analysis.

Results Insert the results here, but do not discuss them or draw any conclusions. This may include a maximum of three illustrations, in a combination of figures and tables. Be sure that each is clear enough to stand alone, with one or two sentences of explanation. Include a table caption at the top of the table, and a figure caption at the bottom of the figure. The caption should include a number and a word description.

Be aware tables and figures should illustrate *different* ideas. Do *not* include large tables of raw data or *every* graph generated. This section should be a *sample* of those items, used to illustrate the points of your discussion.

Table 4-2 Example of a properly formatted table

Section	Instructor	E-mail	Time
–030	Dr. Stephan	beths	M 8:00–9:55
–031	Dr. Park	wpark	M 12:20–2:15
–032	Dr. Sill	sillb	W 8:00–9:55

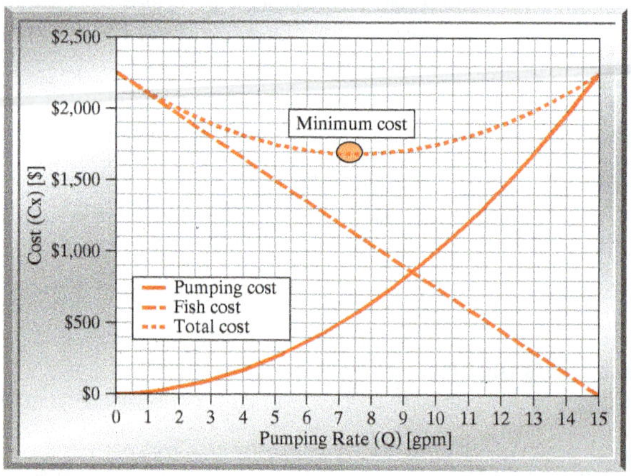

Figure 4-7 Example of a properly formatted figure.

Discussion Explain your results here. This can be up to a maximum of one page. Refer to the table and figure shown in results by name (Table 1 . . .). Be sure to include the items specifically requested in the original project description.

Summary What is the final conclusion? This should be four to five sentences long, and answer the initial questions asked in the introduction and summarize any important findings.

References: List any sources you use here. Any reference format is acceptable; Modern Language Association (MLA) citation style is preferred.

4.5 A ROBOT ARM

Figure 4-8 A robot says "hello!"

Robotic arms have found several applications in our everyday lives. One obvious use is on the factory floor on an assembly line. Manufacturing robots are responsible for performing an action that contributes to the construction of some product. For example, when a robotic arm is used on an assembly line at an automotive factory, then the product is a car. In a general sense, robots have ushered in an age of automation, the use of systems to operate equipment with limited to no human interaction – the robot-equipped assembly line being the most concrete example.

Figure 4-9 Robotic arms are used to assemble most automobiles.

Plenty of robots still require a human operator. Mechanized arms have found their way into medicine as surgeons operate mechanical arms that are incredibly precise; these arms and instruments replace the surgeon's hands inside a patient. Advances in robot-aided surgery allow for numerous patient benefits, including shorter hospital stays, decreased post-surgery pain, and less scarring.

The concepts of robots and artificial intelligence permeated our culture long before the advent of the first real android. Science fiction novelists and screenwriters have toyed with the ideas of the modern day robot – mainly with what could go wrong. For instance, the science fiction writer Isaac Asimov created the "three rules of robotics" which have been modified throughout the years since their original formulation:

- A robot may not injure a human being or, through inaction, allow a human being to come to harm.
- A robot must obey the orders given it by human beings except where such orders would conflict with the first law.
- A robot must protect its own existence as long as such protection does not conflict with the first or second laws.

Robots are currently engrained in popular culture, but may be further integrated into our lives in the future. Typical robots aren't electrically-powered human-like machines (these are often called *androids* or *droids*); plenty of household objects can fall under the classification of a robot. Domestic vacuum cleaners that independently clean the floor are indeed robots, but are only programmed to do one simple task. The continual improvement of technology could bring about new luxuries, perhaps a robot able to clean our entire house automatically.

Figure 4-10 Robotic vacuum cleaner.

The displacement of workers is an often-cited economic consequence of robots. When companies can replace an unskilled worker with a machine able to perform the same task without needing to pay it a salary, the working world must demand a different type of employee.

Although robots are not paid a salary, they still require regular maintenance. In addition, robots have lifespans even though they are not biologically "living" beings. Without proper maintenance, the expected lifespan of the robot can decrease and it can have reduced efficiency – meaning it performs its job worse than before. Moreover, if the robot is improperly installed, it will undergo similar losses. As a result, instead of doing the robot's job, robotics engineers will be designing it and robotics technicians will be performing the necessary maintenance. Robotics is considered a branch of mechanical engineering, computer science, and electrical engineering.

SETTING THE SCENE: A CENTURY OF AUTOMOBILE MANUFACTURING

In the early twentieth century, Henry Ford used new technology to manufacture cars in a very short time. He set up one of the first assembly lines. People were trained to work in his factory. They learned how to put together the different parts that made up every car.

Even though technology has changed dramatically since then, the basic concept behind the Ford assembly line is still used.

- Parts are moved along a belt.
- Parts are put together one piece at a time.
- Many identical products are made.

What is most different today is who—or what—is assembling the parts. A century ago, people worked on assembly lines. Today, more and more of the manufacturing process has been taken over by robots. Have people been replaced? No, but the work they do has changed. Now, people program the computers and make sure all the machinery is running properly. They manage the entire manufacturing system.

DESIGN ACTIVITY 4

ROBOTIC ARM DESIGN

Problem Situation

Robots are replacing people in many factories. They accurately perform repetitive tasks, such as welding car frames, assembling components, and packaging the completed product. Robots are often controlled by computers but can also be controlled by people operating them from a distance. Beanbag production is increasing, and you are directed to develop a robotic arm that can place them in containers. The first step in the development process is to design and construct a pneumatically controlled robotic arm that can grasp a beanbag and place it in a shipping container.

MATERIALS

You will need:

- 1/8" dowel rod
- 2 plungers
- 6 pneumatic syringes and tubing
- 12 centimeter sticks, 60 cm long
- cardboard
- cardboard strips
- glue
- plywood base, 12" square, 3/4" thick
- rubber bands
- small beanbag

Engineering Notebook:

Enter the following information in your engineering notebook. Throughout the entire project, document all steps, ideas, sketches, etc. in the notebook.

Class and period: ______________________________ Date:____________

List Students in Group:

______________________ ______________________

______________________ ______________________

______________________ ______________________

Materials: List the required materials.

Your Challenge

You and your teammates are to construct a pneumatically controlled robotic arm.

1. Clarify the Design Specifications and Constraints

To solve the problem, your design must meet the following specifications and constraints:

- The robotic arm must be able to grasp the beanbag.
- The robotic arm must be able to place the beanbag in a container 6 inches high and located 6 inches to the right or left of where it was grasped.

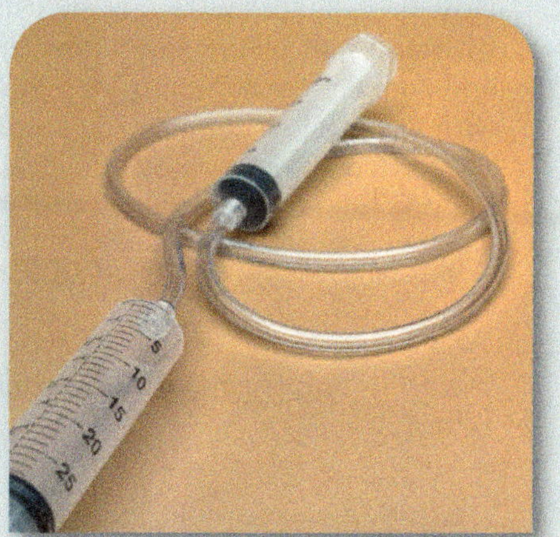

In your engineering notebook:

- State what the design challenge is.
- What are the specifications and constraints the design must meet?

2. Research and Investigate

In order to better complete the design challenge, you need to first gather information to help you build a knowledge base.

Knowledge and Skill Builder 1: **Pneumatics and Hydraulics**

Investigate pneumatics. Take two syringes of the same size, with the plungers partially withdrawn, and connect them with a piece of tubing. This is a pneumatic system. Experiment with them.

When you push one plunger in, does the other plunger move out the same amount?

Does the movement of the second plunger start immediately, or is there a delay? Why?

Look at your syringes. How precise is each mark? To what precision can you accurately measure the volume of liquid?

Can you hold the second plunger so it does not move or moves only a slight amount? Why?

Now investigate hydraulics by filling the syringes and the tubing connecting them with water and then eliminating the air from the system. Fill the syringes with water and connect the tubing to one of the syringes. Depress the plunger, filling the tube with water. Now connect the second syringe. There should be no air bubbles in the system. You have created a hydraulic system. Perform the same experiments as with the pneumatic system.

When you push one plunger in, does the other plunger move out the same amount?

Does the movement of the second plunger start immediately, or is there a delay? Why?

Can you hold the second plunger so it does not move or moves only a slight amount? Why?

What differences did you observe between the pneumatic and hydraulic systems?

What are some of the advantages and disadvantages of using each system?

Knowledge and Skill Builder 2: **Levers and Linkages**

A pivot point is a fundamental part of levers and linkages. You will need to use a piece of cardboard as a surface to which levers can be connected. Take an 18" cardboard strip and punch holes in it every 3", including one at the midpoint (9" from each end). Mount the cardboard strip on the larger piece of cardboard, using the midpoint as the pivot. (You can use paper fasteners as pivots. Inserting a piece of plastic straw in the hole will allow smoother rotation.) Pull down on one side, as shown in the illustration, and trace the path of both ends with a pencil. Notice the arc and the vertical distance moved. When the fulcrum is in the middle, both ends move the same amount, vertically and in an arc.

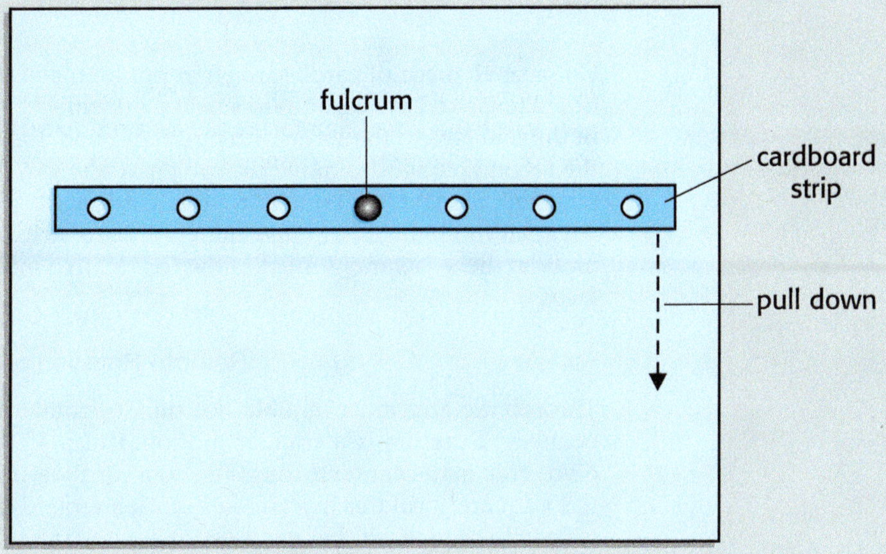

Now, relocate the pivot to the other holes and perform the same experiment. Record your measurements in the table. (Hint: You can use a string to measure the curved distance.)

Pivot Location	Left side vertical distance moved in inches	Left side distance of arc path in inches and degrees	Right side vertical distance moved in inches	Right side distance of arc path in inches and degrees
Center				
First left hole from center				
First right hole from center				
Second left hole from center				
Second right hole from center				

What did you find out about the lengths of the paths?

Cut a small piece of cardboard strip, put a hole in it, and place it between the cardboard lever and the cardboard surface, attaching the pivot through both. Attach one of the plungers to the lever with a rubber band. Instead of moving the lever, depress the second plunger to move the attached plunger. Notice that you have to hold the attached plunger or tape it to the cardboard surface. Observe the connection between the plunger top and the cardboard strip. As the strip moves in a circular motion, the attachment must allow for relative motion between these two flat surfaces.

Knowledge and Skill Builder 3: **Remote Rotation**

The robotic arm must be able not only to rotate, but also to lower and grasp the beanbag. Rotation is a circular motion. Cut a 4" square of cardboard and place a pivot hole in the center. Mount this to a cardboard surface. Notice that even though it is a square, it rotates in a circular motion around the pivot. Glue a small cardboard strip to the square so that it sticks out beyond the square as shown in the illustration. Pulling or pushing the strip will cause the square to rotate around its pivot.

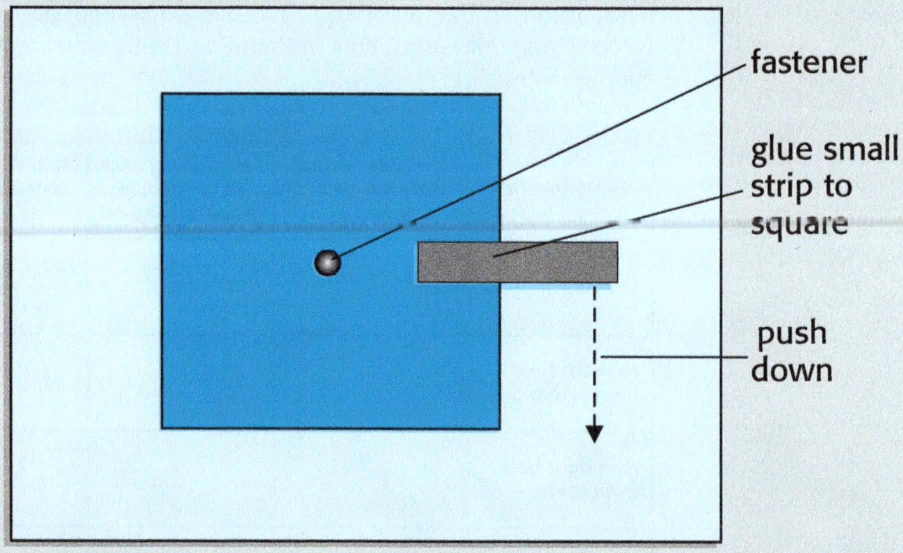

As you did in Knowledge and Skill Builder 2, attach one of the plungers to the cardboard strip with a rubber band. Instead of moving the strip, depress the second plunger to move the attached plunger. Notice that you have to hold the attached plunger or tape it to the cardboard surface. Observe the connection between the plunger top and the cardboard strip. As the strip moves in a circular motion, the attachment must allow for relative motion between these two flat surfaces. What are other ways you can create a flexible connection?

Where should you position the plunger to cause the square to rotate a greater amount? (Note: It is possible to relocate the strip.) Why?

Knowledge and Skill Builder 4: **Hinges**

Using two centimeter sticks, construct a frame with legs. The dimensions of the frame are 10 cm × 5 cm, with each leg measuring 10 cm long. The illustration shows what the structure will look like. Drill 1/8" holes as shown. Be sure to wear your safety glasses when you are drilling.

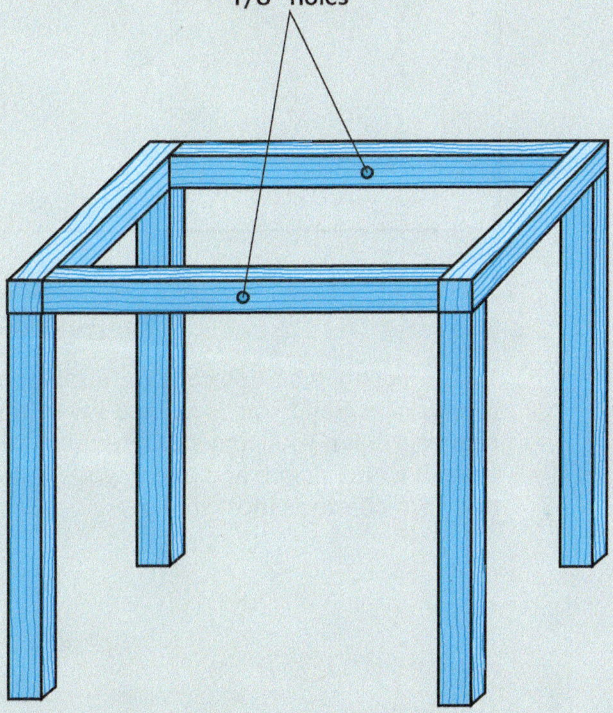

Cut a centimeter stick so that you have 2 sticks, each 15 cm long. Drill a 1/8" hole about 5 cm from the end in each stick. Take a piece of stick about 2 cm long and drill a hole in it. Take a piece of dowel and slide it through all three pieces of wood. This will be your pivoting arm.

Mount the pivoting arm on the dowel in the frame; this is a hinge. Notice that you can increase the space between the longer arms by including two or more drilled sticks. To prevent the arm from sliding in the frame, you can use drilled sticks to fill the space. Join the arms together by gluing a piece of wood between them. You have created a simple pivot arm. Where would you place the plunger to move the pivot arm? You can create support structures to hold the plunger. Indicate this in a sketch.

Knowledge and Skill Builder 5: **Getting a Grip**

Take a beanbag and place it on a table. Pretend that your forearm and hand are a robotic arm. Bend your wrist and lower your forearm until your fingers are above the beanbag. Open your fingers. Determine how you want to grasp the beanbag. Will you grasp it with a thumb and one finger? A thumb and two fingers? A different combination? Which way is most stable?

Now pick up the beanbag. Rotate your elbow, and place the beanbag in a different location. This is what the robotic arm will do. Is it necessary to have all your fingers move to grasp the beanbag? Try grasping the beanbag by just moving your thumb and holding your finger(s) stationary. What did you find out?

In your engineering notebook or digital portfolio, make a sketch of a gripper using the hinges and centimeter sticks. Show the location of the pivot and where the plunger would be attached.

3. Generate Alternative Designs

Describe two of your possible solutions to the problem. Remember to consider the specifications and constraints. In your description, indicate what you consider each solution's strengths and weaknesses. Include descriptions and sketches. Include such considerations as size, plunger location, hydraulic or pneumatic systems, and the mechanisms for causing each type of movement.

4. Choose and Justify the Optimal Solution

Choose your preferred solution. Explain why your solution meets the specifications and constraints. Why is this the better alternative?

What trade-offs, if any, did you make in selecting this alternative?

5. Develop a Prototype

Make a scale drawing, either a side view or isometric view, of the robotic arm and then construct it. If possible, include a photo or technical drawing of your prototype.

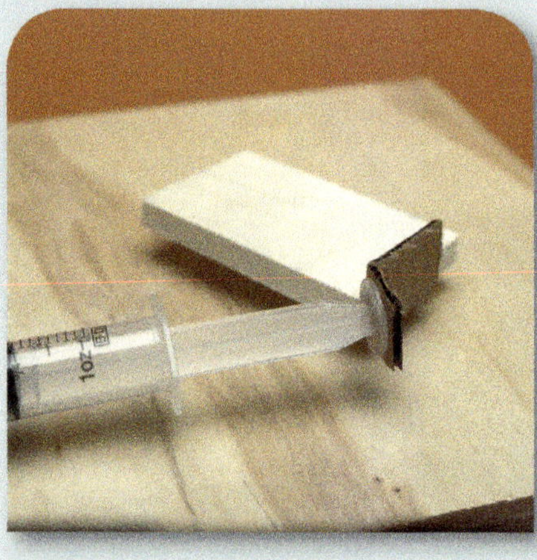

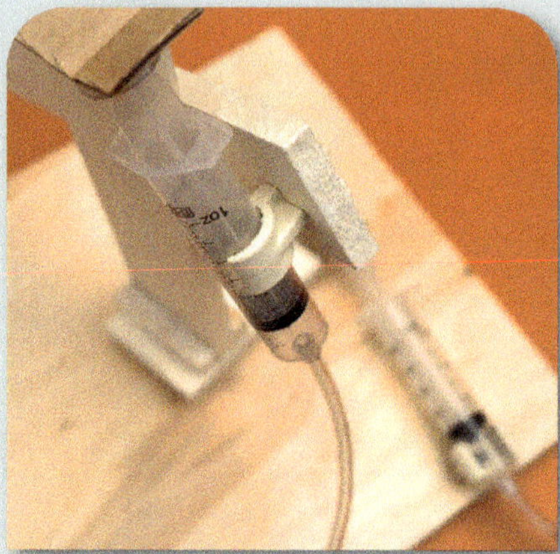

6. Test and Evaluate

Did your design meet the initial specifications and constraints? Indicate the tests you conducted and the experiments you performed to verify this.

Did you add any additional specifications? Describe the testing procedure and explain how the design meets these specifications.

7. Redesign the Solution

If you had to redesign your solution, what problems did you encounter? Did you change your original design concept? Why? If you had to redesign your model, what changes would you recommend in your new design? Explain your reasoning. What additional trade-offs, if any, would you have to make?

8. Communicate Your Achievements

Describe the plan you will use to present your solution to your class. If you work as an individual, you may use PowerPoint slides. A group may use a shared document (such as Google doc), PowerPoint slides, or other presentation mode as described by your teacher.

Chapter 4 ASSESSMENT QUESTIONS

1. If you are doing a design at school and your design will probably be very similar to other designs, why should you document your work in an engineering notebook?

2. Describe the key differences between an engineering notebook and a digital portfolio.

3. Describe how negligent and improper maintenance affects the performance of robots.

4. Research and describe two applications of robots and automation.

5. Research to identify an emerging trend in robotics and automation. Why is it considered an emerging trend? What types of engineers would be involved in this area?

6. Research: Find an advertisement for a job involving robotics and automation on the Internet. What type of engineer is invited to apply? What are some of the responsibilities of that position?

7. Research: Identify five inventions credited to Thomas Edison.
   - **(a)** How did one of these inventions change our lives (if it did)?
   - **(b)** Can you find any evidence of Edison's documentation on each invention? If a picture from the design process exists, capture the image.
   - **(c)** If you were working on one of these inventions, how would the entry for this invention begin in your engineering notebook?
   - **(d)** Select one design on which to focus: Describe specifically how this invention affected the following:
     - i. Individuals
     - ii. Society
     - iii. Economics within our society
     - iv. Culture: your own culture or a different culture
     - v. The environment
   - **(e)** Consider these five inventions. Can you generate an idea of an invention you would like to see, or perhaps would like to invent? Describe the invention as it would exist in 10 years. How will this affect society? On your own or with a partner, design a presentation to present this idea to your class.

8. Research: Consider the hydraulic experiment from the project. Investigate an area in which hydraulics are used to do incredible amounts of work. Find three examples; describe these applications and sketch each system.

9. Research: Find an engineer who made a significant contribution to society before the year 1999. Write a one-page paper with the following information:
   - **(a)** Name, birth year (and year of death, if applicable), and the engineer's hometown or place he or she lived while making the contribution.
   - **(b)** The significant contribution he or she made. Describe this engineer's role and how the invention affected society.
   - **(c)** Find how this engineer documented his or her design process. Is there a record of engineering notebooks or digital portfolios available?

10. Congratulations! You have been invited to participate in the design of a car that would not touch the pavement: Strong magnets would be used to "drive" over roads in which magnets would be imbedded. Please write a paragraph on each of the following aspects:

(a) How would the project be initiated? What are the first few steps after initiating the design?
(b) Describe the design process your team might implement.
(c) Would a prototype need to be developed? Why (or why not)?
(d) How would the team plan the project?
(e) How would the team execute, control, and monitor the project? What does it mean to execute the project? How is it different to control vs. to monitor a project?
(f) How would the project close? What would signify that the project was complete? What needs to be produced in order to close the project?

11. The purpose of this activity is to discuss the graphical representation of various phenomena, such as "A total eclipse" or "Using a toaster." The presentations should last 1 *minute*, 30 *seconds*.

Later in this text, much time is spent on graphing mechanics and interpretation. You are to choose a topic from the following list and represent it graphically. A variety of presentations of the available data that tell different things are possible. Do not forget to use all your senses and imagination.

You must prepare a proper graph, incorporated in PowerPoint, and use it during your presentation. You are not allowed to copy a premade graph from the Internet! In your speech, you should discuss the process, explain how information is shown on the graph, and be prepared to answer questions about your process. *You are limited to three slides*: (1) a title slide, (2) an introductory slide, and (3) a graph. If needed, you will be allowed a single 3×5 inch index card with notes on one side only.

Presentation Topic Suggestions

- Moving a desk down a set of stairs
- Letting go of a helium balloon
- Feedback from an audio system
- A cow being picked up by a tornado
- A kangaroo hopping along
- A glass of water in a moving vehicle
- The flight of a hot air balloon
- Using a toaster
- A train passing
- Dropping ice in a tub of warm water
- Slipping on a banana peel
- Sound echoing in a canyon
- Detecting a submarine using sonar
- Pouring water out of a bottle
- Hammering nails
- Playing with a yo-yo
- Climbing a mountain
- Pumping air into a bicycle tire
- The plume from a smokestack
- Formation of an icicle
- A rabbit family
- Eating a stack of pancakes
- A solar eclipse
- Traffic at intersections
- River in a rainstorm
- Hair

- Forest fire
- Cooking a Thanksgiving turkey
- Brushing your teeth
- Baking bread
- Eating at a fast-food restaurant
- Studying for an exam
- Power consumption of your laptop
- Oil supply
- Air temperature
- Student attention span during class
- Baseball
- A diet
- Sleeping
- Bird migration
- Strength of concrete
- Human population
- A mosquito
- A video game
- Taking a test
- A trip to Mars
- The North Pole
- Space elevators

12. Choose a major transportation structure failure to research from the list of potential topics that follows, and write a memo or presentation that includes the following:

- The name and location of the structure, when it was constructed, and other significant attributes.
- If possible, a photo or sketch.
- A summary of the event(s) surrounding the structure's failure, including dates. Discuss aspects of the design that contributed to its failure (materials, structural flaws, misuse, improper maintenance, climatic conditions, etc.). If it was rebuilt, repaired, or replaced, how was the design modified or improved?

(a) Angers Bridge over the Maine River in Angers, France
(b) Autoroute 19 Overpass, Quebec
(c) Arroyo Pasajero Twin Bridges in Coalinga, California
(d) Banqiao Dam, China
(e) Charles de Gaulle Airport, France
(f) Hartford Civic Center, Connecticut
(g) Hyatt Regency hotel walkway in Kansas City, Missouri
(h) Kemper Arena, Missouri
(i) L'Ambiance Plaza in Bridgeport, Connecticut

13. Write a 1-page memo to your teacher proposing that the class have an extra project dealing with robotics. Properly format the memo, and make a convincing argument.

14. List five of the most important factors in giving a successful speech in your opinion. Why are these the most important?

15. Research: Select two materials used in the robotics project.

(a) Identify two physical properties of each of the two materials that were important to the success of the project. Why were they important?

(b) Identify two mechanical properties of the grabbing mechanism. Why were these properties important?

(c) Write a paragraph on the difference between physical and material properties.

CHAPTER 5
ENGINEERING GRAPHICS AND CAD

computer-aided design
CAD

Engineering designs start as images in the mind's eye of an engineer. Engineering graphics has evolved to communicate and record these ideas electronically or on paper in both two and three dimensions. In the past few decades, the computer has made it possible to automate the creation of engineering graphics using software for **computer-aided design**, or **CAD**. Today engineering design and CAD are inextricably connected. Engineering design is communicated visually using engineering graphics produced with CAD software.

5.1 THE IMPORTANCE OF ENGINEERING GRAPHICS

"Visualizing" a picture or image in your mind is a familiar experience. The image can be visualized at many different levels of abstraction. Think about light and you might see the image of a light bulb in your "mind's eye." Alternatively, you might think about light versus dark. Or you might visualize a flashlight or table lamp. Such visual thinking is necessary in engineering and science. Albert Einstein said that he rarely thought in words. Instead, he laboriously translated his visual images into verbal and mathematical terms.

Visual thinking is a foundation of engineering. Walter P. Chrysler, founder of the automobile company, recounted his experience as an apprentice machinist where he built a model locomotive that existed "within my mind so real, so complete, that it seemed to have three dimensions there." Yet, the complexity of today's technology rarely permits a single person to build a device from his own visual image. The images must be conveyed to other engineers and designers. In addition, those images must be constructed in such a way that they are in a readily recognizable, consistent, and readable format. This ensures that the visual ideas are clearly and unambiguously conveyed to others. **Engineering graphics** is a highly stylized way of presenting images of parts or assemblies.

engineering graphics

A major portion of engineering information is recorded and transmitted using engineering graphics. To demonstrate the effectiveness of engineering graphics compared with that of a written description, try to visualize an ice scraper on the basis of this word description:

> An ice scraper is generally in the shape of a 140 × 80 × 10 mm rectangular prism. One end is beveled from zero thickness to the maximum thickness in a length of 40 mm to form a sharp edge. The opposite end is semicircular.

A 20-mm-diameter hole is positioned so that the center of the hole is 40 mm from the semicircular end and 40 mm from either side of the scraper.

It is evident immediately that the shape of the ice scraper is much more easily visualized from the graphical representation shown in Figure 5-1 than from the word description. Humans grasp information much more quickly when that information is presented in a graphical or visual form rather than as a word description.

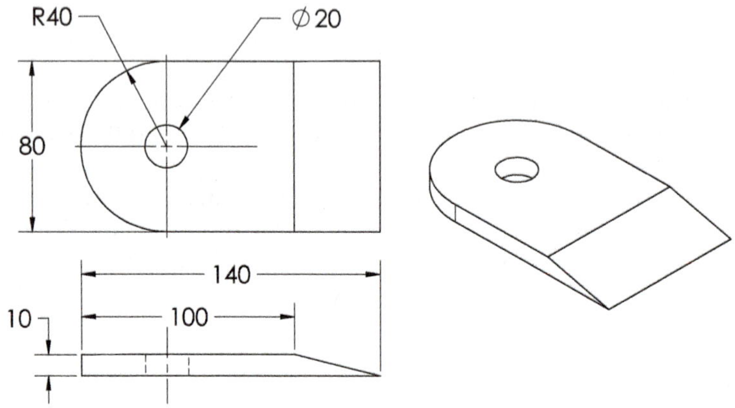

Figure 5-1 Graphical representation of an ice scraper.

Engineering drawings, whether done using a pencil and paper or a computer, start with a blank page or screen. The image in the engineer's mind must be transferred to the paper or computer screen. The creative nature of this activity is similar to that of an artist. Many designs begin with a pencil and paper sketch (Figure 5-2).

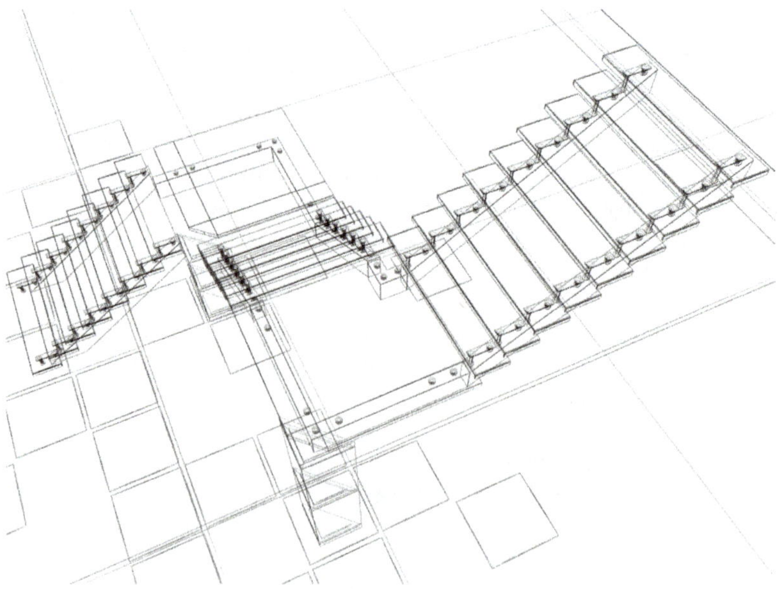

Figure 5-2 A sketch of a staircase.

5.2 ENGINEERING GRAPHICS

The first authentic record of engineering graphics dates back to 2130 B.C. based on a statue now in the Museum of the Louvre, Paris. The statue depicts an engineer and governor of a small city-state in an area later known as Babylon. At the base of the statue are measuring scales and scribing instruments along with a plan of a fortress engraved on a stone tablet.

pictorial perspective

Except for the use of pen and paper rather than stone tablets, it was not until printed books appeared around 1450 that the techniques of graphics advanced. Around the same time, **pictorial-perspective** drawing was invented by artist Paolo Uccello. This type of drawing presents an object much as it would look to the human eye or in a photograph, as shown in Figure 5-3. The essential characteristic of a perspective drawing is that parallel lines converge at a point in the distance as parallel railroad tracks seem to converge in the distance. Copper-plate engravings permitted the production of finely detailed technical drawings in large numbers using pictorial perspective. The pictorial-perspective drawings were crucial to the advancement of technology through the Renaissance and until the beginning of the Industrial Revolution. Similar to the artist version of pictorial perspective is the engineer's, architect's, and technology's version: the **2D drawing**, also called the two-dimensional drawing. A 2D drawing depicts all of the detail of an object without it popping out at you! Specifically, it is a drawing that has length and width but not depth. A great example of a 2D drawing that you would commonly find is a blueprint. A sample of a blueprint

Figure 5-3 An example of pictorial perspective.

can be seen in figure 5-4 below. Architects commonly used blueprints to plan and show how a new building should be laid out. But these drawings could not convey adequately details of the construction of an object. One solution to this problem was the use of

exploded view

the **exploded view** developed in the 15th century and perfected by Leonardo da Vinci. The exploded view of an assembly of individual parts shows the parts spread out along a common axis, as shown for the water pump in Figure 5-5. The exploded view reveals details of the individual parts, along with showing the order in which they are assembled.

The Industrial Revolution brought with it the need to more closely tie together the concept of a design with the final manufactured product using technical drawing. The perspective drawing of a simple object in Figure 5-6a shows pictorially what the object

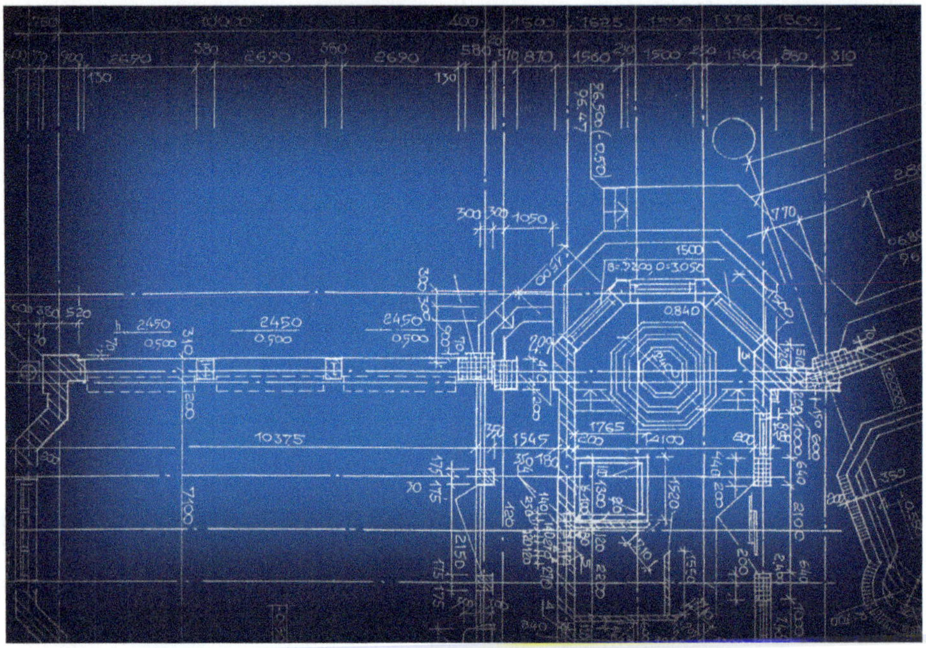

Figure 5-4 A blueprint is one kind of 2D drawing.

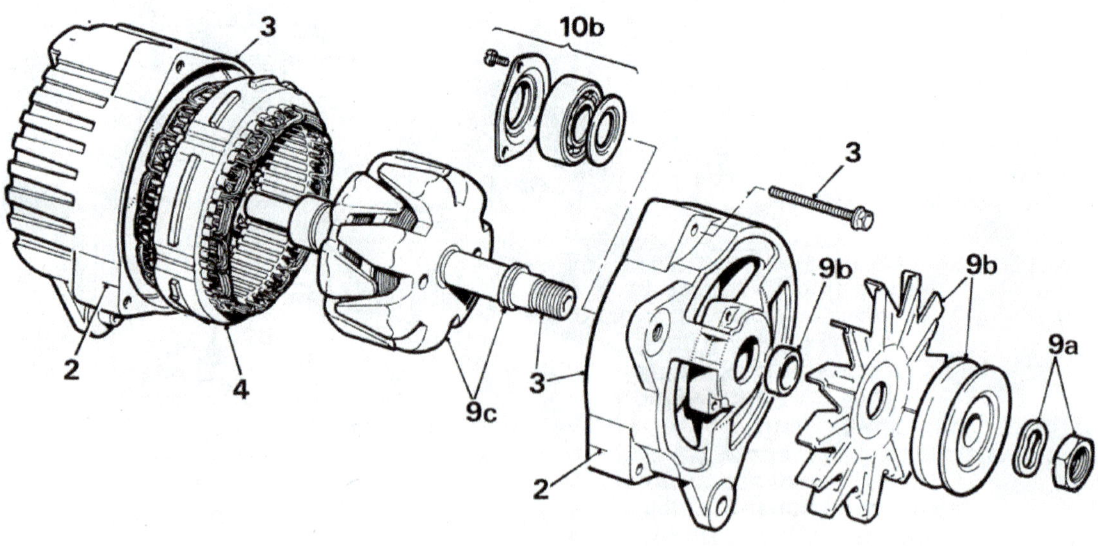

Figure 5-5 Assembled and exploded view of a water pump.

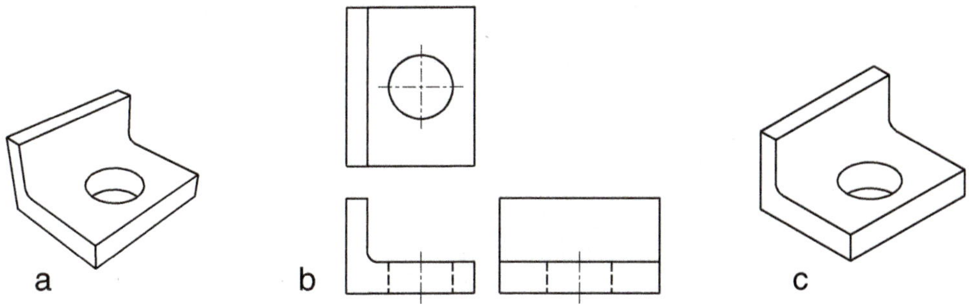

Figure 5-6 (a) Perspective projection (b) Orthographic projection (c) Isometric projection.

orthographic projections

descriptive geometry

isometric projection

metric

looks like. However, it is difficult to represent accurately dimensions and other details in a perspective drawing. **Orthographic projections**, developed in 1528 by German artist Albrecht Dürer, accomplish this quite well. An orthographic projection typically shows three views of an object. Each view shows a different side of the object (say the front, top, and side). An example of an orthographic projection is shown in Figure 5-6b. Orthographic projections are typically easy to draw, and the lengths and angles in orthographic projections have little distortion. As a result, orthographic drawings can convey more information than a perspective drawing. But their interpretation takes more effort than a pictorial perspective, as is evident from Figure 5-6. French philosopher and mathematician René Descartes laid the foundation for the mathematical principles of projections by connecting geometry to algebra in the 17th century. Much later Gaspard Monge, a French mathematician, "invented" the mathematical principles of projection known as **descriptive geometry**. These principles form the basis of engineering graphics today. But because those principles were thought to be of such strategic importance, they remained military secrets until 1795. By the 19th century, orthographic projections were used almost universally in mechanical drawing, and they are still the basis for engineering drawings today.

The **isometric projection** is used more often today than the pictorial-perspective projection. Historically, it was used for centuries by engravers. Then, in the early 19th century, William Farish, an English mathematician, formalized the isometric projection and introduced it to engineers. The isometric projection simplifies the pictorial perspective. In an isometric projection, parallel lines remain parallel, as shown in Fig. 5.6c, rather than converging to a point in the distance. In addition, the three coordinate axes have the same **metric** (hence the name ***iso-metric***). This means that a length unit (say, a centimeter) along one axis has identical length in the drawing along the other two axes. Keeping parallel lines parallel and retaining the same metric along all axes in an isometric projection distorts the appearance of the object slightly. But the distortion in an isometric projection is negligible for objects of limited depth. For situations in which the depth of the object is large, such as an architectural view down a long hallway, the pictorial perspective is preferable. The advantage of the isometric projection, though, is that it is much easier to draw than a pictorial-perspective projection.

5.3 PROJECTIONS

3-D projection
multiview projection

The goal in engineering graphics, whether it is freehand sketching or CAD, is to represent a physical object or the mind's eye image of an object so that the image can be conveyed to other persons. Objects can be shown as **3-D projections** or **multiview projections**. Figure 5-7 shows the handle of a pizza cutter both ways. The 3-D projection

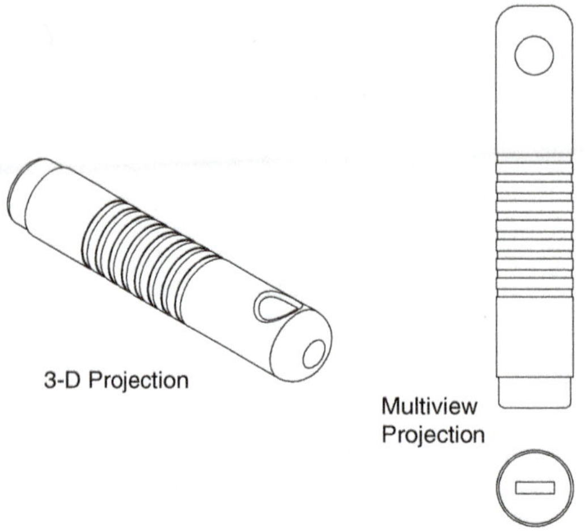

Figure 5-7 3-D projection and multiview projection of a handle.

clearly suggests the three-dimensional character of the handle, even though it is displayed on a two-dimensional medium (the page). 3-D projections are useful in that they provide an image that is similar to the image in the designer's mind. But 3-D projections are often weak in providing adequate details of the object, and there is often some distortion of the object. For instance, the circular hole through the right end of the handle becomes an ellipse in an isometric 3-D projection.

Multiview projections are used to overcome the weaknesses of 3-D projections. Multiview projections are a collection of flat 2-D drawings of the different sides of an object. For instance, the side and bottom end of the pizza cutter handle are shown in the multiview projection in Figure 5-7. Because there are two views, it is quite easy to depict the details of the object. In addition, taken together, multiview projections provide a more accurate representation of the object than the 3-D projection—a circular hole appears as a circle in a multiview projection. On the other hand, multiview projections require substantial interpretation, and the overall shape of an object is often not obvious upon first glance. Consequently, the combination of the overall image provided by 3-D projections and the details provided by multiview projections yields a representation of an object that is best. The shape of the object is immediately evident from the 3-D projection, and the detail needed for an accurate description of the object is available from the multiview projection.

5.3.1 3-D PROJECTIONS

Two types of 3-D projections are available in most CAD software: isometric and perspective. Three projections of a house are shown in Figure 5-8; these 3-D projections represent all three dimensions of the house in a planar image. Although it is clear in all three cases that the object is a house, each type of 3-D projection has its advantages and disadvantages.

projector
construction line
parallel projection

The isometric projection has a standard orientation that makes it the typical projection used in CAD. In an isometric projection, the width and depth dimensions are sketched at 30° above the horizontal as shown in Figure 5-9. This results in the three angles at the upper front corner of the cube being equal to 120° in the plane of the paper. Lines along the edges of the cube, known as **projectors** or **construction lines**, are parallel, so this type of projection is known as a **parallel projection**. Notice that the three sides of the cube are equal in length in the drawing, leading to the term isometric

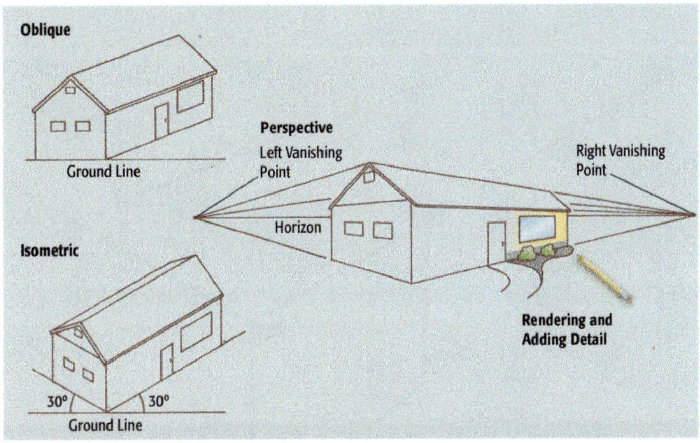

Figure 5-8 Isometric, oblique, and perspective projections.

(equal measure). Isometric drawings work quite well for objects of limited depth. However, an isometric drawing distorts the object when the depth is significant. In this case, a pictorial-perspective drawing is better.

perspective

foreshortening

A pictorial-perspective, or simply **perspective**, projection is drawn so that parallel lines converge in the distance, unlike isometric or trimetric projections, where parallel lines remain parallel. As shown in Figure 5-10, the projectors converge toward two points, known as vanishing points. This results in **foreshortening**—lines of an object are changed to produce the illusion of the part extending into the third dimension. Perspective projections can also be drawn with a single vanishing point, which is useful for showing interiors of buildings, or with three vanishing points, which can display perspective for very large objects. Perspective projections are most useful in providing a realistic image of an object when the object spans a long distance, such as the view of a bridge or aircraft from one end or of a tall building viewed from near its base. Generally, small manufactured objects are adequately represented by isometric or trimetric projections.

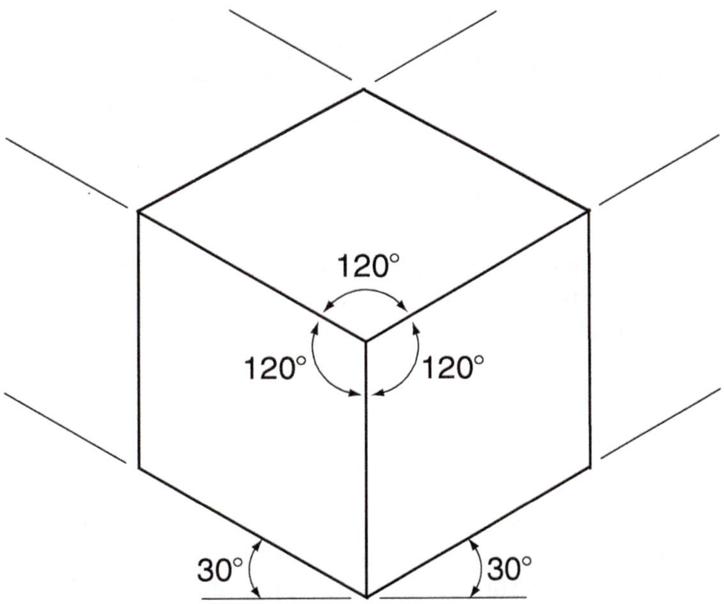

Figure 5-9 Isometric projection of a cube showing parallel projectors.

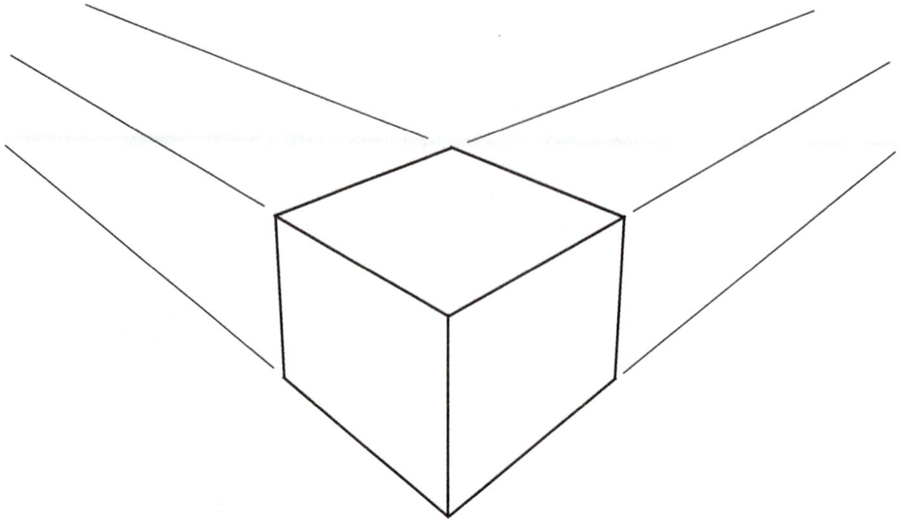

a

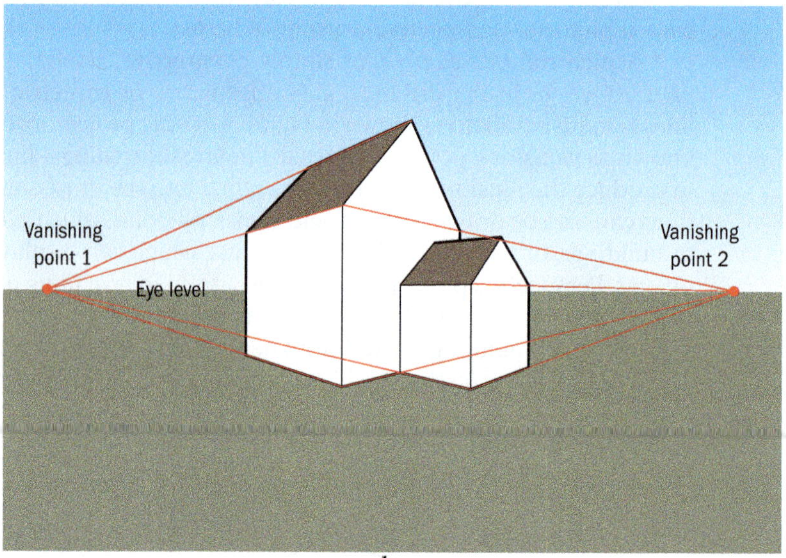

b

Figure 5-10 Pictorial-perspective projection of (a) a cube and (b) a building showing converging projectors.

oblique projection

Two types of pictorial sketches are used frequently in freehand sketching: isometric and oblique. The isometric projection is often used in freehand sketching because it is relatively easy to create a realistic sketch of an object. But the **oblique projection** is usually even simpler to sketch. The oblique projection places the principal face of the object parallel to the plane of the paper with the axes in the plane of the paper perpendicular to one another. The axis into the paper is at an arbitrary angle with respect to the horizontal. Figure 5-11 compares an isometric projection of a cube with a hole in it with an oblique projection of the same object. The advantage of the oblique projection is that details on the front face of the object retain their true shape. For instance, the circle on the front face is circular in the oblique projection, while it is elliptical in the isometric projection. This feature often makes oblique freehand sketching somewhat easier than isometric sketching.

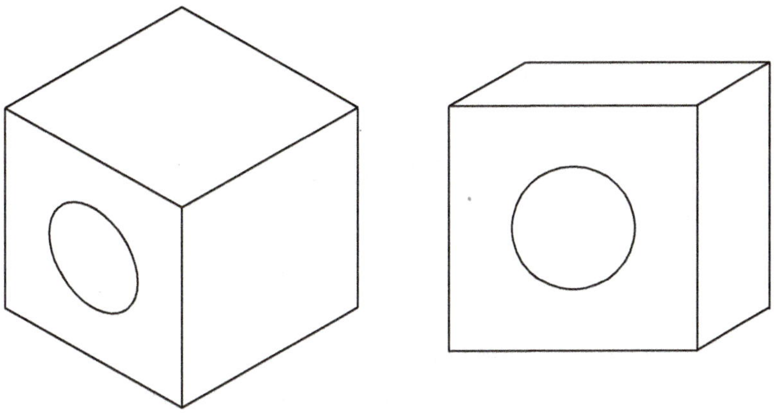

Figure 5-11 Isometric and oblique projections of a cube with a hole.

5.3.2 MULTIVIEW PROJECTIONS

The standard means of multiview projection in engineering graphics is the orthographic projection. Although 3-D projections provide a readily identifiable visual image of an object, multiview projections are ideal for showing the details of an object. Dimensions can be shown easily, and most features remain undistorted in multiview projections.

An orthographic projection is most easily thought of as a collection of flat 2-D views of different sides of an object—front, top, side, and so forth. For instance, two orthographic projections could be used for a coffee mug. The front view would show the sidewall of the mug along with the loop forming the handle. The top view would show what one would see looking down into the mug—a circular rim of the mug, the bottom of the inside of the mug, and the top of the handle that sticks out of the side of the mug. Dimensions of the mug could easily be added to the projections of each side of the mug to create an engineering drawing.

One useful way of looking at multiview projections is to imagine a glass box surrounding the object as shown in Figure 5-12. The image of each side of the object can be projected onto the wall of the glass box. Now an observer on the outside of the box can see each side of the object as projected onto each of the six walls of the box. Solid lines show the edges evident in the projection, and dashed lines show lines that are hidden by the object. Now imagine unfolding the glass box as if each of the edges of the glass box were a hinge, so that the front view is in the middle. Now the unfolded glass box represents all six sides of the object in a single plane as shown in Figure 5-13. In unfolding the glass box, the top view is positioned above the front view, the bottom view is below the front view, the right-side view is to the right of the front view, and so on. The dimensions of the object remain the same in all views. For example, the horizontal dimension (the width of the object) in the front view is identical to that same dimension in the top view and the bottom view. The views also remain aligned so that the bottom edge in the front view is even with the bottom edge in the right-side, left-side, and rear views. Likewise, the top edges remain aligned. Finally, the same edges in adjacent views are closest together. For instance, the same edge of the object is at the left side of the front view and the right side of the left-side view. This edge in the front view is closest to the same edge in the left-side view. See Figure 5-14 for an example using a cell phone.

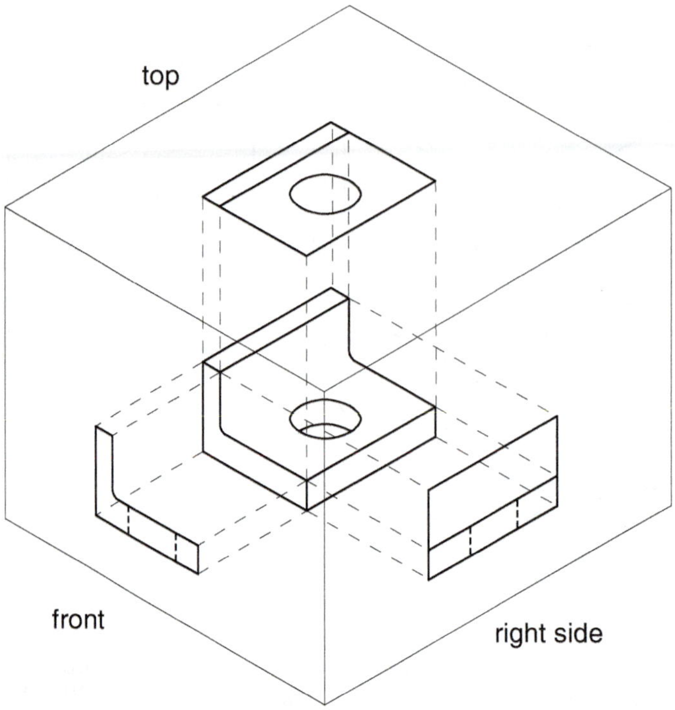

Figure 5-12 Sides of an object can be projected onto an imaginary glass box surrounding the object.

In many cases, three views are needed to represent an object accurately, although in some cases (like a coffee mug) only two views are necessary, and in other cases more than three views are needed to show complex features of the object. It is helpful to

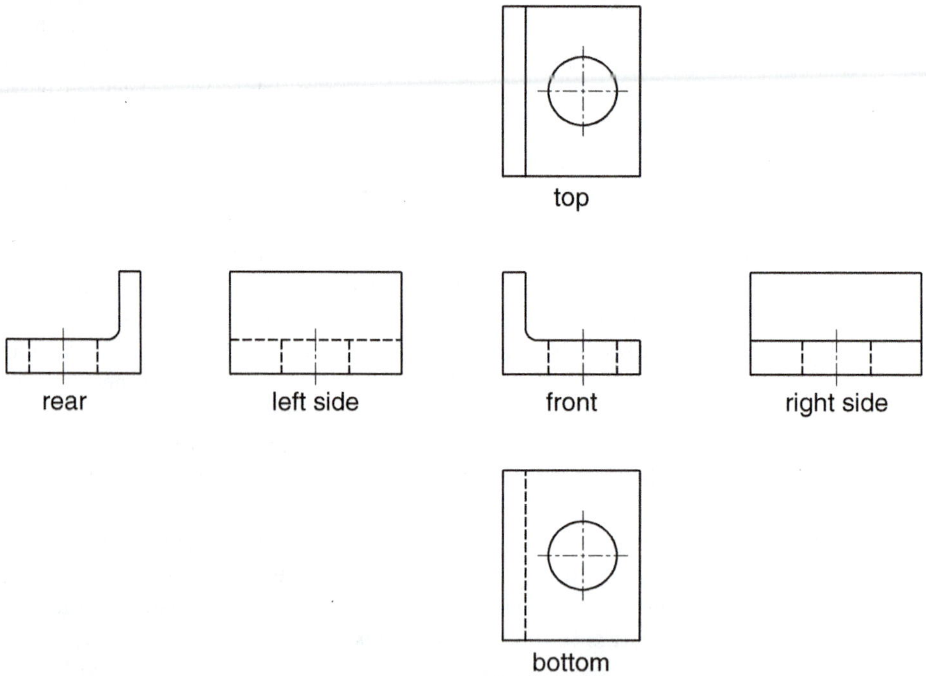

Figure 5-13 Third-angle projection resulting from unfolding the imaginary glass box surrounding the object.

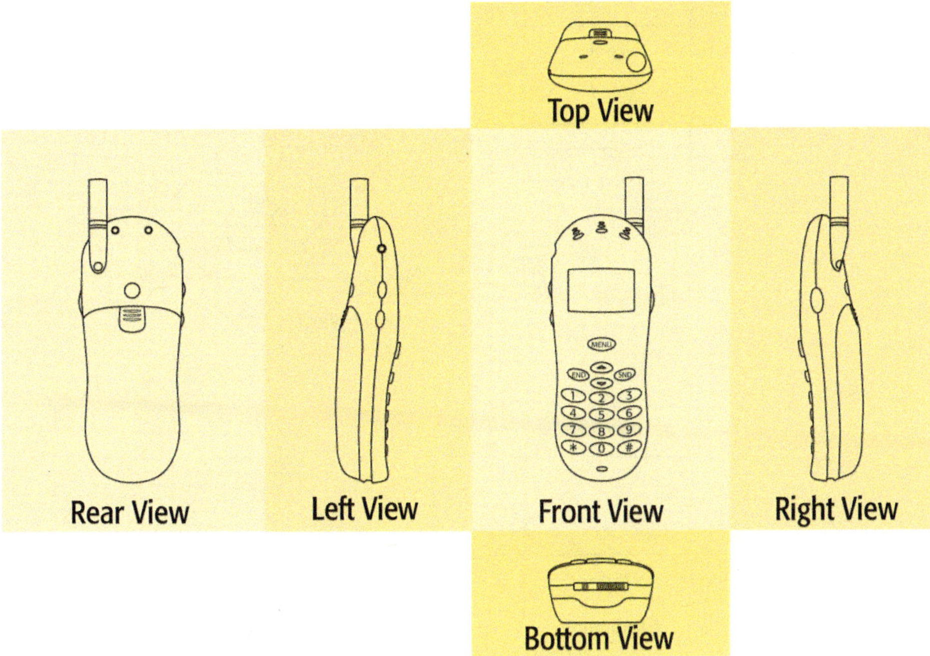

Figure 5-14 Orthographic views show only two dimensions at a time.

select the side of the object that is most descriptive of the object as the front view. Sometimes this may place an object so that what is normally thought of as the front of the object is not shown in the front view of the multiview projection. For example, what is usually described as being the side of a car should be chosen as the front view, because this view is probably most descriptive and easily recognizable as a car. A view of the front of a car (grille, bumper, and windshield) is not as descriptive or as obvious as a view of the side of the car. Furthermore, the object should be properly oriented in the front view. For instance, a car should be shown with its wheels downward in their normal operating position for the front view. The other views that are shown in addition to the front view should be views that best represent features of the object. Normally the minimum number of views necessary to accurately represent the object is used. The standard practice is to use the front, top, and right-side views. But the choice of which views to use depends on the object and which details need to be shown most clearly.

A problem that frequently occurs in orthographic projections is that one of the faces of the object is at an angle to the orthographic planes that form the imaginary glass box. An example is the object shown in Figure 5-15. The circular hole with a keyway (the rectangular cutout on one side of the hole) that is perpendicular to the angled face appears in both the top view and the right-side view. However, it is distorted in both of these views, because it is on an angled plane of the object. An **auxiliary view** is used to avoid this distortion. In this case, a view of the object is drawn so that the angled face is parallel to the auxiliary view plane. The view is based on the viewer looking at the object along a line of sight that is perpendicular to the angled face. Seen from this direction, the circular hole in the auxiliary view appears as a circle without any

auxiliary view

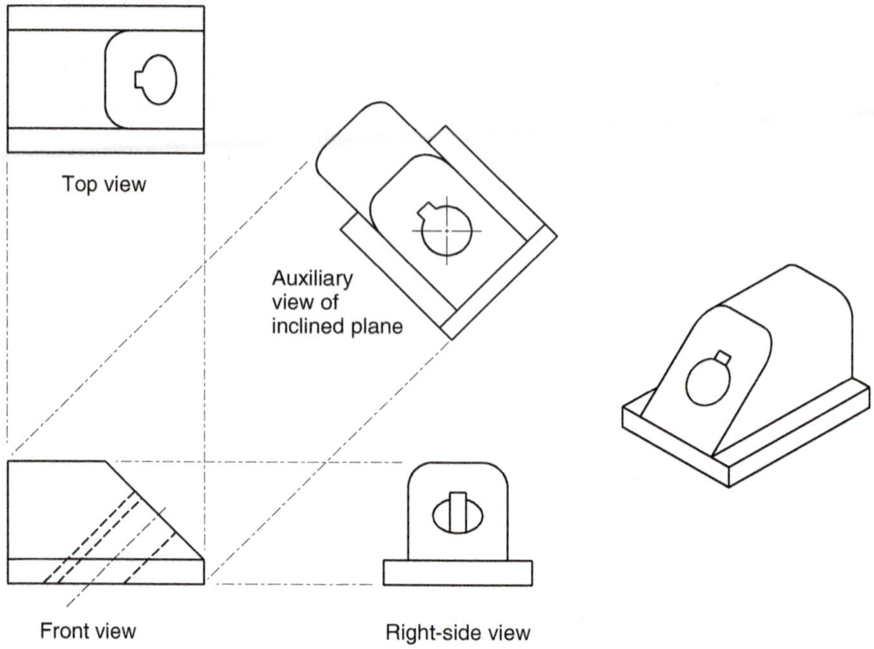

Figure 5-15 Auxiliary view of an inclined face.

distortion. As suggested by the projection lines (dash-dot lines) in Figure 5-15, the auxiliary view is projected from the front view in the same way as the top and right-side views are projected. Thus, its position with respect to the front view depends on the orientation of the angled face. It is normal practice not to include hidden lines or other features that are not directly related to the angled surface in the auxiliary view.

5.3.3 DETAIL DRAWINGS

detail drawings

Detail drawings document the detailed design of individual components using orthographic views. The detail drawing is the final representation of a design that is specific enough so that all of the information necessary for the manufacture of the part is provided. As a result, it is imperative that the detail drawing include the necessary views, dimensions, and specifications required for manufacturing the part. Figure 5-16 shows an example of a detail drawing of a part of a pizza cutter. The detail drawing includes fully dimensioned orthographic views, notation of the material that the part is to be made from, information on the acceptable tolerances for the dimensions, and a title block that records important information about the drawing. Often an isometric projection is included in the detail drawing to further clarify the shape of the part. Detail drawings provide sufficient detail so that the part can be manufactured based on the drawing alone.

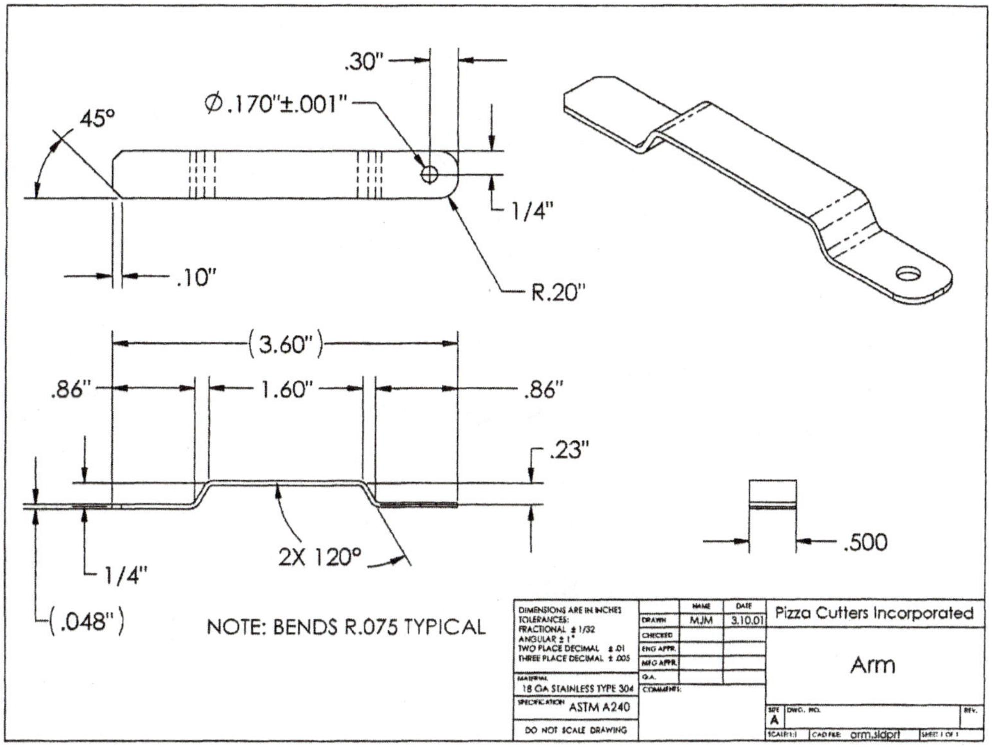

Figure 5-16 Detail drawing of a sheet metal part for a pizza cutter.

5.3.4 ASSEMBLY DRAWINGS

assembly drawing

Assembly drawings show how the components of a design fit together. Dimensions and other details are usually omitted in assembly drawings to enhance clarity. Several styles of assembly drawings are commonly used. Sometimes the assembly drawing is just an isometric projection of the fully assembled device. But an exploded isometric projection is often helpful to show how the individual parts are assembled, as shown in Figure 5-17 for a pizza cutter. In some cases, a sectioned assembly, or cut-away view, shows how complicated devices are assembled. A cutting plane passes through the assembly, and part of the device is removed to show the interior of the assembly. Numbers or letters can be assigned to individual parts of the assembly on the drawing and keyed to a parts list.

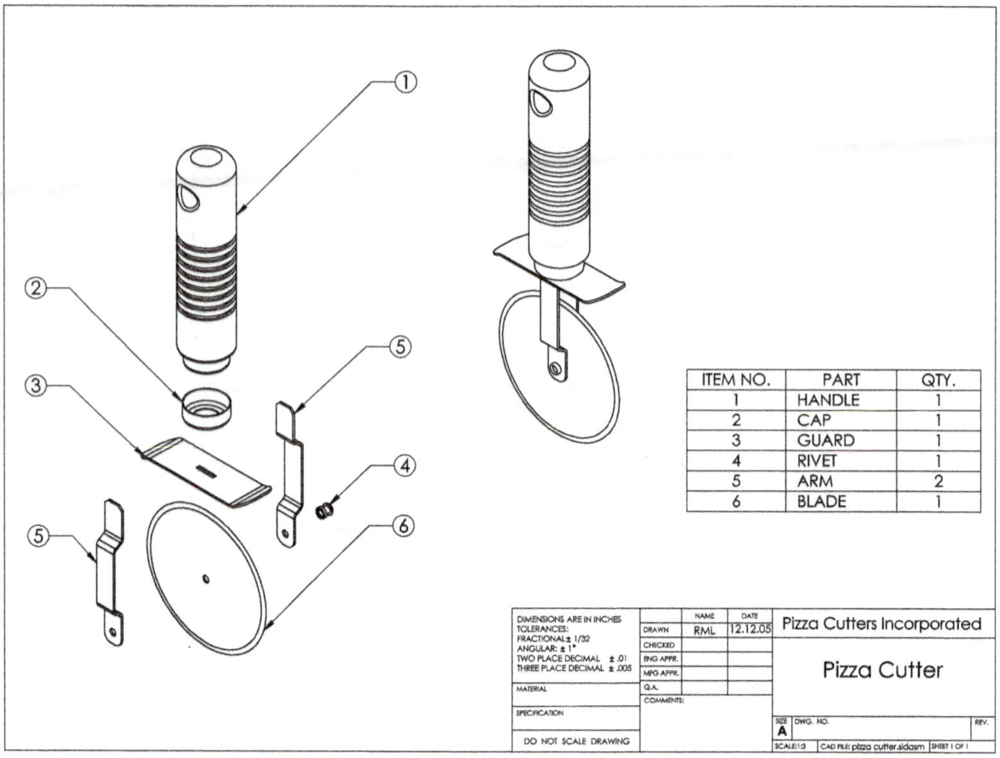

Figure 5-17 Assembly drawing of a pizza cutter.

ITEM	QTY.	PART NO.	DESCRIPTION
1	1	52806	HANDLE
2	1	52825	CAP
3	1	42886	GUARD
4	1	97511	RIVET
5	2	55654	ARM
6	1	56483	BLADE

Figure 5-18 A bill of materials for a pizza cutter.

parts list
bill of materials

Finally a **parts list**, or **bill of materials**, must be included with a set of working drawings. The parts list includes the part name, identification number, material, number required in the assembly, and other information (such as a catalog number for standard parts like threaded fasteners). An example is shown in Figure 5-18 for a pizza cutter. The parts list is used to ensure that all parts are ordered or manufactured and brought to the central assembly point.

Taken together, the detail drawings of each individual part, the assembly drawing, and the bill of materials provide a complete set of working drawings for the manufacture of a part.

5.4 FREEHAND SKETCHING

ideation

An integral part of the creative design process is **ideation**, the generation of concepts or ideas to solve a design problem. Often freehand sketching can be used to explore and communicate concepts that come about in the mind's eye. The process of sketching can solidify and fill out rough concepts. Furthermore, sketching captures the ideas in a permanent form that can be used to communicate the concept to others. In this way, sketches often act as stepping stones to refine and detail the original concept or generate new ideas. Many great design ideas are first sketched on the back of an envelope or in a lab notebook.

Freehand sketching requires few tools: just a pencil and paper. It may be tempting to use straightedged triangles or rulers for drawing straight lines and a compass to draw circles. But these instruments often slow down the process and distract from the purpose of sketching, which is to create a quick, rough graphical representation of the image in the mind's eye.

Two types of pictorial sketches are used frequently in freehand sketching: oblique and isometric. The oblique projection places the principal face of the object parallel to the plane of the paper. The isometric projection tilts the part so that no surface of the part is in the plane of the paper. The advantage of the oblique projection is that details on the front face of the object retain their true shape. This often makes oblique freehand sketching easier than isometric sketching, where no plane is parallel to the paper. The disadvantage of the oblique projection is that it does not appear as "photorealistic" as an isometric projection. In other words, an isometric projection is similar to what a photograph of the object would look like.

5.4.1 OBLIQUE SKETCHING

Often freehand sketching begins with light thin lines called construction lines that define enclosing boxes for the shape that is being sketched. Construction lines are used in several ways. First, the construction lines become the path for the final straight lines of the sketch. Second, the intersections of construction lines specify the length of the final lines. Third, points marked by the intersection of construction lines guide the sketching of circles and arcs. And finally, construction lines guide the proportions of the sketch. This last item is of crucial importance if the sketch is to clearly represent the object. For example, if an object is twice as wide as it is high, the proportions in the sketch must reflect this. Proper proportions of the boxes defined by the construction lines will result in proper proportions of the sketch.

An oblique freehand sketch is easy, since it begins with a two-dimensional representation of the front face of the object. Figure 5-19 shows the steps involved in quickly sketching a part with a circular hole.

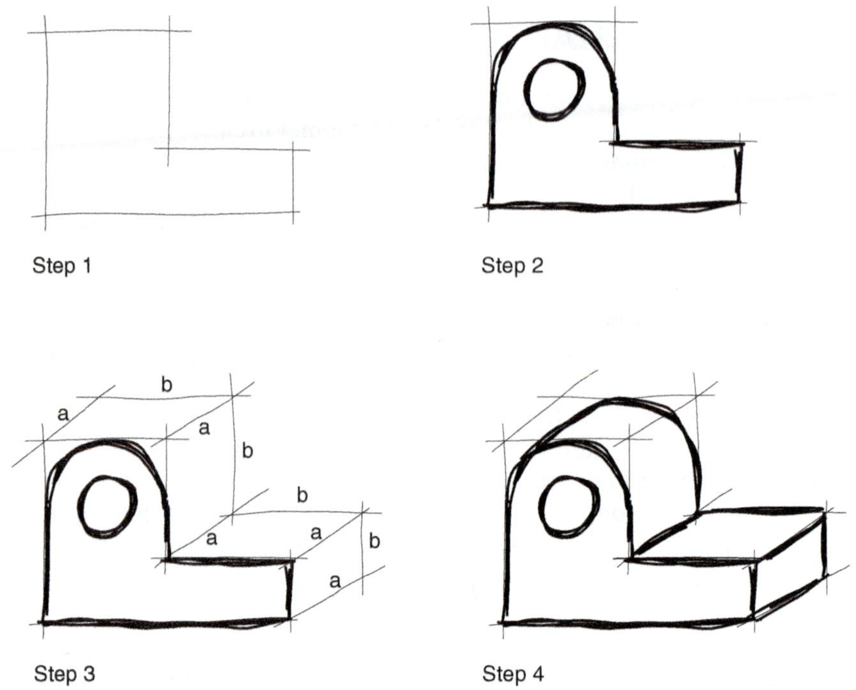

Figure 5-19 Creating a simple oblique freehand sketch.

Step 1: Horizontal and vertical construction lines are lightly drawn to outline the basic shape of the main face of the part. This is known as blocking in the sketch. If you are using a pencil or felt-tip marker, press lightly when drawing the construction lines to produce a thin or light line. If you are using a ball-point pen, draw a single light line.

Step 2: Sketch in the face of the part using the construction lines as a guide. How you sketch the outline of the part depends on the type of pen or pencil that you are using. The idea is to thicken the lines of the part, compared with the construction lines. If you are using a pencil or a felt-tip marker, pressing hard for the outline of the part will result in heavy or dark lines. If you are using a ball-point pen, the line width does not depend much on how hard you press. In this case, the outline of the part is sketched with a back-and-forth motion of the pen to thicken the lines of the part, compared with the construction lines as shown in Figure 5-19. The straight lines are usually sketched first, followed by the arcs. The circle for the hole in the part is added last to complete the face of the part.

Step 3: Sketch receding construction lines (lines labeled *a* extending into the plane of the paper, as shown in Figure 5-19) at a convenient angle. All of the receding lines must be parallel to each other and are usually at an angle of 30° to 45° with respect to the horizontal. The receding lines end at the appropriate depth for the object. Then add the vertical and horizontal lines (lines labeled *b* in the figure) at the back plane of the part. This blocks in the three-dimensional box enclosing the object.

Step 4: Sketch in and darken the lines outlining the part. Again, it is usually easiest to sketch in the straight lines first, then the arcs, and finally any details. Because the construction lines are light compared with the outline of the part, they are not erased.

The final sketch, while rough and lacking detail, clearly shows the design intent for the part.

5.4.2 ISOMETRIC SKETCHING

Isometric freehand sketches are somewhat more difficult to master than oblique sketches because no face is in the plane of the paper for an isometric projection. The steps to construct a simple freehand isometric sketch are shown in Figure 5-20.

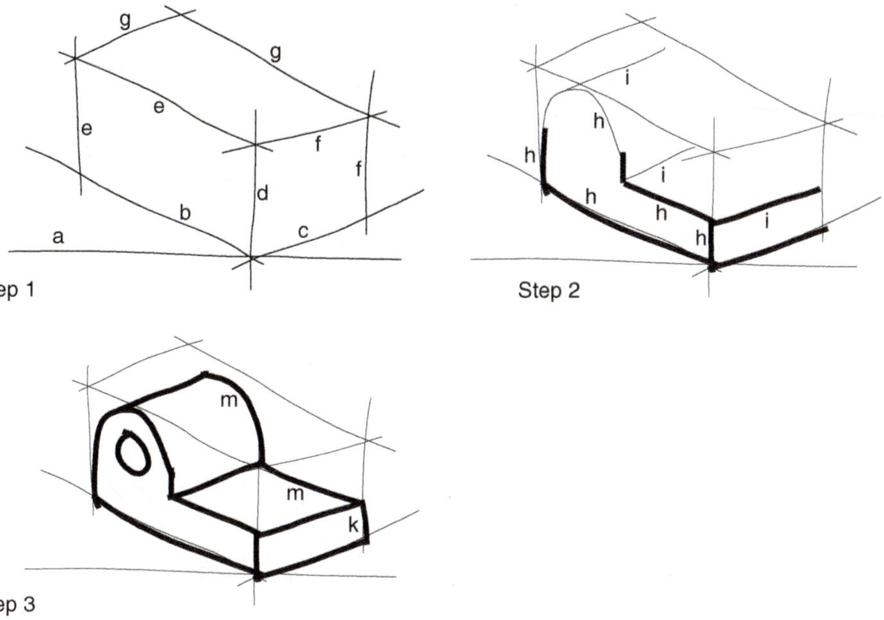

Figure 5-20 Creating a simple isometric freehand sketch.

Step 1: Sketch a light horizontal line (*a*). From this line, draw two intersecting lines at an angle of approximately 30° to the horizontal (*b* and *c*). Then draw a vertical line (*d*) through the intersection of the previous three lines. The three lines labeled *b*, *c*, and *d* form the isometric axes of the sketch. Next sketch the box to block in the front face of the part (*e*). These lines should be parallel to axes *b* and *d*. Similarly, sketch the lines to block in the right face (*f*), making sure that the lines are parallel to axes *c* and *d*. Finish this step by sketching lines parallel to the axes to complete the box that encloses the part (*g*).

Step 2: The outline for the front face is added by sketching in lines and curves (*h*). Then outline the front face using heavy lines. In this case, a single heavy line that can be produced from pressing hard on a pencil or felt-tip marker is used. Next, lines are sketched to indicate the depth of the features of the front view (*i*). These lines should be parallel to axis *c*. They can be darkened after they are drawn lightly.

Step 3: Finally, a line is added to complete the back corner of the part (*k*). Lines and arcs are added to complete the back face of the part (*m*). Then the hole detail is added. Circular holes appear as ellipses in isometric projections, as discussed in the next section.

The choice of whether to use an oblique projection or an isometric projection is often arbitrary. Because the oblique projection is easier to sketch, it is sometimes preferred. On the other hand, an isometric projection provides a more photorealistic image of the object.

5.4.3 SKETCHING AUXILIARY VIEWS

line of sight

When one of the faces of the part is at an angle to the orthographic planes, it may be necessary to sketch an auxiliary view so that the features on the angled face are not distorted. The view is based on the observer looking at the object along a **line of sight** that is perpendicular to the angled face. Sketching an auxiliary view seems quite challenging at first, but following a step-by-step approach makes it nearly as easy as sketching a standard orthographic view. Consider the object shown in Figure 5-21, having an angled face with a circular hole and keyway (rectangular cutout). Only the auxiliary view shows the circle without any distortion.

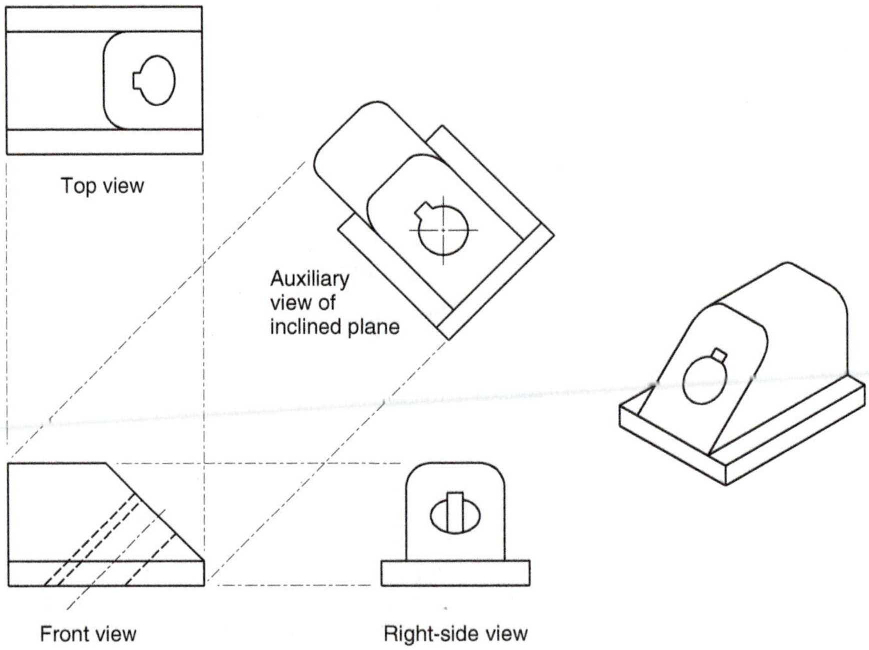

Figure 5-21 Auxiliary view of an inclined face.

Details of sketching an auxiliary view vary, depending on the nature of the object to be sketched. However, Figure 5-22 shows the steps that would usually be used for the object shown in Figure 5-21, as an example.

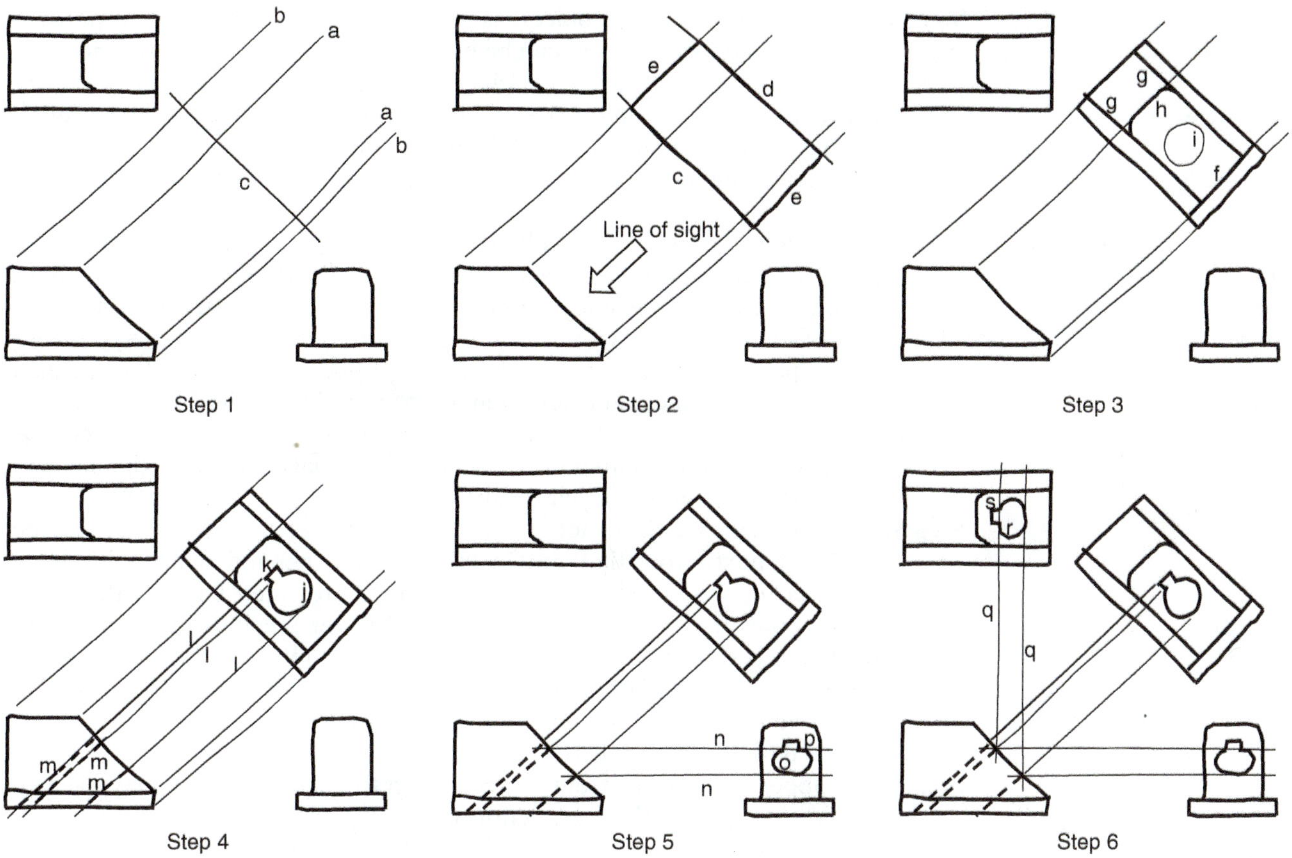

Figure 5-22 Creating an auxiliary view in a freehand sketch.

Step 1: The sketch begins with the three standard orthographic views with a few differences from the way they would normally be done. First, details on the inclined face are omitted from the views until later in the sketching process. In this case, the hole and keyway on the inclined face are not sketched. Second, the three views are separated by a substantial distance. The side view is spaced about twice its width from the front view. Likewise, the top view is spaced about two times its vertical dimension above the front view. This is done to provide space for the auxiliary view. Once these views are sketched, the sketch of the auxiliary view can be done. This begins by sketching two construction lines (*a*) that are perpendicular to the inclined face in the front view and are located at the upper and lower edges of the inclined face. Next, two additional construction lines (*b*) are added. These lines are parallel to the first construction lines and are located at the extreme upper left and lower right edges of the object. These lines will be used to block in the view in the auxiliary plane. The first edge of the object (*c*) is sketched perpendicular to the construction lines (*a*) and (*b*). This line can be drawn any convenient distance from the front view, but usually it is drawn to approximately line up with the lower edge of the top and the left edge of the side view, as shown.

Step 2: Now the outline of the object, as viewed along the line of sight perpendicular to the inclined face, can be blocked in by adding a construction line (*d*). Then the boundaries of the object are added by darkening lines (*c*), (*d*), and (*e*). These are sometimes called **silhouette lines** because they define the outer edges, or silhouette, of the part in the image plane as viewed looking toward the inclined surface in the direction of the arrow indicating the line of sight.

silhouette lines

Step 3: The details of other edges evident in the auxiliary view can be added beginning with line (*f*) for the lower edge of the inclined face, followed by lines (*g*) that indicate the inside edges of the flanges, which protrude from either side of the main body of the part (most easily seen as the widest portions of the part in the side view). The curved upper edge (*h*) of the inclined surface can be sketched next. Finally, the circle for the hole (*i*) can be sketched. The hole appears circular in the auxiliary view, which, of course, is the reason for sketching the view in the first place—to show the inclined face and its features without distortion.

Step 4: The hole can be darkened (*j*) and the rectangular keyway (*k*) can be sketched next. Now that the features of the inclined plane have been sketched in the auxiliary view, they can be projected back to the other views. Lines (*l*) that bound the upper and lower portions of the hole and the upper extent of the keyway are drawn parallel to the other construction lines from the auxiliary view so that they extend across the front view. In the front view, these lines correspond to the hidden lines of the hole, which can be darkened as dashed lines (*m*).

Step 5: The upper and lower extent of the hole in the front view can be projected to the side view with construction lines (*n*). Once this is done, it is relatively easy to position and sketch the hole (*o*), which is elliptical in this view, and the keyway (*p*).

Step 6: In a similar way, construction lines (*q*) represent the leftmost and rightmost extents of the hole in the front view that can be projected to the top view. With these lines drawn, the elliptical hole (*r*) and keyway (*s*) are sketched to complete the drawing.

Notice how the auxiliary view is used to more accurately sketch the features (hole and keyway) that are on the inclined surface when they are projected to the front, side, and top views after the auxiliary sketch has been created. Of course, the technique for sketching auxiliary views depends on the nature of the object that is being sketched. For instance, if the inclined surface were on the left edge of the front view instead of the right edge, the auxiliary view would be sketched to the left and above the front view instead of to the right. Likewise, the examples shown in this chapter for oblique, isometric, and orthographic sketches only demonstrate the typical steps that can be used to generate freehand sketches. The exact steps that are used will vary with the details of the object that is being sketched. The steps in freehand sketching become more obvious with practice and experience.

PROFESSIONAL SUCCESS: WILL FREEHAND SKETCHING EVER BECOME OBSOLETE?

CAD has almost totally eliminated pencil-and-paper drawings. But what about pencil and paper freehand sketching? Although many computers offer "paint" or "draw" programs, it is unlikely that freehand sketching will disappear soon. Just as it is easier to do a calculation in your head or on a piece of scratch paper rather than finding a calculator and punching in the numbers, it is easier to sketch an image on a piece of paper (or a napkin!) than to find a computer, log in, and start the appropriate program. Pencil-and-paper freehand sketches are quick, efficient, easily modified, and easily conveyed to others. And all that is needed is a pencil and a scrap of paper.

Even if the pencil and paper are totally replaced by laptop or handheld computers someday, freehand sketching skills will still be useful. Instead of using a pencil on a piece of paper, a stylus can be used on a touch screen. The only difference is the medium. The freehand sketching techniques themselves are unlikely to change much, although sketching software that automatically assists in generating oblique or isometric sketches as they are drawn is a logical enhancement to freehand sketching.

5.5 CAD—COMPUTER-AIDED DRAFTING

The introduction of the computer revolutionized engineering graphics. Computer-aided drafting is usually termed CAD, an acronym for computer-aided design, computer-assisted drafting, or computer-aided design and drafting. Computer-aided engineering (CAE) is used to refer to the broader range of computer-related design tools.

Several different models for representing a part's geometry have been used in CAD, including 2-D models and 3-D wireframe and surface models. However, solid modeling is the current state of the art in CAD. Solid modeling has the inherent advantage of more accurately representing the "design intent" of the part that is modeled.

model

A CAD **model** is a computer representation of an object or part. It can be thought of as a "virtual" part in that it exists only as a computer image. The model is an engineering document of record. It contains all of the design information, including the geometry, dimensions, tolerances, materials, and manufacturing information. A CAD model replaces the paper blueprints and engineering drawings used a few decades ago.

2-D model

The simplest model used in CAD is a **2-D model**. This model is essentially the computer graphics equivalent of an orthographic projection created using a pencil and paper, as shown in Figure 5-23. A 2-D model represents a three-dimensional object with several views, each showing one side of the object projected onto a plane. On the computer, a 2-D model appears as lines and curves on a flat surface, and it is up to the engineer or designer to interpret several views to create a mind's eye image of the three-dimensional object. The CAD software electronically stores the 2-D model as several views but does not inherently "know" that the views can be connected to one another to form a three-dimensional object.

2 1/2-D model

A **2 1/2-D model** has a third dimension that is recognized by the CAD software, but the third dimension is simply an extrusion of a two-dimensional shape. In this case, the CAD software electronically stores the cross-sectional shape of the object and the depth that it is **extruded**, or stretched, in the third dimension. Thus, the only objects that can be represented using a 2 1/2-D model are those that have a constant cross section.

extruded

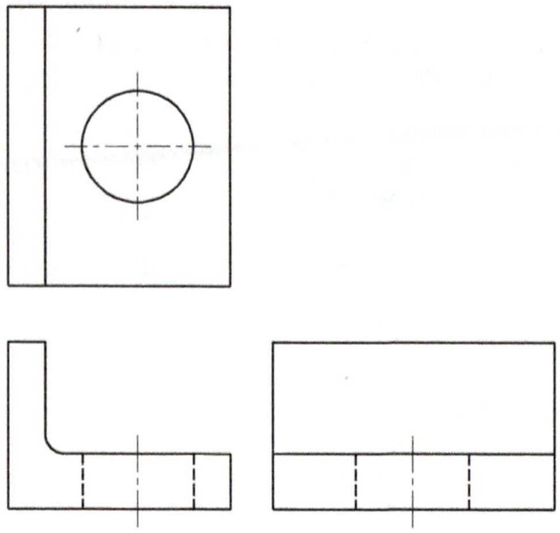

Figure 5-23 2-D model of an object.

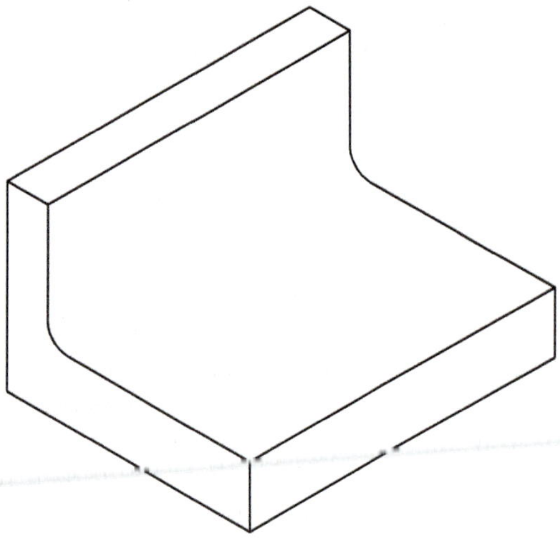

Figure 5-24 $2\frac{1}{2}$-D model of an object formed by extruding a 2-D shape.

For instance, the simple shape shown in Figure 5-24 is the 2 1/2-D model of the object shown using a 2-D model in Figure 5-23 without the hole. The object, including the hole, cannot be represented as a 2 1/2-D model because the hole results in a change in the cross section. A 2 1/2-D model could be used to represent a shape such as an I-beam or C-channel.

3-D model

A **3-D model** is the most general model used in CAD software. In this case, the CAD software electronically stores the entire three-dimensional shape of the object. Most simply, one can think of the software storing the three-dimensional coordinates of various points on the object and then defining how these points are connected. Of course, a two-dimensional drawing representing the orthographic views of the object can be automatically created from the 3-D model. The current generation of CAD software is based on 3-D models.

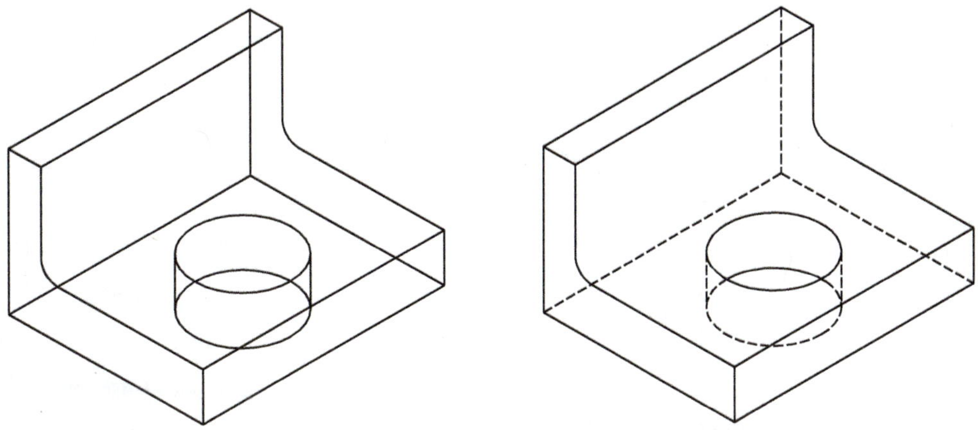

Figure 5-25 (a) 3-D wireframe model. (b) 3-D wireframe model with hidden lines shown as dashed lines.

wireframe model

The simplest 3-D model is a **wireframe model**. Figure 5-25a shows a wireframe representation of the object that we have been considering. In a wireframe model, only edges of the object are represented. Thus, the CAD software needs to store the locations of the vertices (intersections of lines) and information about which vertices are to be connected to each other by lines. A circle or curved section can be represented by a series of closely spaced vertices with linear edges connecting them or as an arc defined by a center, radius, and end points. Before more powerful computers were available, the low storage requirements of a wireframe model made it quite popular. The problem inherent in a wireframe representation is quite clear in Figure 5-25a—it is difficult to interpret the drawing because all of the edges are visible. Figure 5-25b shows the object with the hidden lines shown as dashed lines to indicate that they are behind a surface. 3-D models not based on wireframes are able to deal with hidden lines more effectively and offer other advantages.

surface model

A 3-D **surface model** defines the object in terms of surfaces such as plates (flat) and shells (curved) in addition to edges. This makes it easy to determine whether a line or surface is hidden, because the surfaces are defined. Using a surface model also permits the construction of smoothly curved surfaces, such as a circular hole, rather than using many line segments to form a curved line as with some wireframe models. It is also possible to "fair" or smooth one surface into another surface to provide a sculpted shape. Figure 5-26a shows the simple object with hidden lines removed altogether. Although a surface model can be displayed to look like a wireframe, either with or without hidden lines, it can also be displayed by assigning different degrees of shading to the surfaces. A virtual light source is assumed to be near the object to provide a three-dimensional lighting effect as shown in Figure 5-26b. This is known as **shading** or **rendering**. Although a variety of shading techniques are available, all depend on determining how much light from a virtual light source strikes each portion of the virtual object. The use of shading provides a very realistic image of the object, which permits

shading
rendering

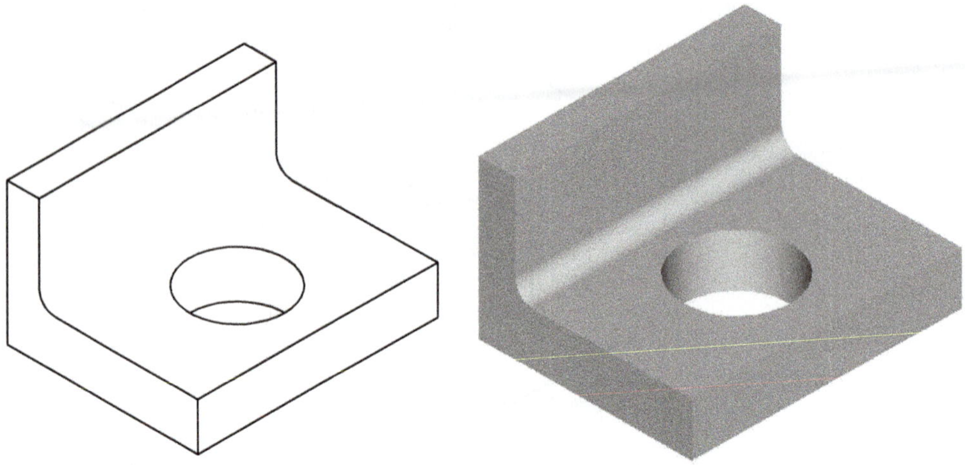

Figure 5-26 (a) 3-D wireframe model with hidden lines removed. (b) 3-D model with shading.

a much more vivid communication of the nature of the object. However, the problem that remains with a surface model is that the interior of the object remains undefined. The surface model just represents the shell of the object.

5.5.1 CAD AND SOLID MODELING

solid modeling

Solid modeling, the current state of the art in CAD, is the most sophisticated method of representing an object. Unlike wireframe or surface models, a solid model represents an object in the virtual environment just as it exists in reality, having volume as well as surfaces and edges. In this way, the interior of the object is represented in the model, as well as the outer surfaces.

constructive solid geometry
primitive modeling

The first attempt at solid modeling was a technique known as **constructive solid geometry** or **primitive modeling**, which is based on the combination of geometric primitives such as right rectangular prisms (blocks), right triangular prisms (wedges), spheres, cones, and cylinders. Each primitive could be scaled to the desired size, translated to the desired position, and rotated to the desired orientation. Then one primitive could be added to or subtracted from other primitives to make up a complex object. It is easy to see how two blocks minus a cylinder (hole), as shown in Figure 5-27, could represent an object similar to the object in Figure 5-26.

constraint-based solid modeling

Constraint-based solid modeling overcomes the weakness of constructive solid geometry modeling by making the modeling process more intuitive. Instead of piecing together geometric primitives, the constraint-based modeling process begins with the creation of a 2-D sketch of the profile for the cross section of the part. The initial sketch need not be particularly accurate; it need only reflect the basic geometry of the part's cross-sectional shape. Details of the cross section are added later. The next step is to constrain the 2-D sketch by adding enough dimensions and parameters to completely define the shape and size of the 2-D profile. The name constraint-based modeling

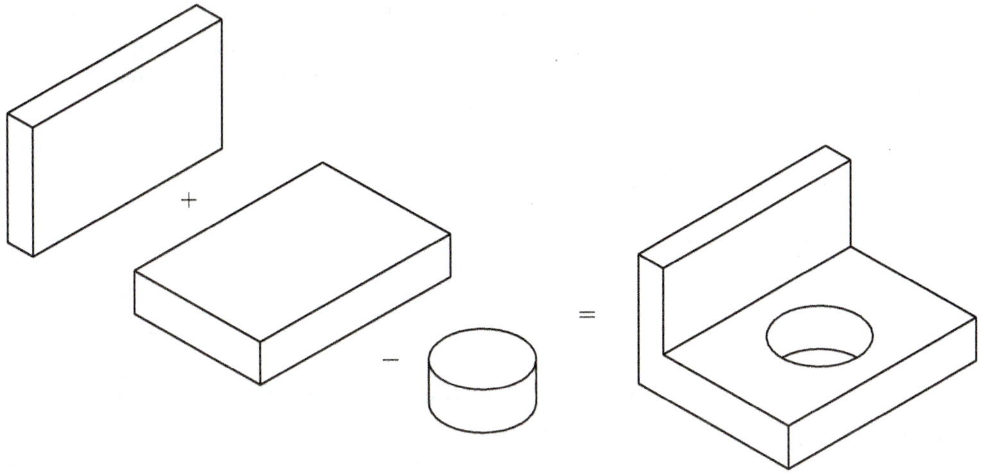

Figure 5-27 Combining geometric primitives to represent an object.

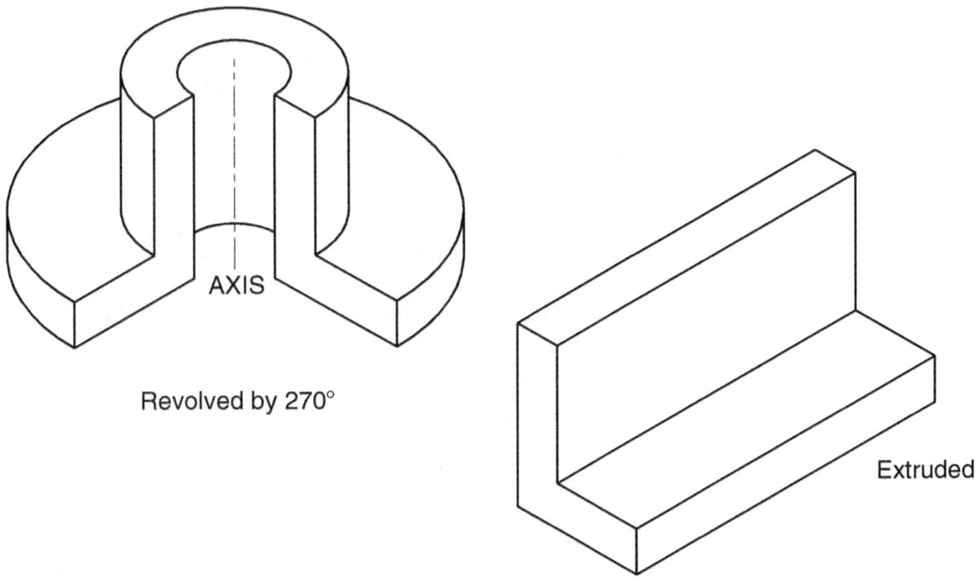

Figure 5-28 Revolving by 270° and extruding an L-shaped cross section.

revolving

arises because the shape of the initial 2-D sketch is "constrained" by adding dimensions to the sketch. Finally, a three-dimensional object is created by **revolving** or extruding the 2-D sketched profile. Figure 5-28 shows the result of revolving a simple L-shaped cross section by 270° about an axis and extruding the same L-shaped cross section along an axis. In either case, these solid bodies form the basic geometric solid shapes of the part. Other features can be subsequently added to modify the basic solid shape.

Once the solid model is generated, all of the surfaces are automatically defined, so it is possible to shade it in the same way as a surface model is shaded. It is also easy to generate 2-D orthographic views of the object. This is a major advancement from other modeling schemes. With solid modeling, the two-dimensional drawings are produced as views of the three-dimensional virtual model of the object. In traditional CAD, the three-dimensional model is derived from the two-dimensional drawings. One might look at solid modeling as the sculpting of a virtual solid volume of material. Because the volume of the object is properly represented in a solid model, it is possible to slice through the object and show a view of the object that displays the interior detail. Once several solid objects have been created, they can be assembled in a virtual environment to ensure that they fit together properly and to visualize the assembled product.

5.5.2 THE NATURE OF SOLID MODELING

feature-based modeling
features

base feature

Feature-based modeling attempts to make the modeling process more efficient by creating and modifying geometric **features** of a solid model in a way that represents how geometries are created using common manufacturing processes. Features in a part have a direct analogy to geometries that can be manufactured or machined. A **base feature** is a solid model that is roughly the size and shape of the part that is to be modeled. The base feature is the 3-D solid created by revolving or extruding a cross section, such as those shown in Figure 5-28. It can be thought of as the initial work block. All subsequent features reference the base feature either directly or indirectly. Additional features shape or refine the base feature. Examples of additional features include holes or cuts in the initial work block.

cross section

The analogy between feature-based modeling and common manufacturing processes is demonstrated in Figure 5-29 for making a handle for a pizza cutter. Beginning at the top of the figure, we follow the steps that an engineer would use to create a solid model, or virtual part, on the left and the steps that a machinist would take to create the same physical part in a machine shop on the right. The engineer, using solid modeling software, begins by creating a two-dimensional profile, or **cross section**, of a part, in this case a circle (shown in isometric projection). The analogous step by a machinist is to choose a circular bar stock of material with the correct diameter. Next the engineer extrudes, or stretches, the circular cross section along the axis perpendicular to the plane of the circle to create a three-dimensional base feature (a cylinder in this case). The equivalent action by a machinist is to cut off a length of bar stock to create an initial work block. Now the engineer adds features by cutting away material on the left end to reduce the diameter and by rounding the right end of the cylinder. The machinist performs similar operations on a lathe to remove material from the cylinder. Next the engineer creates a circular cut to form a hole through the cylinder on the right end. The machinist drills a hole in the right end of the cylinder. Finally, the engineer creates a pattern of groove cuts around the handle. Likewise the machinist cuts a series of grooves using a lathe. In similar fashion, a geometric shape could be added to the base feature in the solid model, analogous to a machinist welding a piece of metal to the work block. Feature-based techniques give the engineer the ability to easily create and modify common manufactured features. As a result, planning the manufacture of a part is facilitated by the correspondence between the features and the processes required to make them.

design intent

Constraint-based modeling permits the engineer or designer to incorporate "intelligence" into the design. Often this is referred to as **design intent**. Unlike traditional CAD software, the initial sketch of a two-dimensional profile in constraint-based solid modeling does not need to be created with a great deal of accuracy. It just needs to represent the basic geometry of the cross section. The exact size and shape of the profile is defined through assigning enough parameters to fully "constrain" it.

dimensional constraints

There are two primary types of constraints. **Dimensional constraints** are used to specify distances between items in a solid model. For example, a hole is prescribed by the location of its center, typically in terms of horizontal and vertical distances from features of a part such as edges or other holes. In addition, the radius and depth of the hole need to be specified. Given these four dimensional constraints, the hole size, depth, and location are fully defined.

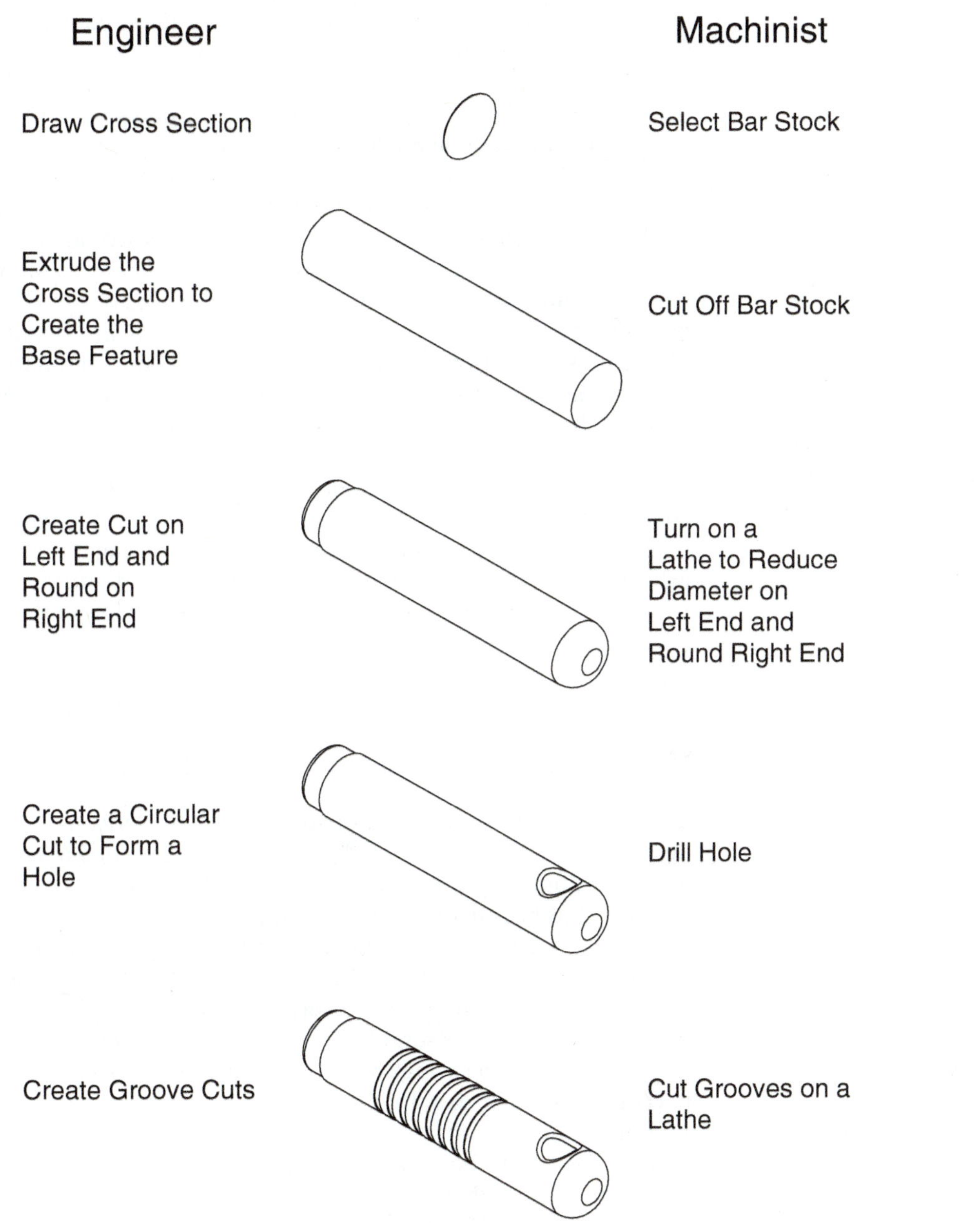

Figure 5-29 The analogy between feature-based modeling done by an engineer or a designer and common manufacturing processes done by a machinist.

geometric constraints

Instead of specifying numerical dimensions, **geometric constraints** define positional relationships between entities in the model in terms of the geometry. For instance, a pipe can be formed from a solid circular rod by creating a hole that is concentric with the circumference of the rod. The radius and depth of the hole require dimensional constraints, but the position of the hole does not. Instead, the position can be prescribed purely by a geometric constraint, specifically that the hole is concentric with the rod. Often geometric constraints are particularly useful in creating assemblies of individual parts, as demonstrated in Section 5.6.

parametric

Another aspect of solid modeling is that the model is **parametric**. This means that parameters of the model may be modified to change the geometry of the model. A dimension is a simple example of a parameter. When a dimension is changed, the geometry of the part is updated. Thus, the *parameter drives the geometry*. For example, the position of the hole could be specified parametrically, so that the center of the hole is located at a position that is one-half of the total length of the part. If the total length was specified as 80 mm, then the hole would be 40 mm from either side. However, if the length was changed to 100 mm, then the hole would automatically be positioned at half of this distance, or 50 mm from either side. Thus, no matter what the length of the part is, the hole stays in the middle of the part. The power of this approach is that when one dimension is modified, all linked dimensions are updated according to specified mathematical relations, instead of having to update all related dimensions individually.

history-based modeling

parent–child relation

The last aspect of solid modeling is that the order in which parts are created is critical. This is known as **history-based modeling**. For example, a hole cannot be created before a solid volume of material in which the hole occurs has been modeled. If the solid volume is deleted, then the hole is deleted with it. This is known as a **parent–child relation**. The child (hole) cannot exist without the parent (solid volume) existing first. Parent–child relations are critical to maintaining design intent in a part. Most solid modeling software recognizes that if you delete a feature with a hole in it, you do not want the hole to remain floating around without being attached to a feature.

5.6 VIRTUAL ASSEMBLY OF PARTS

One of the great advantages of solid modeling software is that it allows engineers or designers to assemble modeled parts to create a virtual assembly. This helps the designer to work out problems with part interactions early in the design phase, before manufacturing has begun. Often it is possible to check automatically for clearances or interferences between parts. A virtual assembly also permits the engineer or designer to get an idea of what the assembled product looks like before it is fabricated.

In the case of the pizza cutter, the assembly begins by bringing two parts of the pizza cutter into the assembly window of the CAD software, in this case the rivet and one arm, as shown in Figure 5-30. Geometric constraints are used to orient the two parts properly. For example, a geometric constraint is used to make the shaft of the rivet and the cylindrical surface of the hole concentric with one another, as shown in Figure 5-31. Next, the surface of the arm that the arrow points to in Figure 5-31 must be constrained so that it coincides with the underside of the head of the rivet that the arrow points to in Figure 5-32. The blade can be added by making the hole in the blade concentric with the shaft of the rivet and by making the two surfaces with arrows in

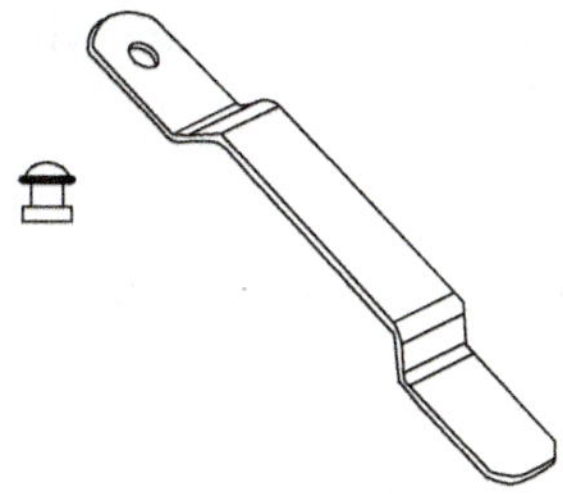

Figure 5-30 The pizza cutter assembly begins as two parts, the rivet and an arm, oriented randomly in space.

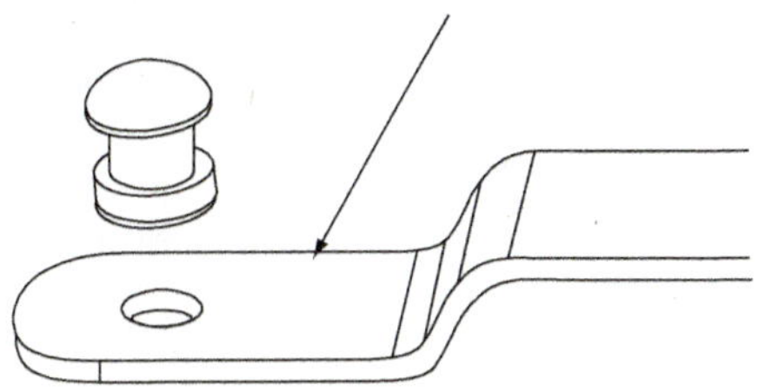

Figure 5-31 The shaft of the rivet is constrained to be concentric with the hole in the arm.

Figure 5-33 coincident. Likewise, the hole in the second arm is concentric to the shaft of the rivet, and the surfaces indicated in Figure 5-34 are made coincident. At this point, the two arms can still rotate with respect to one another; hence, a geometric constraint is added so that the surfaces indicated in Figure 5-35 are parallel. Similar geometric constraints can be used to assemble the remaining parts to create the finished assembly, shown in Figure 5-36.

interference

Once the pizza cutter is assembled, it is possible to check for **interference** between parts. In this case, there was an error in the metal guard that is at the end of the handle to prevent the user's hand from slipping down the handle onto the sharp blade. Notice in Figure 5-37 that the rectangular assembly that holds the blade cannot fit through the

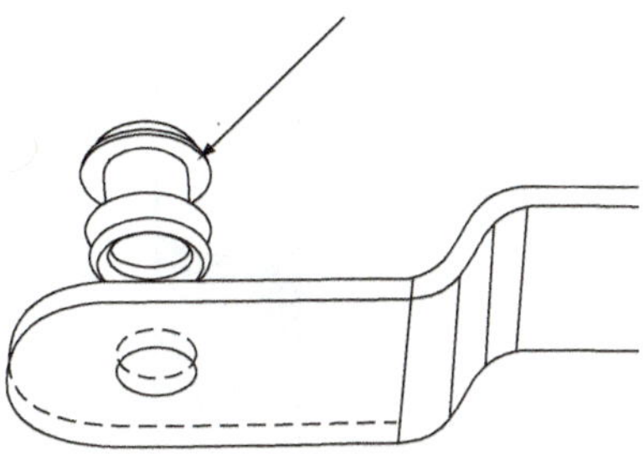

Figure 5-32 The underside of the head of the rivet (see arrow) is constrained to be coincident with the upper plane of the arm (indicated by the arrow in Figure 5-31).

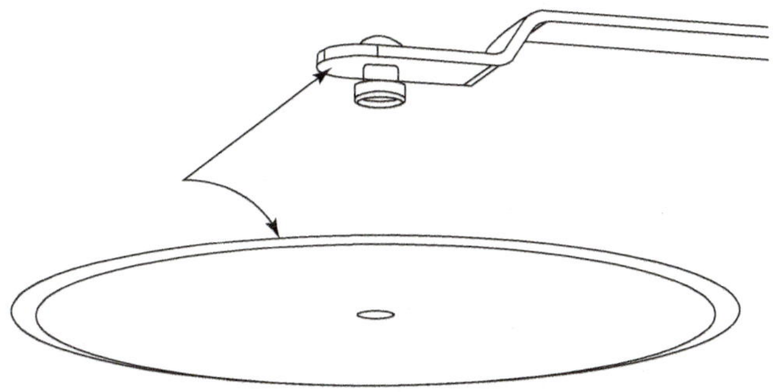

Figure 5-33 The lower surface of the arm and the upper surface of the blade (indicated by arrows) are constrained to be coincident.

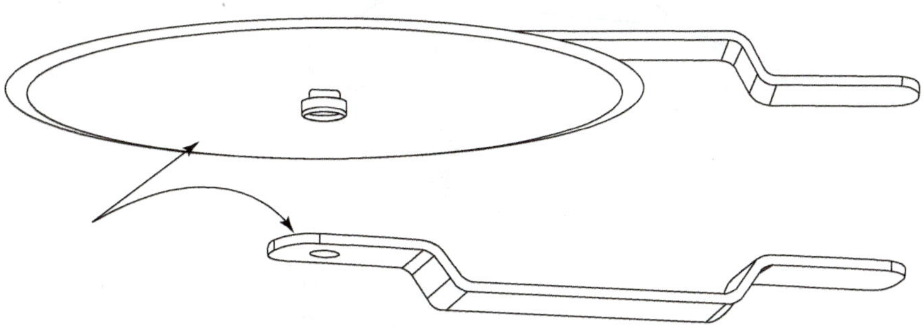

Figure 5-34 The lower surface of the blade and the upper surface of the second arm (indicated by arrows) are constrained to be coincident.

circular hole in the guard. The solid modeling software will detect this problem, which can be fixed by returning to the CAD file of the guard and changing the hole feature from a circular hole to a rectangular hole. Because of the associative nature of the CAD software, this modification in the part will be automatically reflected in the assembly, resulting in the modified guard in Figure 5-38, shown in an exploded view.

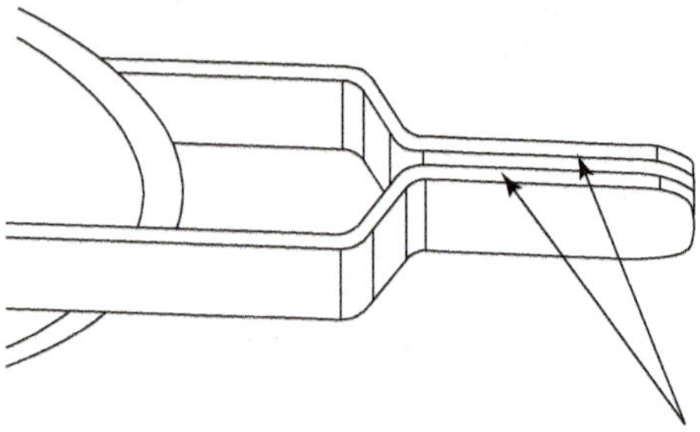

Figure 5-35 The sides of the two arms (indicated by arrows) are constrained to be parallel so that the arms are aligned with each other.

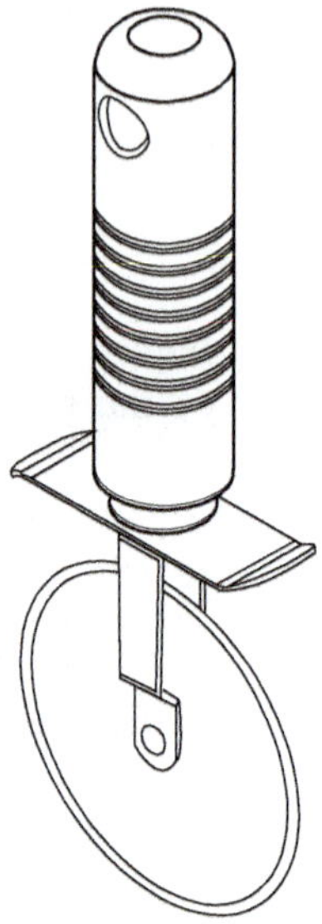

Figure 5-36 The assembled pizza cutter.

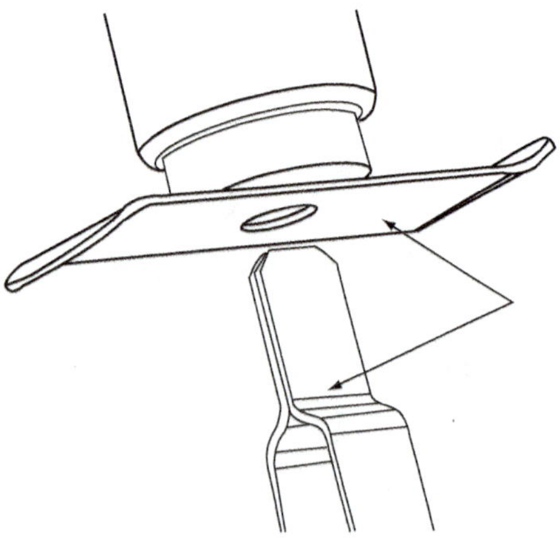

Figure 5-37 The rectangular ends of the arms cannot fit through the circular hole of the guard, a problem that is automatically detected by the CAD software as interference between the parts.

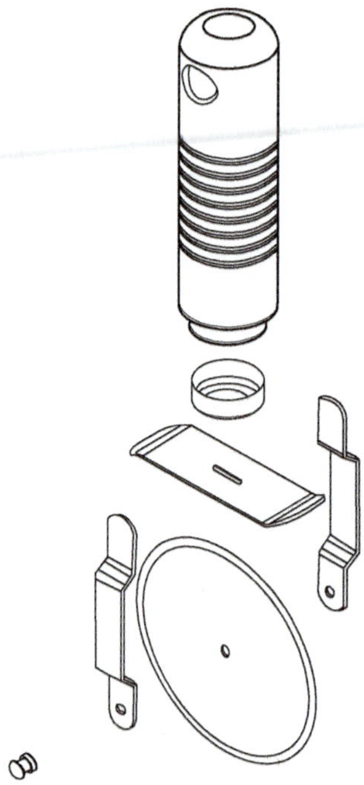

Figure 5-38 An exploded view of the pizza cutter showing the guard with a rectangular hole instead of a circular hole.

The steps that have been shown in creating the handle and the pizza cutter assembly are merely examples of how solid-modeling CAD software can be used in the design of parts. There are many alternative methods of accomplishing the same thing, such as changing the order in which the features are added to the base feature or changing the order in which the parts are assembled. Furthermore, many details have been left out, particularly with regard to differences in how various CAD software packages accomplish the creation of parts and assemblies. However, the steps used to create the handle and the pizza cutter assembly demonstrate some key characteristics of solid modeling: starting with a base feature and adding other features to it; using constraints, both dimensional and geometric, to incorporate "design intent"; and using the associative capability of the software to permit changes to ripple through associated objects such as the assembly.

5.7 DRAWING STANDARDS

Not only is the accurate and clear depiction of details of a part necessary in an engineering drawing, but it is also necessary that the drawing conform to commonly accepted standards or conventions. There are two reasons for the existence of standards and conventions. First, using standard symbols and projections ensures a clear interpretation of the drawing by the viewer. For example, the symbol ∅ associated with a dimension indicates that the dimension is a diameter of a circular feature. Without this convention, it would be more difficult to unambiguously represent diameter dimensions on a drawing.

Most CAD programs automatically use a standardized presentation of drawings, usually based on standards from the American National Standards Institute (ANSI; see www.ansi.org) in the United States and the International Standards Organization (ISO; see www.iso.org) in the rest of the world. Figure 5-39 shows the dimensions of a part in one view using both the ANSI and ISO standards. Notice that the numbers for the dimensions in the ANSI standard are **unidirectional**, having the same orientation, whereas the values for the dimensions in the ISO drawing are **aligned** with the dimension line.

unidirectional
aligned

Even with the use of CAD software to ensure standard presentation in drawings, it is up to the engineer to implement drawing standards in some cases. An example is in the placement of dimensions in a drawing. The user controls where dimensions appear on the drawing. Furthermore, even in cases where the CAD software automatically implements drawing standards, it is still necessary to understand what the conventions mean so that they are properly interpreted.

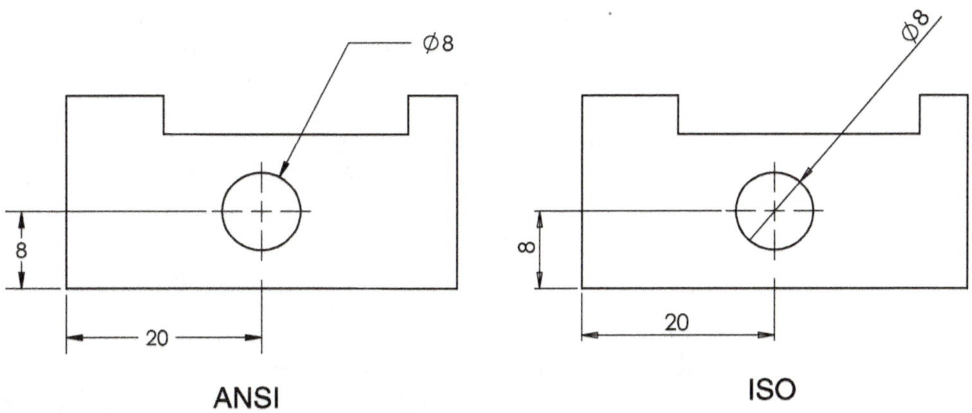

Figure 5-39 Drawings of the same object using the ANSI and ISO standards.

There is some variability among different CAD software packages in how various details are shown. For instance, the length of centerlines, the type of line used for hidden lines, and the shape of arrowheads may differ from one CAD software package to another. An example is the gray color sometimes used for hidden lines. Although hidden lines are generally shown as dashed lines in final drawings, they may be shown as gray or other colors while the part is being modeled.

5.7.1 SHEET LAYOUTS

Engineering drawings are created on standard-size sheets that are designated by the code indicated in Table 5-1. Engineering drawings are almost always done in "landscape" orientation so that the long side of the drawing is horizontal. The choice of drawing size depends upon the complexity of the object depicted in the drawing. The drawing should be sized so that the projections of the part, dimensions, and notes all fall within the borders of the drawing with adequate spacing, so that the drawing is not cluttered. The drawing should be large enough so that all details are readily evident and readable. As a result, regardless of the physical size of the part, simple parts are usually drawn on smaller sheets because it is not necessary to show much detail. Complex parts are usually drawn on larger sheets so that there is room to show adequate detail. When using CAD software, it is necessary to consider the printed drawing size in addition to the drawing size on the computer screen. On the computer screen, it is possible to zoom in to read detailed items, but this cannot be done for a printed drawing. Consequently, the sheet size should be chosen carefully.

Table 5-1 Standard Sheet Sizes.

ANSI	ISO
A – 8.50" × 11.00"	A4 – 210 mm × 297 mm
B – 11.50" × 17.00"	A3 – 297 mm × 420 mm
C – 17.00" × 22.00"	A2 – 420 mm × 594 mm
D – 22.00" × 34.00"	A1 – 594 mm × 841 mm
E – 34.00" × 44.00"	A0 – 841 mm × 1189 mm

title block

The **title block** on a drawing records important information about the drawing. Normally the title block of a drawing is in the lower right corner of the drawing. ANSI standard title blocks can be used, but often individual companies use their own standard title block. The title block, such as the one shown in Figure 5-40, is used primarily for drawing control within a company. The title block includes information regarding the part depicted in the drawing, such as its name and part number; the person who created the drawing; persons who checked or approved the drawing; dates the drawing was created, revised, checked, and approved; the drawing number; and the name of the company. This information allows the company to track the drawing within the company if questions arise about the design.

Other information can appear in the title block to provide details about the manufacture of the part, such as the material for the part, the general tolerances, the surface finish specifications, and general instructions about manufacturing the part (such as "remove all burrs"). Many times, though, this information is shown as notes on the drawing rather than in the title block.

Finally the title block provides information about the drawing itself, such as the units used in dimensioning (typically inches or millimeters) and the scale of the drawing. The

scale

scale of the drawing indicates the ratio of the size of the drawing to the size of the actual

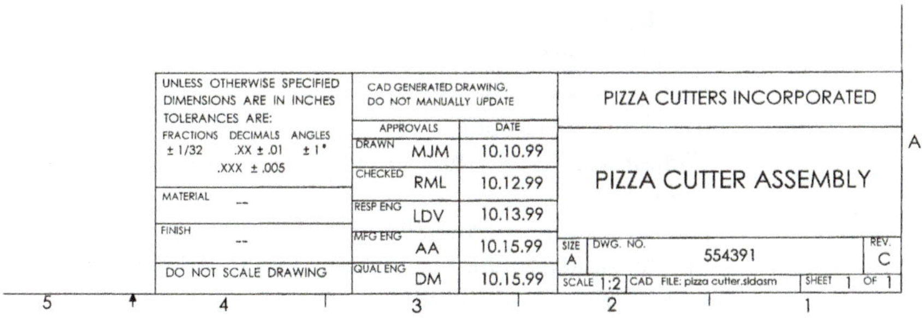

Figure 5-40 A typical title block.

object. It can be thought of as the degree to which the object is enlarged or shrunk in the drawing. Scales on drawings are denoted in several ways. For instance, consider an object that is represented in the drawing as half of its actual size. Then the drawing is created such that 1 inch on the drawing represents 2 inches on the physical object. This can be reported in the Scale box in the title block as HALF SCALE, HALF SIZE, 1 = 2, 1:2, or 1/2" = 1". In the numerical representations, the left number is the length of the drawing and the right number is the length of the actual object. A part that is drawn twice as large as its physical size would be denoted as DOUBLE, 2 × (denoting "2 times"), 2 = 1, or 2:1. In some cases, such as architectural drawings, the scale can be quite large. For instance, 1/4" = 1' – 0" corresponds to a scale of 1:48, where " denotes inches and ' denotes feet. Common scales used in engineering drawing are 1 = 1, 1 = 2, 1 = 4, 1 = 8, and 1 = 10, for English units and 1:1, 1:2, 1:5, 1:10, 1:20, 1:50, and 1:100 for metric units. A 1:1 scale is often specified as FULL SCALE or FULL SIZE.

5.7.2 LINES

The types of lines used in engineering drawing are standardized. Different linetypes have different meanings, and understanding these different types of lines makes interpreting drawings easier. The linetypes vary in thickness and in the length and number of dashes in dashed lines. Most CAD software automatically produces linetypes that correspond to the way the line is used in the drawing. Figure 5-41 shows some of the more commonly used linetypes:

visible lines

- **Visible lines** (thick, solid lines) represent the outline of the object that can be seen in the current view.

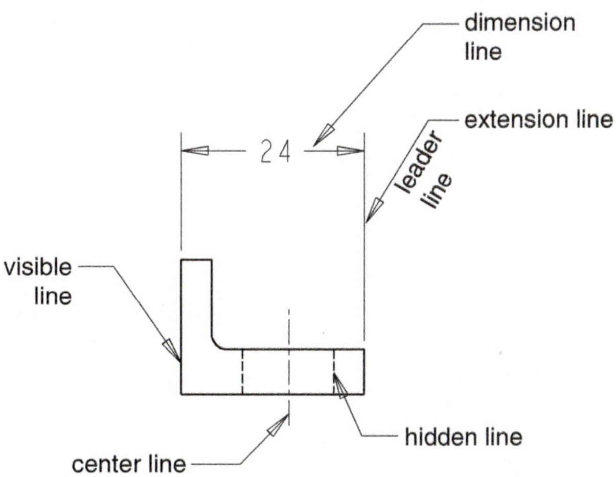

Figure 5-41 Types of lines used in drawings.

- **Hidden lines** (thin, dashed lines) represent features that are hidden behind surfaces in the current view.
- **Centerlines** or **symmetry lines** (thin, long-dash/short-dash lines) mark axes of rotationally symmetric parts or features.
- **Dimension lines** (thin lines with arrowheads at each end) indicate sizes in the drawing.
- **Extension lines** or **witness lines** (thin lines) extend from the object to the dimension line to indicate which feature is associated with the dimension.
- **Leader lines** (thin, solid lines terminated with arrowheads) are used to indicate a feature with which a dimension or note is associated.

The arrowheads for dimension lines, leader lines, and other types of lines may be filled or not filled, depending on the CAD software.

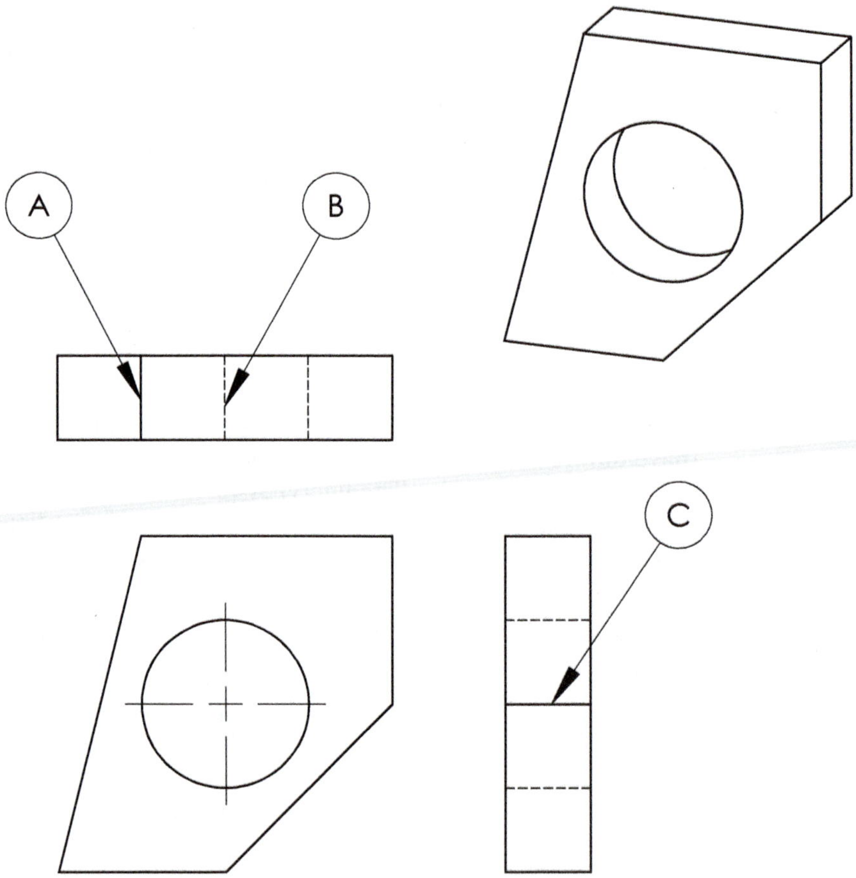

Figure 5-42 An example showing the precedence of lines.

It is common in technical drawings that two lines in a particular view coincide. When this occurs, a convention known as the **precedence of lines** dictates which line is shown: visible lines have precedence over hidden lines, which have precedence over centerlines. An example is shown in Figure 5-42. In the top view, line A is shown as a solid visible line, since the visible line that forms the top left edge has precedence over the dashed hidden line for the side of the hole. Line B of the top view is shown as a dashed hidden line, since it has precedence over the centerline of the hole. Likewise,

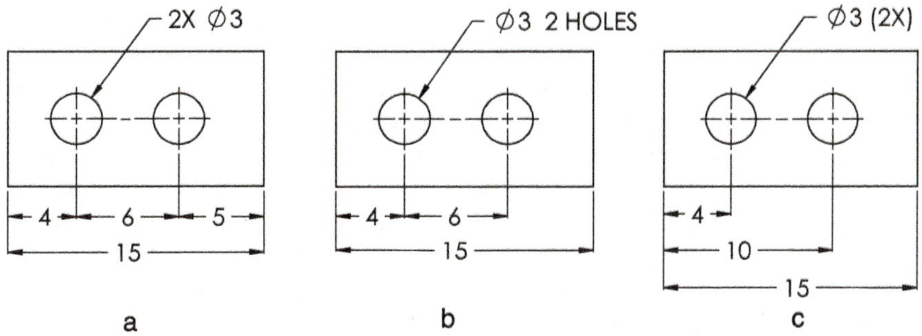

Figure 5-43 (a) An overdimensioned drawing; (b) A dimensioning scheme that emphasizes the 6-mm distance between the holes; (c) A dimensioning scheme that emphasizes the distance from the left edge to each hole.

line C in the side view is shown as a visible line, since the edge on the right side has precedence over the centerline of the hole. Visible lines, hidden lines, and centerlines all have precedence over extension lines.

Conventions also exist for the intersection of lines in a drawing. For instance, an extension line is always drawn so that there is a slight, visible gap between its end and the outline of the object, as shown in Figure 5-41. Similarly, center marks appearing at the center of a circle are drawn so that the short dashes of each line intersect to form a small cross, as shown in Figure 5-43. Fortunately, CAD software automatically takes care of most of the details related to lines, as well as their length and position. The engineer or designer need only know how to properly interpret the various linetypes.

5.7.3 DIMENSION PLACEMENT AND CONVENTIONS

overdimensioned

Engineering drawings are usually drawn to scale, but it is still necessary to specify numerical dimensions for convenience and to ensure accuracy. In general, sufficient dimensions must be provided to define the geometry of the part precisely, but redundant dimensions should be avoided. Furthermore, dimensions on the drawing should reflect the way the part is made or the critical dimensions of the part. For example, Figure 5-43 shows three drawings of a plate with two holes. Figure 5-43a is **overdimensioned**, having one redundant dimension. It is clear that, given the distances from the left edge to the left hole, the distance from the right edge to the right hole, and the overall width of the plate, the distance between the holes has to be 6 mm. Thus, the 6-mm dimension could be omitted. However, the dimensions that are shown in the drawing often suggest which dimensions are most important. If the 6-mm dimension in Figure 5-43a were omitted, it would suggest that the distance between each hole and the nearest edge of the plate is more important than the distance between the holes. But suppose, instead, that the distance between the holes is critical. This might be necessary if the holes in the plate are supposed to align with another part. Then the dimensions shown in Figure 5-43b would clearly indicate that the distance between the holes is the most important dimension by omitting the dimension from the right hole to the right edge. Figure 5-43c indicates that the distance from the left edge to each hole, not the distance between the holes, is critical. The engineer or designer must consider the optimal way to display dimensions to clearly convey which dimensions are most important.

The position of the dimensions is also important. Figure 5-44 shows several examples of incorrect ways of dimensioning a drawing. It is best to avoid having dimensions inside the boundary of the object, as is the case in Figure 5-44a. Dimensions should be kept outside the boundaries of the object whenever possible, unless clarity is improved in some way by having them inside the boundary. Dimension lines should not be too close to the boundaries of the object, as is the case with the 40-mm dimension in Figure 5-44b. A typical spacing is 10 mm (3/8 inch) for the distance of the first dimension from the object and 6 mm (1/4 inch) for the spacing between subsequent dimensions. It is also difficult to interpret the drawing when the dimension line coincides with a visible line of the object, as is the case in Figure 5-44b for the 20-mm dimension (the dimension

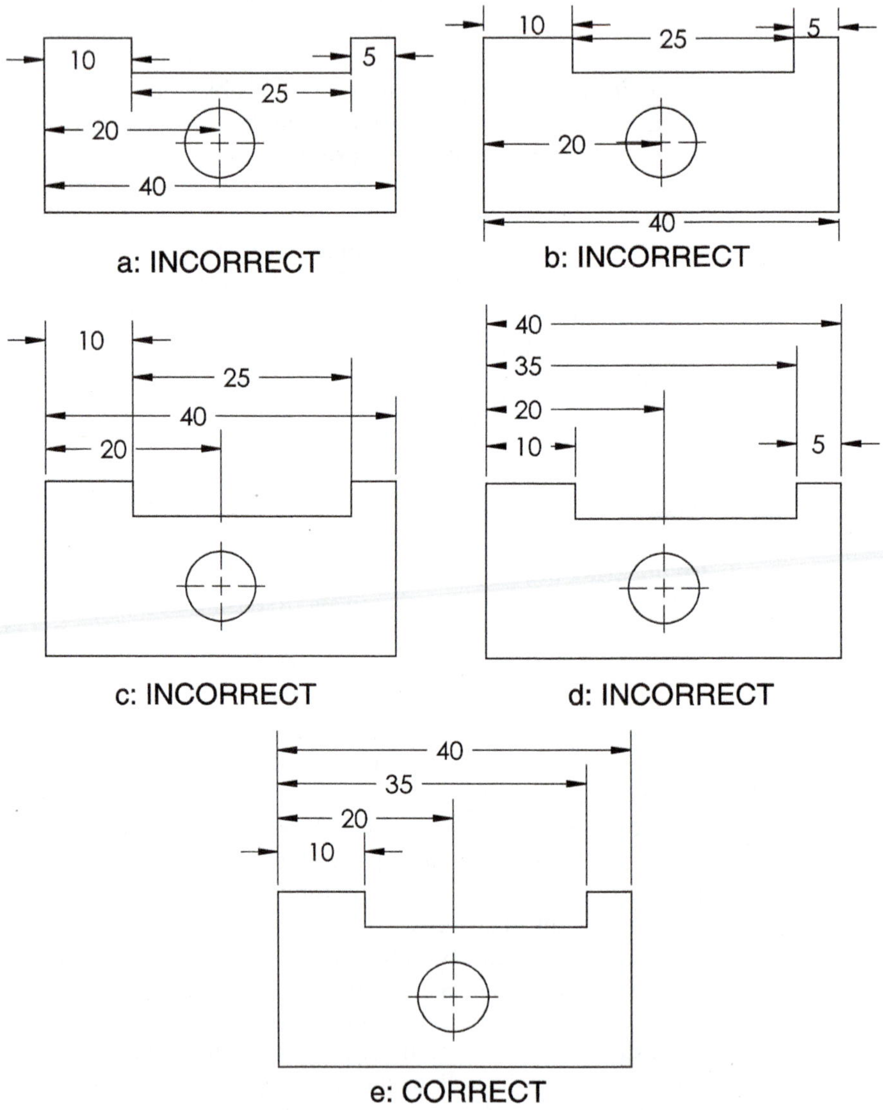

Figure 5-44 (a) Placing dimensions on the object is incorrect; (b) Placing dimensions so that dimension lines align with visible lines of the object or are very close to the object is incorrect; (c) Placing dimensions in random order is incorrect; (d) Overdimensioning is incorrect; (e) A properly dimensioned drawing.

line coincides with the center mark line) and the 25-mm dimension (the dimension line looks like a continuation of the upper boundary of the part). In Figure 5-44c, the dimensions are in a random order with some smaller dimensions farther from the boundary of the object (such as the 10-mm dimension) than larger dimensions (such as the 40-mm dimension). This results in dimension lines crossing extension lines, which should be avoided. It is often helpful to stack the dimensions in order of increasing length as shown in Figure 5-44d. However, there are three problems in this case. First, placing the dimension numbers right above one another is somewhat difficult to read. Second, there is a redundant dimension—the 5-mm dimension is unnecessary. Third, the 10-mm dimension is so small that there is no dimension line, and only the arrowheads appear. This is readily corrected by placing the arrowheads outside of the extension lines instead of inside, as shown in Figure 5-44e. In addition, in this case the dimension numbers are staggered, or offset, by a small amount to make them more easily readable. All the dimensioning conventions are aimed at making the drawing as easily readable as possible so that the information on the drawing is conveyed clearly to the reader.

contour dimensioning

It is natural for a person reading the drawings to look for the dimensions of a feature in the view where the feature occurs in its most characteristic shape and where it is visible (as opposed to hidden). This is known as **contour dimensioning**. For example, the location and size of a hole should be dimensioned in the view where the hole appears as a circle. Likewise, it is best to show the overall dimensions of the object in a view that is most descriptive of the object. If possible, both the horizontal and vertical location of a feature should be dimensioned in the same view. The upper part of Figure 5-45 shows an object properly dimensioned in millimeters to show the relation of the dimensions to the features in the most obvious manner. In the lower part of Figure 5-45, the dimensioning has several problems. The overall vertical dimensions of the L-shaped profile, the 8- and 35-mm dimensions, are shown on the right view where it is not clear that the object is L-shaped. The diameter of the hole is shown in a view where the hole is shown only as a pair of parallel hidden lines, not as a circle. And the vertical location of the hole is shown in one view, while the horizontal location is shown in another view. Finally, the vertical location of the hole is shown as a distance to its edge (the 7-mm dimension). Generally, the locations of holes should be dimensioned to the center of the hole.

The conventions for millimeter and inch dimensioning are different. For millimeter dimensioning, dimensions less than 1 mm have a zero preceding the decimal point, and integer dimensions may or may not include a decimal point and following zeros. Thus, legitimate millimeter dimensions are 0.8, 6, 8.00, and 1.5. Inch dimensions do not include a zero before the decimal point for values less than 1 inch. Zeros to the right of the decimal point are usually included. Legitimate inch dimensions are .800, 6.00, 8.00, and 1.500. For both millimeter and inch dimensions, the number of digits to the right of the decimal point may be used to indicate the tolerance (acceptable variation). It is standard practice to omit the units for dimensions. However, it is wise to note the units in the title block or as a note on the drawing.

Circles and arcs are dimensioned using special symbols and rules. A radius is denoted using a leader line with an arrow pointing at the arc, as shown in the upper portion of Figure 5-45. Sometimes a small cross is placed at the center of the radius. The position of the center should be dimensioned unless the radius is for a rounded edge, in which case the arc is positioned by virtue of being tangent at its ends to the sides forming the corner that is rounded, as shown in Figure 5-45. The dimension for a radius begins with a capital letter R to indicate radius. Full circles are dimensioned using the diameter with a leader line having an arrow pointing at the circle, as shown

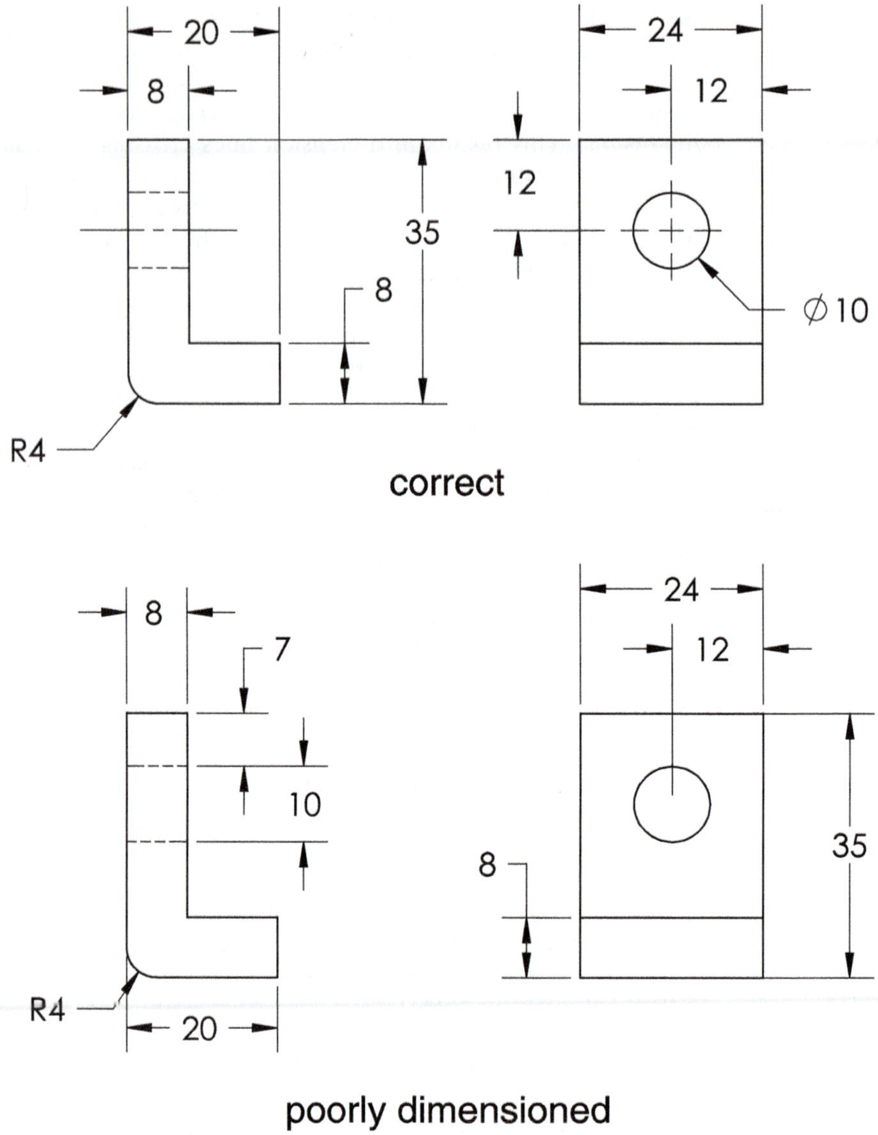

Figure 5-45 Properly and improperly dimensioned drawings.

in Figures 5-39 and 5-45. If space permits, the leader line can extend across a diameter of the circle with arrows at each end of the diameter. The dimension for a diameter begins with the Greek letter phi (Ø)to indicate diameter. The location of the center of the circle must also be dimensioned.

Sometimes maintaining clarity makes it preferable to dimension several concentric circles from a side view, where the circle does not appear as a circle, thus violating the rule to dimension a feature in its most characteristic shape. This also permits the depth of the feature to be indicated along with the diameter in the same view of the object. Figure 5-46 shows how concentric circles can be dimensioned from a side view. The diameter dimensions to the left use (Ø) to indicate diameter. The dimensions above the object indicate the length of the features corresponding to particular diameters along the axis.

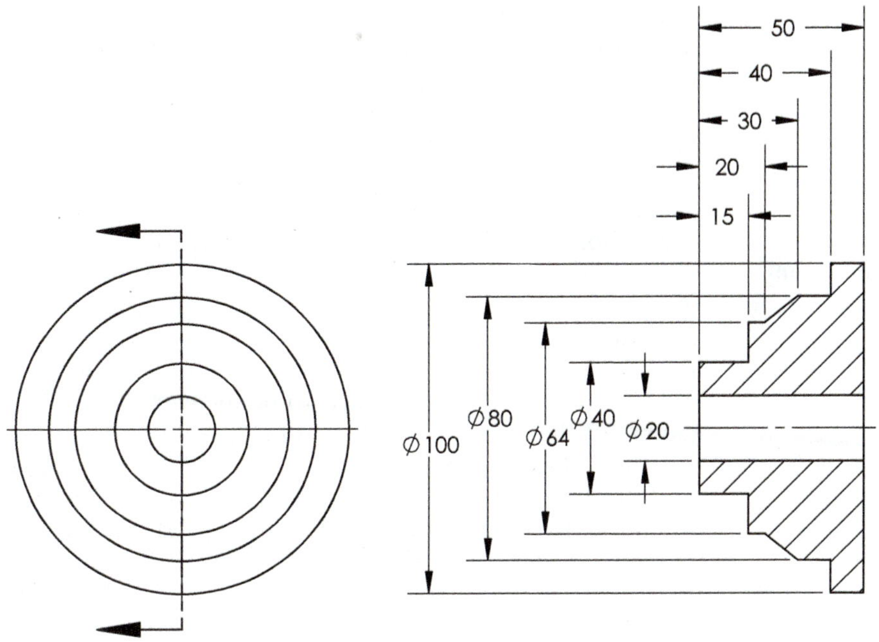

Figure 5-46 Proper dimensioning of concentric circles.

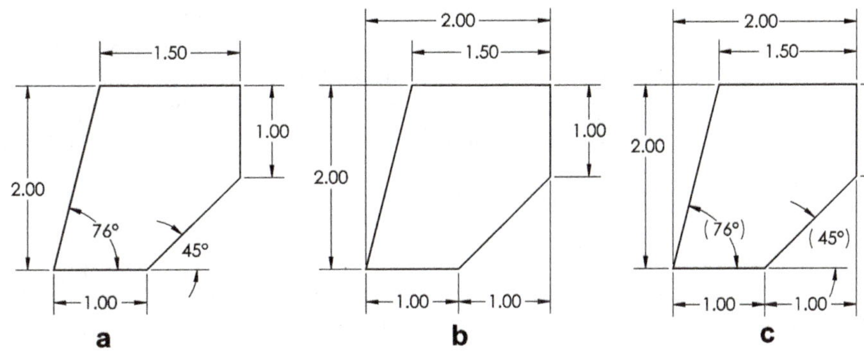

Figure 5-47 (a) Properly dimensioned object using angles; (b) Properly dimensioned object without using angles; (c) Angles are in parentheses to indicate that they are reference dimensions.

Angles are dimensioned as decimals or in degrees and minutes (60 minutes is one degree), so that 32.5 and 32°30' are equivalent, where ' indicates minutes. The dimension line for an angle is drawn as an arc with its center at the apex of the angle, as shown in Figure 5-47a. Here a combination of angles and length dimensions are used to fully dimension the object. Note that the arrows are "inside" for the 76° dimension and "outside" for the 45° dimension. Either type can be used, depending on the space available and readability. Figure 5-47b shows how the same shape can be dimensioned without using any angles. In some ways, this figure is less clear than Figure 5-47a, because some extension lines cross and there are more dimension lines. At first glance, it may appear that Figure 5-47b is overdimensioned, since the 2-inch dimension on the top is simply the sum of the two 1-inch dimensions on the bottom. However, strictly speaking, all of these dimensions are necessary. If the lower right 1-inch dimension were omitted, the upper-right corner of the object would not necessarily be constrained to be 90°. In practice, when lines are drawn at right angles, a 90° angle is usually implied, so the lower right 1-inch dimension could be omitted without creating any ambiguity. Figure 5-47c shows the length dimensions in addition to the angles. While this drawing is overdimensioned, the display of the angles may aid in clarifying the dimensions for the reader of the drawing. Whether to include extra dimensions depends on what critical aspects of the dimensions need to be indicated on the drawing. If it is necessary that the angle in Figure 5-47c be exactly 45°, then this angle should be specified on the drawing and the 1-inch horizontal dimension should be in parentheses to indicate it is a **reference dimension** provided for convenience to the reader. On the other hand, if the 1-inch dimension is critical, it should be indicated on the drawing and the angle could be provided as a reference dimension for convenience, as shown in Figure 5-47c. Reference dimensions are provided only for information to help in interpreting the drawing. They are not intended to be measured or to be used to govern the manufacture of the part. Generally, it is preferable to show the dimensions in the clearest form possible. For this object, Figure 5-47a is superior to the other dimensioning schemes by using the fewest dimensions that overlap the least.

reference dimension

Multiple features that are identical need not be individually dimensioned. For example, the diameter of the two 3-mm holes in Figure 5-43a are dimensioned with a leader line to one hole with the dimension 2 × ∅3. The 2 × indicates "two times." Alternative dimensions are ∅3 (2 ×), ∅3 2 PLACES, or ∅3 2 HOLES.

There are several other "rules" for dimensioning:

- Dimension lines should be outside of the outline of a part whenever possible.
- Dimension lines should not cross one another.
- Dimensions should be indicated on a view that shows the true length of the feature. This is particularly important when dimensioning features that are on a surface that is at an angle to the plane of the orthographic view.
- Each feature should be dimensioned only once. Do not duplicate the same dimension in different views.
- Dimension lines should be aligned and grouped when possible to promote clarity and uniform appearance. See, for example, the placement of the horizontal dimensions in Figure 5.44e.
- The numerical dimension and arrows should be placed between the extension lines where space permits. If there is space only for the dimension, but not the arrows, place the arrows outside of the extension lines. When the space is too small for either the arrows or the numerical value, place both outside of the extension lines. See Figure 5.45 for examples.
- Place the dimension no closer than about 10 mm (3/8 inch) from the object's outline.
- Dimensions should be placed in clear spaces, as close as possible to the feature they describe.
- When dimensions are nested (such as the horizontal dimensions in Figure 5.44e), the smallest dimensions should be closest to the object.
- Avoid crowding dimensions. Leave at least 6 mm (1/4 inch) between parallel dimension lines.
- Extension lines may cross visible lines of the object.
- Dimensions that apply to two adjacent views should be placed between the views, unless clarity is enhanced by placing them elsewhere. The dimension should be attached to only one view. Extension lines should not connect two views.
- Dimension lines and extension lines should not cross, if possible. Extension lines may cross other extension lines.
- A centerline may be extended to serve as an extension line, in which case it is still drawn as a centerline.

- Centerlines should not extend from one view to another view.
- Leader lines are usually sloped at about 30°, 45°, or 60° and are never horizontal or vertical.
- Numerical values for dimensions should be approximately centered between arrowheads unless several dimensions are nested, as in Figure 5-44e. In this case, dimensions should be staggered.
- When a rough, noncritical dimension (such as a round) is indicated on a drawing, add the note TYP to the dimension to indicate that the dimension is "typical" or approximate.
- Sometimes it is helpful to include a reference dimension that is redundant or otherwise unnecessary, but is included for the convenience of the person reading the drawing. Such dimensions are usually in parentheses.

5.8 ASSEMBLY DRAWINGS

An assembly drawing shows all of the components of a design either assembled or in an exploded view. Many times assembly drawings include sections. Most dimensions are omitted in assembly drawings. Individual parts are not dimensioned, but some dimensions of the assembled mechanism may be included. Hidden lines are seldom necessary in assembly drawings, although they can be used where they clarify the design. Leader lines attached to a ballooned letter or detail number, as shown in Figure 5-48, reference the parts of the assembly. The leader lines should not cross and nearby leader lines should be approximately parallel. Sometimes parts are labeled by name rather than number. The parts list may be on the assembly drawing (usually on the right side or at the bottom) or it may be a separate sheet. The assembly drawing may

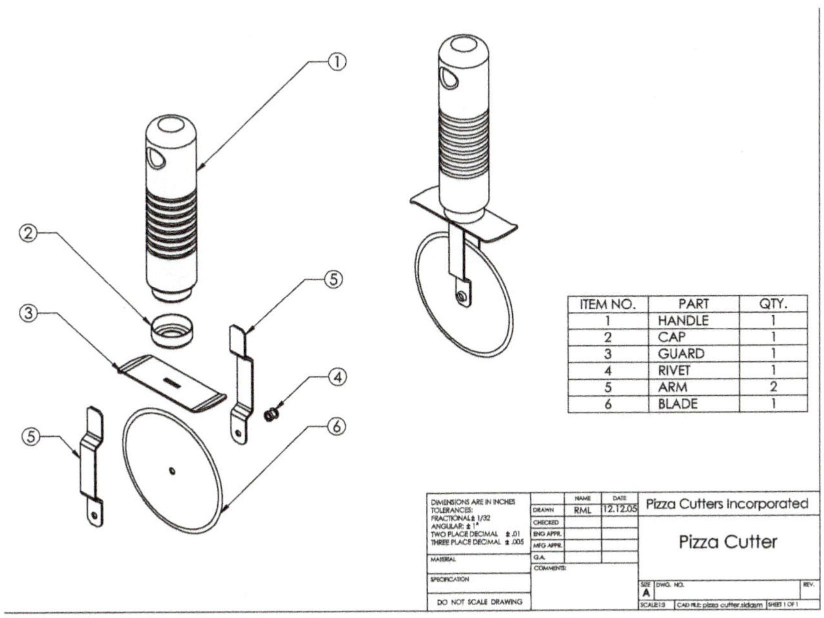

Figure 5-48 Assembly drawing of a pizza cutter.

also include machining or assembly information in the form of notes on the drawing. Often assembly drawings include assembly sections. These are typically orthographic or pictorial section views of parts as put together in an assembly. The assembly cross section on the left side of Figure 5-49 shows the interior structure of the parts of a pizza cutter and how they fit together. The detail view in the upper right of Figure 5-49 shows the cross section through the rivet to indicate how the parts are assembled.

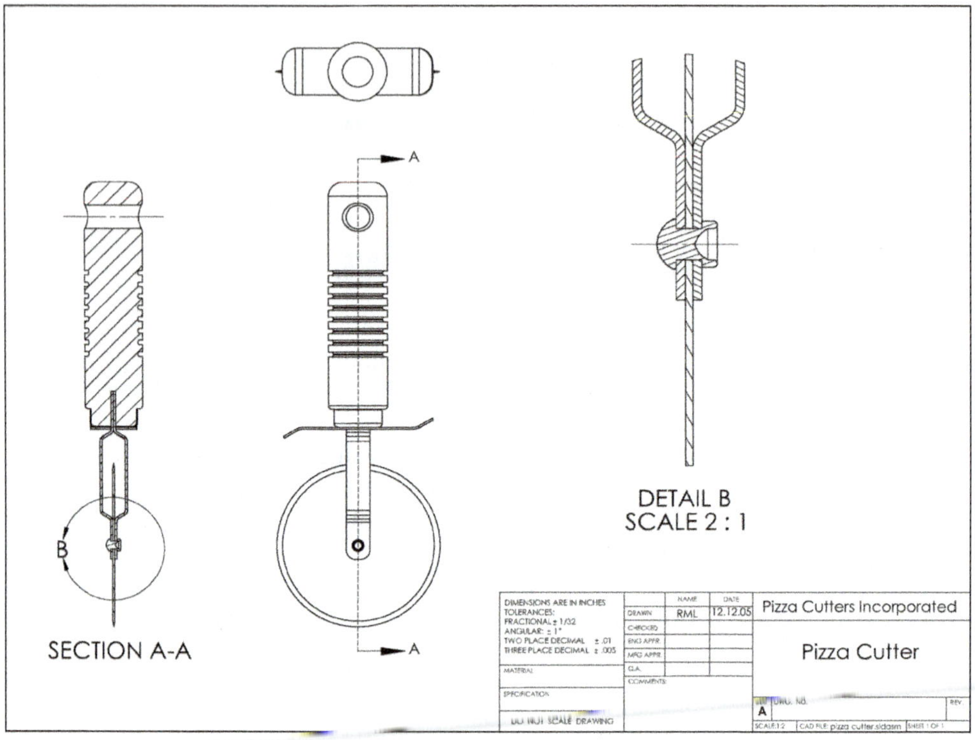

Figure 5-49 Detail view of a small section of an assembly.

Chapter 5 ASSESSMENT QUESTIONS

1. How does freehand sketching differ from technical drawing?
2. Design a poster pointing out three safety hazards in your laboratory or classroom. Include sketches of each as part of the poster.
3. Explain how "paint" or "draw" programs such as Adobe Illustrator, MS paint, or drawing tools in MS Word differ from CAD engineering graphics software.
4. Create the 2D Orthographic Projection for a) through i).

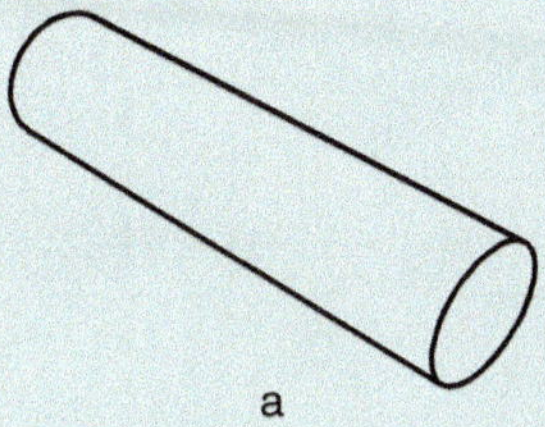
a

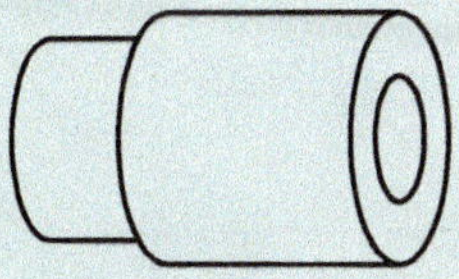
b

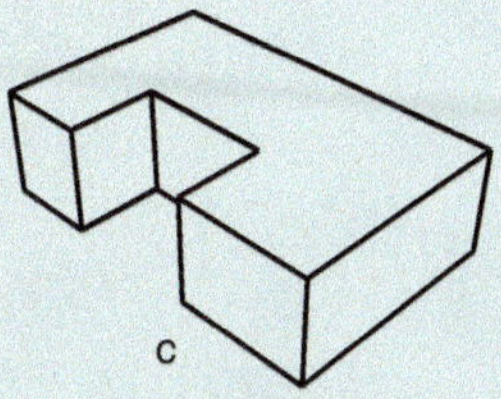
c

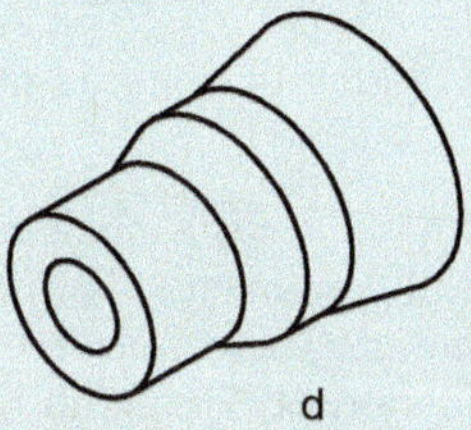
d

e

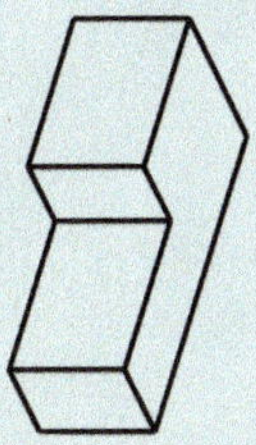
f

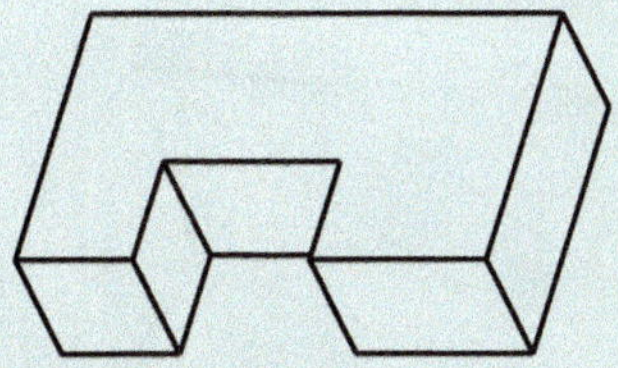
g

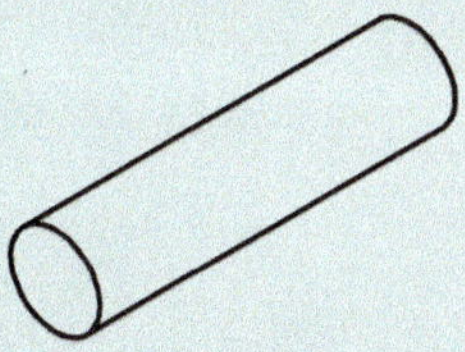
h

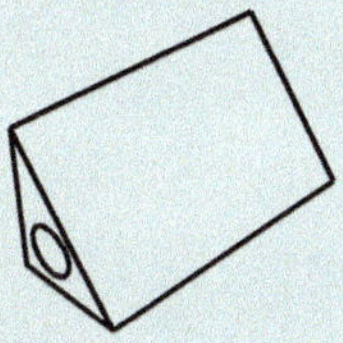
i

5. The items shown are 2 inches wide, 1.5 inches high, and 1 inch deep. All holes are through holes.
   - **(a)** Create freehand orthographic sketches of the objects in the figures on the following page. (The objects are shown as oblique projections.)
   - **(b)** Create freehand isometric sketches of the objects. The objects are shown as oblique projections.)
   - **(c)** For the following diagrams, draw a 2D drawing of the face facing you. (There is no need to draw in any hidden lines.)

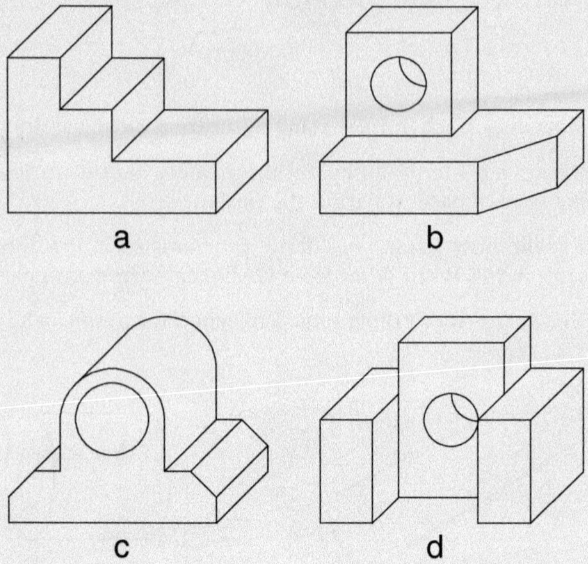

6. Use a CAD package to draw each figure from question 5.
7. Look around your classroom. With a partner create a blueprint of your specific room.
8. Create a freehand orthographic sketch of a coffee mug. Only two views are necessary to fully represent the mug.
9. The wireframe model shown in the figure is a small cube centered within a large cube. The nearest corners of the two cubes are connected. For instance, the lower left front corners of both cubes are connected with a line. This wireframe model is ambiguous; that is, it could represent several different solid bodies. Freehand sketch or trace the figure. Then use dashed lines for hidden lines to show at least two different solid-model configurations that are possible for this wireframe model.

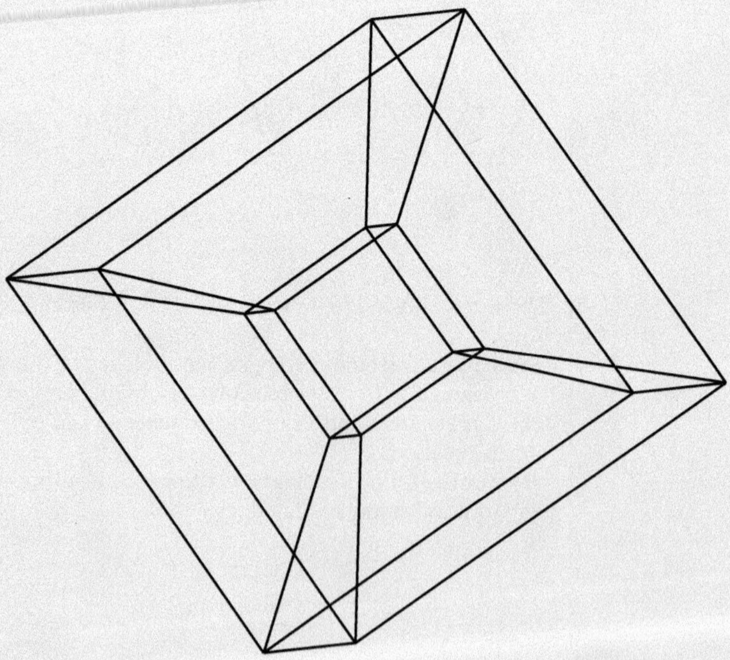

10. Consider the following objects. Sketch what the base feature would look like. List what features would be added to model the object. The type of base feature to be used (extrude, revolve, or both) is noted.

(a) hexagonal cross section of a wooden pencil that is sharpened to a point (do not include eraser)—extrude.

(b) coffee mug without the handle—revolve.

(c) nail—extrude and revolve.

(d) push pin—revolve.

11. Three orthographic views and an isometric projection of an object that has overall dimensions of 4 × 4 × 4 inches are to be shown in a drawing. Determine the best ANSI sheet size for a an object of this size at the scales indicated below:

(a) Half size.

(b) 2:1.

(c) 1 = 1.

12. The drawing shown below has many errors in the dimensioning. Sketch or draw with CAD the orthographic views and show the proper placement of the dimensions.

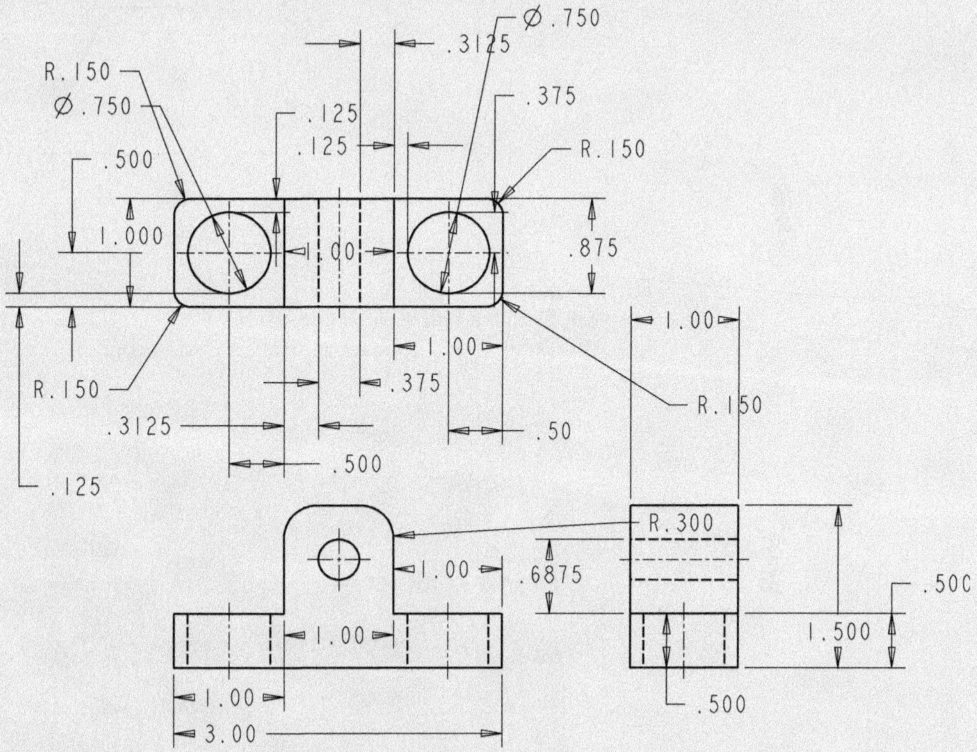

13. The figure below shows the dimensioned front view and several different corresponding right-side views, each with a horizontal width of 1 inch. Using CAD, draw three orthographic views (front, top, and right side) as well as the auxiliary view of the inclined surface for the corresponding right-side view.

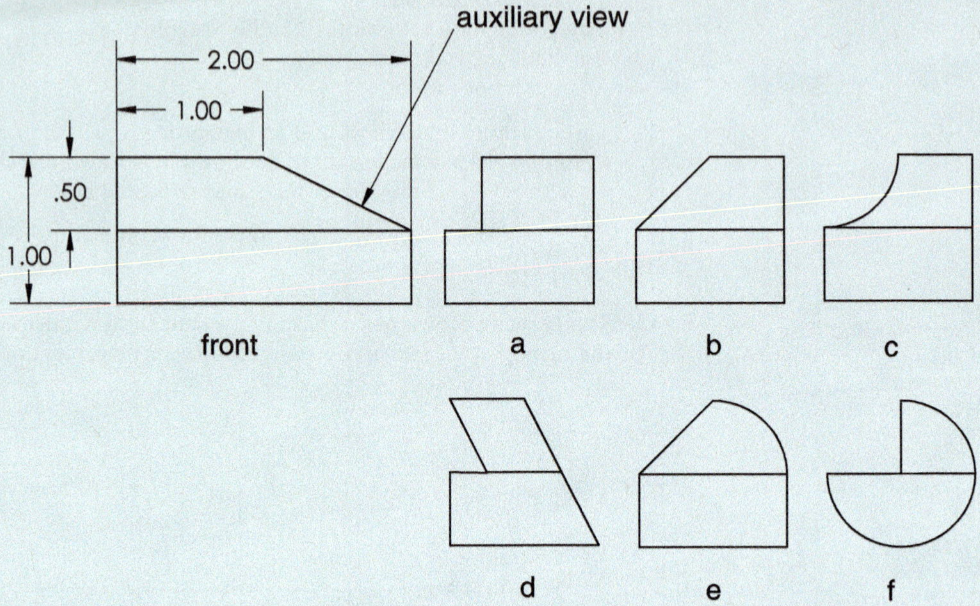

14. Research:
  (a) Sketch a half-door-hinge.
  (b) Draw a half-door-hinge using a CAD program.

15. Find a cellphone:
  (a) Sketch the cellphone.
  (b) Draw the cellphone using a CAD program.

16. Assume we are producing a CAD drawing and the following line types are used: visible, hidden, center, dimension, extension, and leader. Explain how each of the lines help in the creation and interpretation of the drawing.

CHAPTER 6
ELECTRICITY AND ENERGY

All systems in the natural and synthetic worlds require energy to function properly. Energy is used to make products; to move goods and people; and to heat, cool, and light the places where people work and live.

renewable energy sources

nonrenewable energy sources

Like materials, energy sources are either renewable or nonrenewable. **Renewable energy sources** are those that can be replaced. Human and animal muscle power are renewable. So is the energy we get from burning wood. Solar, wind, gravitational, tidal, geothermal, nuclear fusion energies are other renewable types of energy. **Nonrenewable energy sources** are those that cannot be replaced once we use them up. These include coal, oil, natural gas, and nuclear fission (atomic energy).

The original source for nearly all forms of energy on Earth is the sun. Plants receive the power to grow from the sun. People and other animals get energy from the plants and other foods they eat. Coal, oil, and gas come from decayed plant and animal matter.

The following are the seven different kinds of energy sources:

- human and animal muscle power—from food
- solar energy—from the sun
- wind energy—from the airflow across Earth's surface
- gravitational energy—from tides and falling water
- geothermal energy—from heat inside Earth
- chemical energy—from fossil fuels and wood
- nuclear energy—from the conversion of radioactive matter into energy

We can use these sources of energy directly. We also can convert them into other forms of energy, such as mechanical, electrical, and light energy. For example, a dam can convert gravitational energy into electrical energy.

6.1 ENERGY AND WORK

energy

Energy is an abstract quantity with several definitions, depending on the form of energy being discussed. You may be familiar with some of the following types of energy.

work

- **Work** (W) is energy expended by exertion of a force (F) over a distance (d). As an example, if you exert a force on (push) a heavy desk so that it slides across the floor, which will make you more tired: pushing it 5 feet or pushing it 50 feet? The farther you push it, the more work you do.

potential energy

- **Potential energy** (PE) is a form of work done by moving a weight (w)—which is a force—a vertical distance (H). Recall that weight is mass (m) times gravity (g). Note that this is a special case of the work equation, where force is weight and distance is height.

kinetic energy

energy of translational motion

- **Kinetic energy** (KE) is a form of energy possessed by an object in motion. If a constant force is exerted on a body, then by $F = ma$, we see that the body experiences a constant acceleration, meaning the velocity increases linearly with time. Since the velocity increases as long as the force is maintained, work is being done on the object. Another way of saying this is that the object upon which the force is applied acquires kinetic energy, also called **energy of translational motion**. For a nonrotating body moving with some velocity (v) the kinetic energy can be calculated by $KE_T = (½)mv^2$.

  This, however, is not the entire story. A rotating object has energy whether it is translating (moving along a path) or not. If you have ever turned a bicycle upside down, spun one of the wheels fairly fast, then tried to stop it with your hand, you understand that it has energy. This is **rotational kinetic energy**, and for an object spinning in place (but not going anywhere), it is calculated by $KE_R = (½)I\omega^2$. The Greek letter omega (ω) symbolizes angular velocity or the object's rotational speed, typically given in units of radians per second. The moment of inertia (I) depends on the mass and the geometry of the spinning object. The table shown lists the moments of inertia for a few common objects.

rotational kinetic energy

  For an object that is rotating and translating, such as a bowling ball rolling down the lane toward the pins, the total kinetic energy is simply the sum of the two:

$$KE = KE_T + KE_R = (1/2)mv^2 + (1/2)I\omega^2 = (1/2)(mv^2 + I\omega^2)$$

thermal energy

- **Thermal Energy** or heat (Q) is energy associated with a change in temperature (ΔT). It is a function of the mass of the object (m) and the specific heat (C_p), which is a property of the material being heated:

$$Q = mC_p\Delta T$$

IMPORTANT CONCEPT

Work
$W = F\Delta x$

Potential Energy
$PE = mg\Delta H$

Kinetic Energy, translational
$KE_T = 1/2m(v_f^2 - v_i^2)$

Kinetic Energy, rotational
$KE_R = 1/2\,I(\omega_f^2 - \omega_i^2)$

Kinetic Energy, total
$KE = KE_T + KE_R$

Thermal Energy
$Q = mC_p\Delta T$

MOMENT OF INERTIA (I)

m = mass, r = radius

Object	I for KE_R
Cylinder: thin shell	mr^2
Cylinder: solid	$1/2\,mr^2$
Sphere: thin shell	$2/3\,mr^2$
Sphere: solid	$2/5\,mr^2$

ENERGY

J = N m
SI unit of energy = joule

joule

The SI unit of work is **joule**, defined as one newton of force acting over a distance of one meter. The unit is named after James Joule (1818–1889), an English physicist responsible for several theories involving energy, including the definition of the mechanical equivalent of heat and Joule's law, which describes the amount of electrical energy converted to heat by a resistor (an electrical component) when an electric current flows through it. In some mechanical systems, work is described in units of foot pound-force [ft lb_f].

British thermal unit

calorie

For energy in the form of heat, units are typically reported as British thermal units and calories instead of joules. A **British thermal unit** [BTU] is the amount of heat required to raise the temperature of one pound-mass of water by one degree Fahrenheit. A **calorie** [cal] is amount of heat required to raise the temperature of one gram of water by one degree Celsius.

EXAMPLE 6-1

In the morning, you like to drink your coffee at a temperature of exactly 70 degrees Celsius [°C]. The mass of the coffee in your mug is 470 grams. To make your coffee, you had to raise the temperature of the water by 30 degrees Celsius. How much energy in units of British thermal units [BTU] did it take to heat your coffee? The specific heat of water is 4.18 joules per gram degree Celsius [J/(g °C)].

First, you must determine the type of energy we are using. The parameters discussed include mass, temperature, and specific heat. Examining the energy formulas given above, the equation for thermal energy fits. Also, the words "How much energy . . . did it take to heat your coffee" fits with an understanding of thermal energy.

Step One: Convert to Base SI Units	
No conversion needed	
Step Two: Calculate	
Method	**Steps**
(1) Determine appropriate equation	$Q = mC_p\Delta T$
(2) Insert known quantities	$Q = \frac{470\ \text{g}}{} \Big\vert \frac{4.18\ \text{J}}{\text{g°C}} \Big\vert \frac{30\text{°C}}{}$
(3) Calculate	$Q = 59{,}370\ \text{J}$
Step Three: Convert from Base SI Units to Desired Units	
Method	**Steps**
(1) Term to be converted	59,370 J
(2) Conversion formula (3) Make a fraction (4) Multiply	$\frac{59{,}370\ \text{J}}{} \Big\vert \frac{9.48 \times 10^{-4}\ \text{BTU}}{1\ \text{J}}$
(5) Cancel, calculate, be reasonable	56 BTU

6.2 POWER

power
watt
horsepower

Power is defined as energy per time. The SI unit of power is **watt**, named after James Watt (1736–1819), a Scottish mathematician and engineer whose improvements to the steam engine were important to the Industrial Revolution. He is responsible for the definition of **horsepower** [hp], a unit of power originally used to quantify how the steam engine could replace the work done by a horse.

IMPORTANT CONCEPT

Power = energy / time

W = J / s

SI unit of power = watt

NOTE

Power is the RATE at which energy is delivered over time.

To help understand the relationship between energy and power, imagine the following. Your 1,000-kilogram car has run out of gas on a level road. There is a gas station not far ahead, so you decide to push the car to the gas station. Assume that you intend to accelerate the car up to a speed of one meter per second (about 2.2 miles per hour), and then continue pushing at that speed until you reach the station. Ask yourself the following questions:

- Can I accelerate the car to one meter per second in one minute?
- On the other hand, can I accelerate it to one meter per second in one second?

Most of you would probably answer "yes" to the first and "no" to the second, but why? Well, personal experience! But that is not really an explanation. Since the change in kinetic energy is the same in each case, to accelerate the car in one second, your body would have to generate energy at a rate 60 times greater than the rate required if you accelerated it in one minute. The key word is rate, or how much energy your body can produce per second. If you do the calculations, you will find that for the one-minute scenario, your body would have to produce about $1/90$ horsepower, which seems quite reasonable. On the other hand, if you try to accomplish the same acceleration in one second, you would need to generate $2/3$ horsepower. Are you two-thirds as powerful as a horse?

As another example, assume that you attend a class on the third floor of your school. When you are on time, you take 2 minutes to climb to the third floor. On the other hand, when you are late for class, you run up the three flights in 30 seconds.

- In which case do you do the most work (expend the most energy)?
- In which case do you generate the most power?

EXAMPLE 6-2

A 50-kilogram load is raised vertically a distance of 5 meters by an electric motor in 60 seconds. Assume the energy used by the system was found to be 2,450 joules. How much power in units of watts does the motor use, assuming no energy is lost in the process?

Step One: Convert to Base SI Units	
No conversion needed	
Step Two: Calculate	
Method	**Steps**
(1) Determine appropriate equation	$\text{Power} = \dfrac{\text{energy}}{\text{time}}$
(2) Insert known quantities	$\text{Power} = \dfrac{2{,}450\ \text{J}}{60\ \text{s}}$
(3) Calculate, be reasonable	$\text{Power} = 41\dfrac{\text{J}}{\text{s}} \left\lvert \dfrac{1\ \text{W}}{1\ \dfrac{\text{J}}{\text{s}}} \right. = 41\ \text{W}$
Step Three: Convert from Base SI Units to Desired Units	
No conversion needed	

Note that since power = energy/time, energy = power * time. We pay the electric company for energy calculated this way as kilowatt-hours. If power is constant, we can obtain the total energy involved simply by multiplying the power by the length of time that power is applied. If power is *not* constant, we would usually use calculus to determine the total energy, but that solution is beyond the scope of this book.

6.3 EFFICIENCY

efficiency

Efficiency (η, Greek letter eta) is a measure of how much of a quantity, typically energy or power, is lost in a process. In a perfect world, efficiency would always be 100%. All energy put into a process would be recovered and used to accomplish the desired task. We know that this can never happen, so *efficiency is always less than 100%*. If a machine operates at 75% efficiency, 25% of the energy is lost. This means you have to put in "extra" energy to complete the work.

input

output

The use of the terms "input" and "output" require some explanation. The **input** is the quantity of energy or power or whatever required by the mechanism from some source to operate and accomplish its task. The **output** is the amount of energy or power or whatever is actually applied to the task itself by the mechanism. Note that the rated power of a device, whether a light bulb, a motor, or an electric heater, refers to the input power—the power needed to operate the device—not the output power. In an ideal, 100% efficient system, the input and output would be equivalent. In an inefficient system (the real world), the input is equivalent to the sum of the output and the power or energy lost.

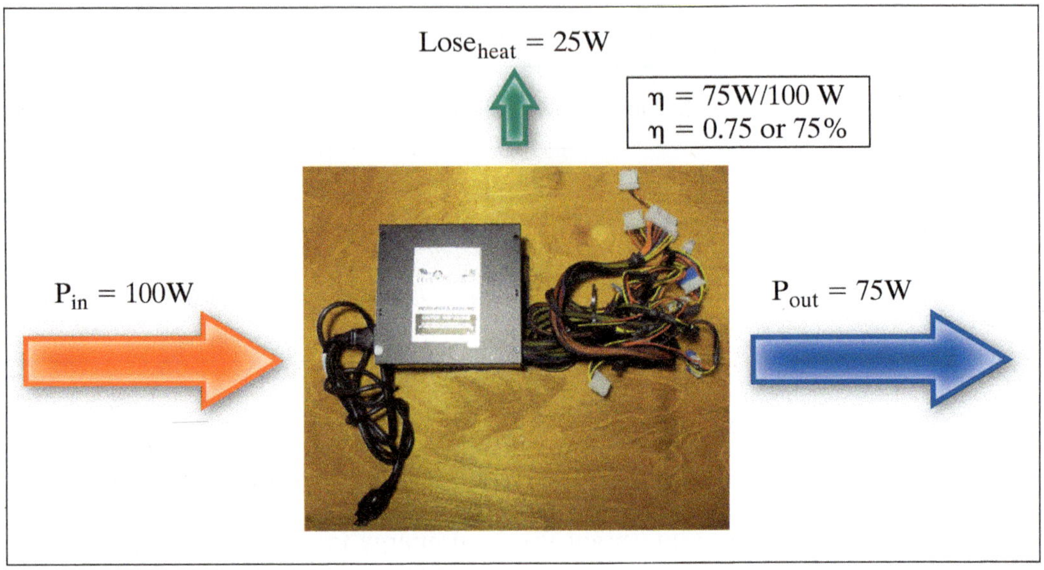

Figure 6-1 Energy is a measure of the output power vs. the input power.

IMPORTANT CONCEPT

Efficiency is always less than 100%.
Efficiency (η) = output/input
Efficiency (η) = output/(output + loss)
Input = quantity required by mechanism to operate
Output = quantity actually applied to task
Loss = quantity wasted during the application

Orders of Magnitude

Table 6-1 gives you an idea of orders of magnitude of power and energy as related to real-world objects and phenomena. All values are approximate and, in most cases, have been rounded to only one significant figure. Thus, if you actually do the calculations from power to energy, you will find discrepancies.

A few things of possible interest:

- U.S. power consumption (all types) is one-fifth of the total world power consumption.
- The Tsar Bomba generated 1.5% of the power of the sun, but only lasted 40 nanoseconds.
- Total human power consumption on the planet is about 0.01% (1/10,000) of the total power received from the sun.

Table 6-1 Order of magnitude for power and energy comparison

Power	"Device"	Energy per Hour	Energy per Year
10 fW	Minimum reception power for cell phone	40 pJ	300 nJ
	Single human cell	4 nJ	30 mJ
10 mW	DVD laser	40 J	300 kJ
500 mW	Cell phone microprocessor	2 kJ	15 MJ
50 W	Modern 2GHz microprocessor	200 kJ	1.5 GJ
100 W	Human at rest	400 kJ	3 GJ
500 W	Human doing strenuous work	2 MJ	15 GJ
750 W	Power per square meter bright sunshine	3 MJ	25 GJ
2 kW	Maximum human power for short period	(NA)	
20 kW	Average U.S. home	80 MJ	600 GJ
100 kW	Typical automobile	400 MJ	3 TJ
150 MW	Boeing 747 jet	500 GJ	5 PJ
1 GW	Large commercial nuclear reactor	4 TJ	30 PJ
20 GW	Three Gorges Hydroelectric Dam (China)	80 TJ	600 PJ
4 TW	U.S. total power consumption	15 PJ	100 EJ
20 TW	Total human power consumption	80 PJ	600 EJ
100 TW	Average hurricane	400 PJ	(NA)
200 PW	Total power received on the Earth from the sun	1 ZJ	6 YJ
5 YW	Largest fusion bomb ever built (Russian Tsar Bomba)	(NA) Total yield 200 PJ	
400 YW	Total power of the sun	10^6 YJ	10^9 YJ

● EXAMPLE 6-3

A standard incandescent light bulb has an efficiency of about 5%; thus, $\eta = 0.05$. An incandescent bulb works by heating a wire (the filament) inside the bulb to such a high temperature that it glows white. About 95% of the power delivered to an incandescent bulb is discarded as heat. Only 5% results in "light" energy.

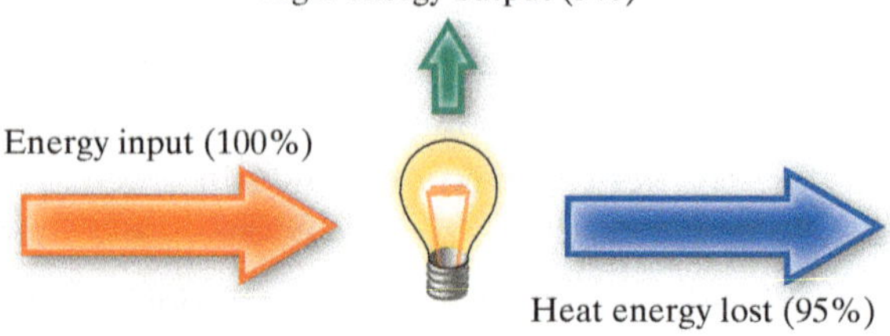

If a 100-watt bulb is turned on for 15 minutes, how much energy is "lost" as heat during the 15-minute period?

Step One: Convert to Base SI Units	
No conversion needed	
Step Two: Calculate	
Method	**Steps**
(1) Determine appropriate equation	Energy = (Power)(Time)
(2) Insert known quantities	Energy = (Power)(15 min)
For Unknown Quantities, Repeat the Process	
Method	**Steps**
(1) Determine appropriate equation	Input power = output power/efficiency "Lost" power = input power − output power
(2) Insert known quantities	100 W = output power/(0.05)
(3) Calculate	Output power = 5 W "Lost" power = 100 W − 5 W = 95 W
(2) Insert known quantities	Energy = (95 W)(15 min)
(3) Calculate, be reasonable	$E = 95\ \frac{\text{J}}{\text{s}} \Big\| \frac{15\ \text{min}}{} \Big\| \frac{60\ \text{s}}{1\ \text{min}}$ Energy lost = 85,500 J
Step Three: Convert from Base SI Units to Desired Units	
No conversion needed	

● EXAMPLE 6-4

Over the past few decades, the efficiency of solar cells has risen from about 10% to the most recent technologies achieving about 40% conversion of solar energy to electricity. The losses are due to several factors, including reflectance and resistive losses, among others.

Assume you have an array of solar cells mounted on your roof with an efficiency of 28%. If the array is delivering 750 watts of electricity to your home, how much solar power is falling on the photoelectric cells?

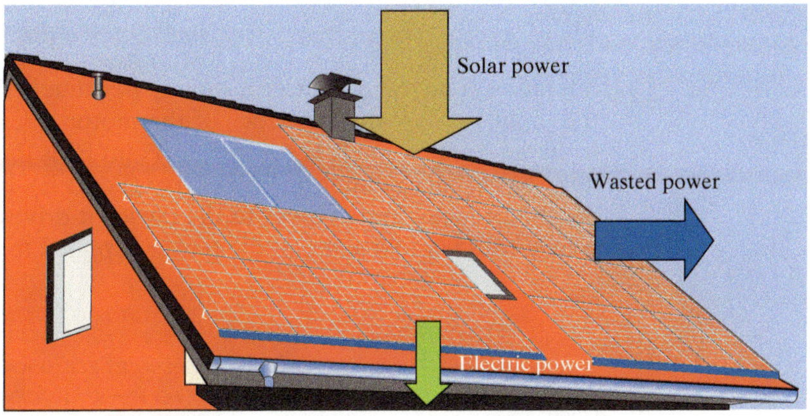

Step One: Convert to Base SI Units	
No conversion needed	
Step Two: Calculate	
Method	**Steps**
(1) Determine appropriate equation	Input power = output power/efficiency
(2) Insert known quantities	Input power = 750 W/0.28
(3) Calculate, be reasonable	Input power = 2,680 W
Step Three: Convert from Base SI Units to Desired Units	
No conversion needed	

NOTE

How much power is wasted by the array of solar cells? If the array received 2,680 watts and delivered 750 watts, then the difference is 2,680 − 750 = 1,930 watts; thus, 1,930 watts are wasted.

● EXAMPLE 6-5

A simple two-stage machine is shown in the diagram below. Initially, an electric motor receives power from the power grid, accessed by being plugged into a standard electrical wall socket. The power received by the motor from the wall socket is the "input" power or the power the motor uses or requires (Point A).

The spinning drive shaft on the motor can then be used to power other devices, such as a hoist or a vacuum cleaner or a DVD drive; the power available from the spinning shaft is the "output" power of the motor (Point B). In the process of making the drive shaft spin, however, some of the input power is lost (Point C) because of both frictional and ohmic heating as well as other wasted forms such as sound.

Since some energy is being lost as heat or other unusable forms every second of operation (remember, power is energy per time!) the output power *must* be less than the input power—the efficiency of the motor is less than 100%. The power available from the spinning shaft of the motor (Point B) is then used to operate some device: the hoist or vacuum cleaner or DVD drive (Point E). In other words, the output power of the motor is the input power to the device it drives. This device will have its own efficiency, thus wasting some of the power supplied to it by the motor (Point D).

NOTE

The overall efficiency of two linked devices is the product of the two efficiencies.

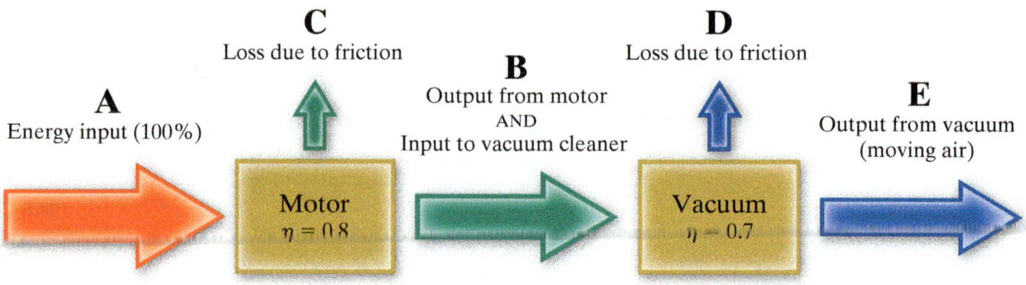

Assume that a motor with an efficiency of 80% is used to power a vacuum cleaner that has an efficiency of 70%. If the input power to the motor (Point A) is one-half horsepower, what is the output power of the vacuum (Point E)?

The output from the motor (Point B) is $P_B = \eta P_A = 0.8\ (0.5\ \text{horsepower}) = 0.4\ \text{horsepower}$*.*

This is the input to the vacuum, so the output from the vacuum (Point E) is

$$P_E = \eta P_B = 0.7\,(0.4\ \text{hp}) = 0.28\ \text{hp}$$

What is the overall efficiency of this machine? The input power (Point A) is 0.5 horsepower, and the output power (Point E) is 0.28 horsepower; thus, the efficiency of this linked system is:

$$\eta = P_A/P_E = 0.28\ \text{hp}/0.5\ \text{hp} = 0.56\text{, or an efficiency of 56\%.}$$

6.4 ELECTRICAL CONCEPTS

The basic concepts of electricity and electrical devices are perhaps less familiar to most students than are many of the other physical phenomena covered previously in this text. This is due partly to lack of practical experience, and partly to the fact that in general these phenomena are themselves invisible, only their effects being perceptible by people. These effects range from receiving an electric shock to the almost magical performance of touch-screen devices. Table 6-2 summarizes the concepts discussed in this section.

WARNING:

Electrical parameters can get confusing at times since V is used as both a variable name for voltage and for the unit volt, and C is used as a variable name for capacitance as well as the unit coulomb.

Table 6-2 Summary of electrical properties

Property	Symbol	Related Equations	Typical Units	Equivalent Units
Charge	Q	Coulomb's Law: $\lvert F\rvert = k_e\dfrac{\lvert Q_1 Q_2\rvert}{r^2}$	coulomb [C]	C = A s
Current	I	$Q = It$	ampere [A]	A = C/s
Voltage	V		volt [V]	V = J/C
Resistance	R	Ohm's Law: $V = IR$	ohm [Ω]	Ω = V/A
Conductance	G	$G = 1/R$	siemens [S]	S = A/V
Electric power	P	$P = VI = I^2 R = V^2/R$	watt [W]	W = V A
Capacitance	C	$Q = CV \quad E_C = \frac{1}{2}CV^2$	farad [F]	F = C/V
Inductance	L	$E_L = \frac{1}{2}LI^2$	henry [H]	H = V s/A

6.4.1 ELECTRIC CHARGE

electric charge

Electrons and protons, as well as some subatomic particles, have a property known as **electric charge**. On a small scale, charge can be measured in terms of the elementary charge (e). The magnitude of the elementary charge on either an electron or a proton is 1, and by convention, the charge of a proton is called positive ($e = +1$) and that of an electron negative ($e = -1$).

A force acts on a charged particle when in the vicinity of another charged particle. If the charges are alike, both positive or both negative, the force is a repulsive force, and the charges tend to accelerate away from each other. If the charges are unlike, one positive and one negative, the force is attractive with the particles tending to accelerate toward each other.

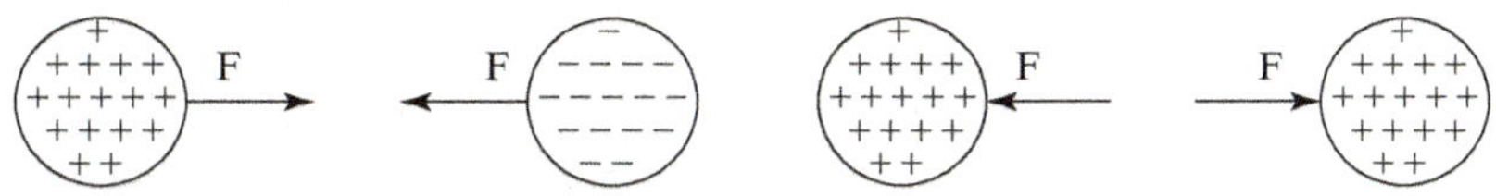

Figure 6-2 Like charges repel; opposite charges attract.

charge
coulomb

The value of the elementary charge (e) is inconveniently small, so **charge** (Q) is generally quantified using the derived unit **coulomb** [C]. A charge of one coulomb represents the total charge on approximately 6.24×10^{18} protons. Another way to say this: the elementary charge of a single electron is $1/6.24 \times 10^{18} = 1.6 \times 10^{-19}$ C.

Coulomb's Law

The actual force exerted on a charged object varies with both the amount of charge on each object (Q_1 and Q_2) and the distance (r) between the charges. This relationship is defined by **Coulomb's Law**, named after the French physicist Charles-Augustin de Coulomb (1736 – 1806) who first described and quantified the attractive and repulsive electrostatic force.

$$|F| = k_e \frac{|Q_1 Q_2|}{r^2}$$

In this equation, k_e is Coulomb's constant, and is approximately equal to 9×10^9 N m^2/C^2.

6.4.2 ELECTRIC CURRENT

Electric current is superficially analogous to a current of water or other fluid. Just as a current of water is a movement of water molecules in a pipe or channel, electric current is a movement of electric charge in a wire or other solid material.

electric current
amperes

Electric current (I) is measured in **amperes** [A], one of the base units in the metric system, and is named for Andre-Marie Ampere, (1775 – 1836), the French physicist who is credited with discovering electromagnetism.

The derived unit coulomb is defined in terms of the ampere as one ampere second: 1 C = 1 A s. This may be easier to understand on an intuitive level by rephrasing this as one ampere equals one coulomb per second: 1 A = 1 C/s. In other words, a current of one ampere represents a movement of one coulomb of charge past any given point in the wire every second.

To put the magnitude of the ampere in context: for those who have received an electric shock by sticking your finger in a light socket, for instance, you realize the sensation is rather unpleasant. In general, in countries that use 120 volts in domestic appliances, such a shock is typically about 5 milliamperes [mA] or one two-hundredth of an ampere. This level of current is small enough that although unpleasant, your muscles will still respond to the commands from your brain, and you can release the live wire or pull your hand away. However, in circuits powering large appliances such as stoves or air conditioners that use 240 volts, the current from a typical shock is roughly twice the above value, or 10 mA. This is very close to the current level that will overload your nervous system so your muscles will no longer obey and you become unable to let go. This is FAR more dangerous.

When denoting a current on a circuit diagram, an arrow is used to indicate the assumed direction. If the current actually flows the other way, the numeric value will be negative.

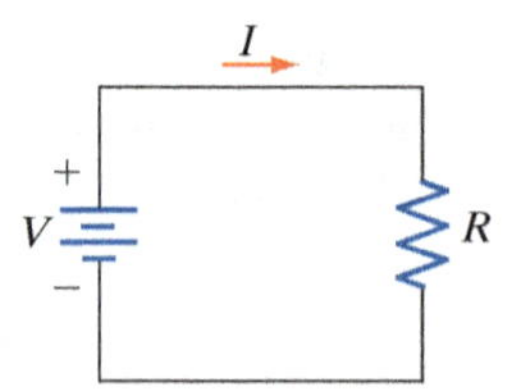

Figure 6-3 A battery and resistor.

A couple of centuries ago when people were just beginning to seriously experiment with electricity, they hypothesized that there was a flow of some substance from one terminal of their primitive devices to the other. Perhaps needless to say, they knew nothing about electrons, since the structure of the atom was completely unknown at that time. They had a 50-50 chance of correctly guessing the direction this "substance" flowed, but luck was against them and they got it wrong. For many decades, scientists assumed that this substance, called charge, flowed from the terminal they called positive (an excess of charge) to the other terminal they called negative (a deficiency of charge). Eventually, the structure of the atom was deciphered and scientists realized that for many years they had been working with the opposite assumption to the correct one, since in most situations it is electrons flowing from the negative terminal to the positive terminal. As a consequence, even to this day, most engineers solve problems using "conventional current," that assumes charge flows from positive to negative. If you want the actual direction of the flow of electrons, you just multiply the current by minus one.

6.4.3 VOLTAGE

To really understand voltage requires knowledge of the concept of the electric field. This is unfortunately a bit too complicated for the limited time and space we have here, so we will merely attempt to help you develop a feel for how voltage affects other electrical parameters. You will study electric fields in some depth in physics and may learn even more in other courses, particularly if you choose to study electrical or computer engineering in college.

A somewhat inaccurate explanation of voltage, although one that can be useful in understanding it, is that voltage is what pushes the charges around to create current. In a sense, it quantifies the amount of force that can be exerted on an electric charge by other accumulated charge.

voltage

Voltage (V) is quantified using units of volts [V], and is a measure of how much work is required to move an electric charge in the vicinity of other electric charges. The unit of volt is named for Italian physicist Alessandro Volta, (1745 – 1827), who possibly invented the first chemical battery, called a voltaic pile.

One volt is defined as one joule per coulomb. In other words, if one joule of energy is required to move one coulomb of charge from one place to another, the voltage between those two points is one volt.

Since work is required to move an electric charge in the vicinity of other charges, we might recall the definition of work in another context. Work equals force times the distance through which that force moves an object: $W = F\,d$. Similarly, work equals charge times the difference in voltage through which that charge moves: $W = Q\,V$.

One specific case of work is potential energy. If a force is used to raise a mass above the surface of the planet, the work done is stored as energy in the mass of the object being raised. When the object is dropped, the energy will convert form from potential energy to kinetic energy. Similarly, if a bunch of charge is moved closer to an accumulation of like charges, the work done to move the packet of charge is stored as energy in the packet of charge. If a path is provided for the packet of charge to move, the stored energy will be converted into another form, often heat, but it might also include light, sound, chemical energy, etc.

Similar to the need to know whether an object is being lifted (storing potential energy) or is falling (converting the stored energy into kinetic energy), we need to know whether charges are moving toward like charges or away from them. Just like we use an arrow to denote the assumed direction of a current, we need some sort of notation to indicate the assumed polarity of a voltage – which end is assumed to be more positive than the other. This allows us to keep track of whether energy is being stored or released. We do this by placing a plus sign on one side of the device through which the current is flowing and a negative sign on the other end.

6.4.4 ELECTRICAL RESISTANCE

Resistance is a measure of how difficult it is to move charges through a material. In some substances, such as many metals, electrons can move quite easily. In other materials such as glass or air, considerable force, thus considerable voltage, is required to make electrons move therein.

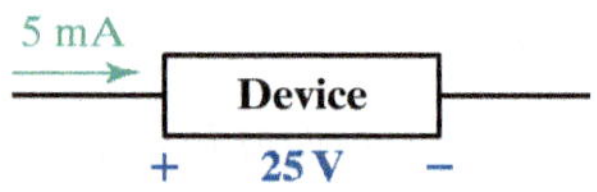

Figure 6-4 A device with voltage and current.

resistance

Resistance (R) is quantified using units of ohms [Ω], where one ohm is defined as one volt per ampere. For example, if a 1 volt battery were connected to a device having a resistance of one ohm, one ampere of current would flow through it, assuming the chemical reaction could replenish the charge rapidly enough to maintain such a current. The ohm is named for Georg Simon Ohm, the German physicist who first described the relationship linking voltage, current, and resistance.

Resistance relates the voltage across a device to the current through the device. **Take particular note of the choice of prepositions – across and through.** Understanding this choice will help you understand voltage and current.

Electric current is the movement of charge, typically electrons moving **THROUGH** a substance. Voltage is to some extent a measure of the force being exerted on the moving charges by forces at either end of the device. *This is where it can be a little confusing, particularly without using electric fields in the discussion.* However, imagine that on each side of a device is an accumulation of charge, each exerting a force on the electrons inside the device. Each of those forces might be a "push" or a "pull," and the total force on the electrons in the device is the difference in these forces. The difference in the forces from one side of the device to the other is referred to as the voltage **ACROSS** the device.

We can talk about a voltage at a point, such as, "the voltage at point A is 15 volts," but such statements are really based on some reference point, often the actual, physical ground itself, so the statement is really equivalent to something like "the voltage across (or between) point A and ground is 15 volts."

As an analogy, if you are trying to push a sofa across the room, but someone else is trying to push the sofa in the opposite direction with the same force, the net force is zero and the sofa does not move. If one person pulls and the other pushes, however, the sofa will move quicker than with either person alone.

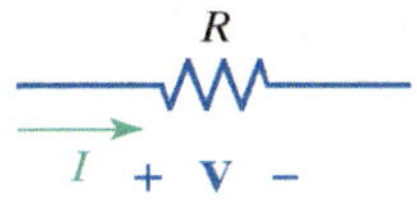

Figure 6-5 Resistor.

Ohm's Law

Resistance is related to current and voltage by **Ohm's Law:** $\boldsymbol{V = IR}$. Note the following implications of Ohm's Law:

- To maintain a specific current through a resistance requires a voltage proportional to the resistance. A larger resistance makes it harder to "push" the electrons through the device, thus a larger voltage is required.
- Similarly, current is inversely proportional to resistance. For a given voltage, if the resistance increases, the voltage cannot "push" as many electrons through the device per second, so the current must decrease.

In some contexts, it is computationally simpler to use conductance instead of resistance. Conductance (G) is measured in siemens [S] and is simply the reciprocal of resistance: $G = 1/R$. An older unit for conductance that you might find, particularly in older references, is the mho (ohm spelled backwards) and is represented by an upside-down omega [℧].

EXAMPLE 6-6

The voltage across a resistor is 15 volts, and the current through it is 6 milliamps [mA]. What is the value of the resistance?

$$V = IR \text{ so } R = \frac{V}{I} = \frac{15\,\text{V}}{6\,\text{mA}}\left|\frac{1000\,\text{mA}}{\text{A}}\right. = 2500\,\frac{\text{V}}{\text{A}}\left|\frac{\Omega\text{A}}{\text{V}}\right. = 2500\,\Omega = 2.5\,\text{k}\Omega$$

6.4.5 ELECTRIC POWER

Conceptually, electric power is perhaps easiest to understand by examining the formula for gravitational potential energy. A mass has its potential energy increased by expending energy to lift it higher above the surface of the planet, since the mass of the object and the mass of the planet are mutually attracting each other.

Similarly, forcing electrons closer to other electrons stores potential energy since they are mutually trying to repel each other. Recall our discussion of batteries. For each electron that is transferred to the negative terminal of the battery by the chemical reaction, a little bit of energy is "stored" in the battery. This is effectively electrical potential energy. The more electrons per second that are jammed together, the more energy per second is stored. Current is measured in charge (electrons) per second, power is proportional to current: $P \propto I$ or $P = XI$, where X is the proportionality constant.

A volt times an ampere is a watt.

$$V = J/C$$
$$A = C/s$$
$$VA = J/s = W$$

Now think back to voltage. Voltage is a measure of how much energy is used to move a given amount of charge: one volt is one joule per coulomb. Therefore, voltage is the proportionality constant and $P = VI$.

EXAMPLE 6-7

A semiconductor diode has 500 millivolts [mV] across it and 700 microamps [μA] of current through it. How much power is the diode absorbing?

$$P = VI = \left(\frac{500\text{ mV}}{}\middle|\frac{1\text{V}}{1000\text{ mV}}\right)\left(\frac{700\ \mu\text{A}}{}\middle|\frac{1\text{A}}{1\times10^{6}\ \mu\text{A}}\right)$$

$$= 3.5 \times 10^{-4}\text{ W} = 350\ \mu\text{W}$$

For resistors, the electrical power absorbed is usually converted to heat, and we can use Ohm's Law to replace either the voltage or the current in this power relationship:

$$P = VI = (IR)\,I = I^2R \quad \text{or} \quad P = VI = V(V/R) = V^2/R$$

Note that these two relationships expressed in terms of resistance are only valid for resistors, not for other electrical components. However, it gives us a means to quickly calculate the power absorbed by a resistor when we know only the voltage or current, but not both.

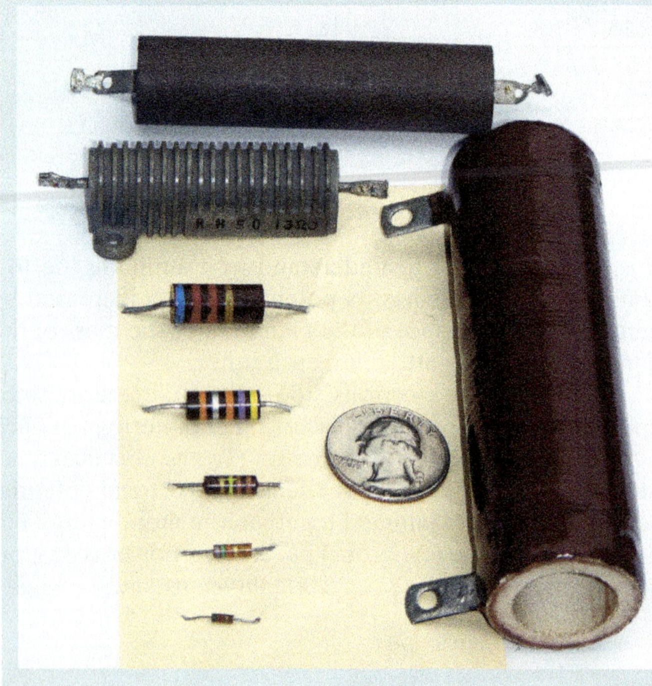

When specifying resistors, both the resistance and the wattage must be given. If you connect a resistor rated at 1 watt in a circuit in which it will have to dissipate 100 watts, it will literally burst into flames, or at least perform an imitation of popcorn by exploding!

From bottom to top, the power rating of these resistors is ⅛ W, ¼ W, ½ W, 1 W, 2 W, 15 W, 25 W, and on the right, 50 W.

Figure 6-6

6.4.6 CAPACITANCE

Another simple electrical device is the capacitor. The capacitor is formed by arranging two conducting, low resistance plates very close together, but separated by an insulator with extremely high resistance. Each plate has a wire connected to it. If a current is run into one of the plates of the capacitor, the charges accumulate on that plate since they cannot cross the insulating barrier to the other plate.

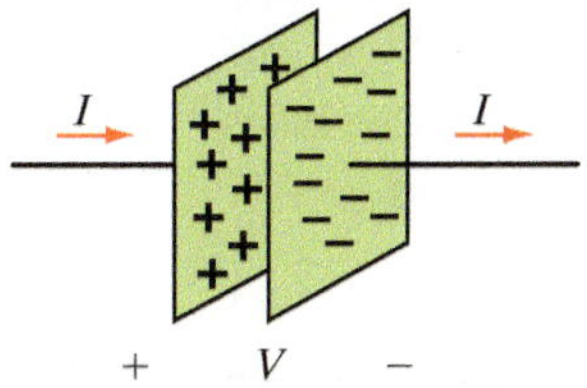

Figure 6-7 Capacitance.

When a current enters one plate of the capacitor, the electrons making up the current begin to accumulate on that plate. Since the two plates are separated by an insulator, the electrons cannot cross over to the other plate. While accumulating negative charge on one plate, the negative electrons in the other plate are repelled, leaving behind an overall positive charge. This superficially gives the appearance that the current is going through the capacitor. However, the electrons entering one plate and leaving the other are different electrons, and a charge separation is accumulated on the plates of the capacitor.

The voltage across the capacitor depends not only on the total charge stored, but also on the physical construction of the device, particularly the surface area of the plates. The charge (Q) stored in a capacitor is proportional to the voltage (V) across it: $Q = CV$, where C is the proportionality constant. Note that C must have units of coulombs per volt, and is called capacitance.

capacitance

farads

Capacitance is measured in units of **farads** [F], where one farad equals one coulomb per volt, or $1\text{ F} = 1\text{ C/V}$. In other words, if a capacitor was storing a charge of one coulomb, and the resulting voltage across its plates was one volt, the capacitance of the device would be one farad.

Since capacitors can contain a separation of charge, they can store energy. The energy stored in a capacitor can be calculated by $E_C = \frac{1}{2}CV^2$. Note that the energy is proportional to the square of the voltage.

NOTE

One farad is a LARGE amount of capacitance. Most capacitors are much smaller than this, and are usually measured in microfarads [μF], nanofarads [nF], or even picofarads [pF].

EXAMPLE 6-8

A 0.01-microfarad [μF] capacitor is initially completely discharged (no stored charge, thus an initial voltage across it of zero). If a constant current of 4 milliamps [mA] charges the capacitor, how long will be required to change the voltage across it to 5 volts [V]?

Since $Q = CV$, then $C = Q/V$.

The total charge delivered by a constant current is $Q = It$. Inserting this into $C = Q/V$ gives $C = It/V$.

Solving for t gives $t = CV/I$.

Therefore,

$$t = \left(\frac{0.01\ \mu\text{F}}{}\left|\frac{1\text{F}}{1 \times 10^{6}\ \mu\text{F}}\right.\right)\frac{5\ \text{V}}{}\left(\frac{}{4\ \text{mA}}\left|\frac{1000\ \text{mA}}{1\text{A}}\right.\right) = 1.25 \times 10^{-5}\,\frac{\text{FV}}{\text{A}}\left|\frac{\text{C}}{\text{FV}}\right|\frac{\text{As}}{\text{C}} = 12.5\ \mu\text{s}$$

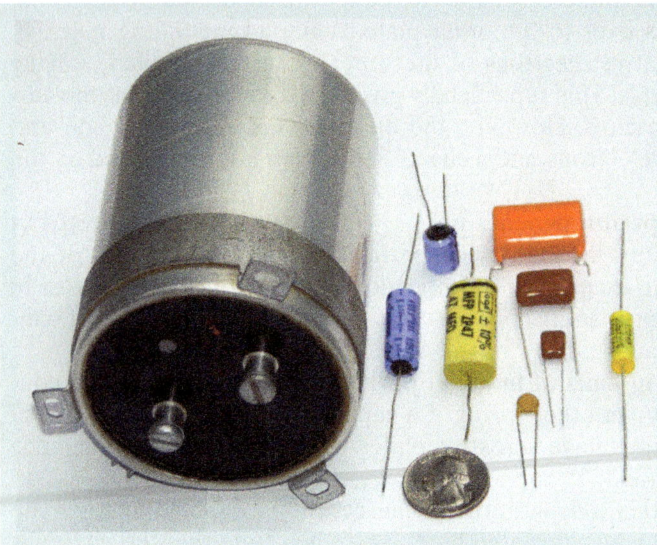

When specifying capacitors, both the capacitance and the maximum voltage must be given. If you connect a capacitor rated at 25 V in a circuit in which it will have 200 V across it, the insulator between the plates will probably fail and the capacitor will be destroyed; it might even explode! The voltage ratings of the capacitors shown range from 25 volts to 500 volts. The big 10,000 μF one on the left is rated at 50 V.

Figure 6-8

6.4.7 INDUCTANCE

If a current is moving through a wire, a magnetic field is generated surrounding that wire. If a wire is placed in a CHANGING magnetic field, a current is induced in the wire. This is the physical basis for the final electrical device we will introduce, the inductor.

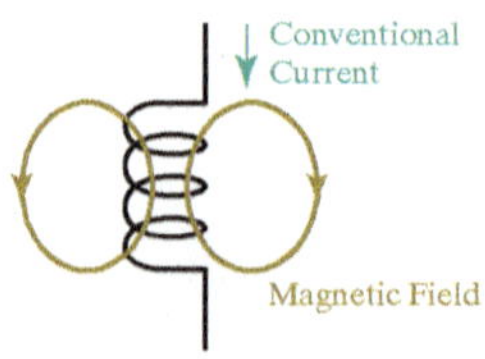

Figure 6-9 Inductance.

inductor

In its simplest form, an **inductor** is just a coil of wire. If a current flows through the coil, the magnetic field generated in each loop adds to the magnetic field generated in every other loop, generating a stronger magnetic field. This is the basis of the electromagnet. Industrial electromagnets can concentrate the magnetic field to a large enough value to pick up huge objects, like cars and trucks. However, in electrical engineering, inductors are more often used for their ability to store energy in the form of a magnetic field.

changing

Note that once a magnetic field has been created in an inductor, if the source of the current that created the magnetic field is removed, the magnetic field will begin to collapse. However, a wire in a **changing** magnetic field such as a collapsing will have a current induced in it, thus allowing the stored energy to be transferred to another place and form.

inductance
henrys

Inductance (L) is measured in units of **henrys** [H], although the millihenry [mH] or microhenry [μH] is often more convenient. The voltage across an inductor is equal to the inductance of the device times the rate of change of current through the inductor; $V = L\frac{dI}{dt}$ for those of you who have begun your study of differential calculus by now. If the current in an inductor changes by one ampere per second, and a voltage is developed across the inductor of one volt, then the inductance is one henry.

Dimensionally, the henry is one volt second per ampere [V s/A]. This can be shown to be dimensionally equal to resistance times time [Ω s] or energy per current squared [J/A^2]. The total energy stored in an inductor can be calculated by $E_L = \frac{1}{2}LI^2$. Note that the energy is proportional to the square of the current. You might note the similarity of this to both the formulae for energy stored in a capacitor $E_C = \frac{1}{2}CV^2$ and kinetic energy $E = \frac{1}{2}mv^2$.

When specifying inductors, both the inductance and the maximum current must be given. If you connect an inductor rated at 5 mA in a circuit which will have 2 amperes through it, the wire will probably melt – a BAD idea!

The inductor shown on the left is wound with many thousands of turns of extremely small wire and its current capacity is rated in microamperes. The one on the right has only ten turns of rather thick wire and can handle about an ampere.

Figure 6-10

Table 6-3 Schematic Symbols

Device	Schematic Symbol	Notes
Battery (Single Cell)	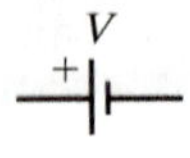	The battery terminal with the longer line next to it is always the positive.
Battery (Multi-cell)		
Resistor		
Capacitor	 If polarized type	For polarized capacitors, the positive terminal is the one with the straight line, not the curved line.
Inductor	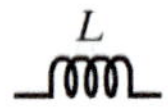	The inductor symbol can also be drawn with curved bumps rather than loops.

6.4.8 RESISTORS IN SERIES

When two resistors are connected in *series*, the *current through both of the resistors is the same*, even though the value of each resistor may be different. The derivation below shows the effective resistance of two resistors connected in series. Note that the voltage is applied to the entire system.

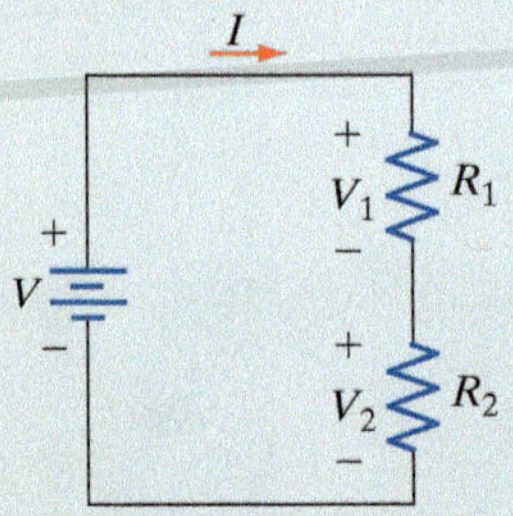

Figure 6-11
Resistors in series have the same current through both of the resistors.

Writing Ohm's Law for two resistors each with the same current:

$$V_1 = IR_1 \tag{1}$$

$$V_2 = IR_2 \tag{2}$$

Solving for V_1 in terms of V_2 since the current is the same:

$$V_1 = R_1 \frac{V_2}{R_2} = V_2 \frac{R_1}{R_2} \tag{3}$$

Writing Ohm's Law as applied to the overall system:

$$V = IR_{\text{eff}} \tag{4}$$

The total voltage applied to the configuration (V) is the sum of the voltage applied to each resistor:

$$V = V_1 + V_2 \tag{5}$$

Eliminating voltage (V) from Equation (5) with Equation (4):

$$IR_{\text{eff}} = V_1 + V_2 \tag{6}$$

Eliminating current (I) from Equation (6) with Equation (2):

$$R_{\text{eff}} \frac{V_2}{R_2} = V_1 + V_2 \tag{7}$$

Substitution for V_1 with Equation (3):

$$R_{\text{eff}} \frac{V_2}{R_2} = V_2 \frac{R_1}{R_2} + V_2 \tag{8}$$

Dividing Equation (8) by V_2:

$$R_{\text{eff}} \frac{1}{R_2} = \frac{R_1}{R_2} + 1 \tag{9}$$

Multiplying Equation (9) by R_2:

$$\boldsymbol{R_{eff} = R_1 + R_2} \tag{10}$$

6.4.9 RESISTORS IN PARALLEL

When two resistors are connected in *parallel*, the *voltage across both of the resistors is the same*. The current through each resistor may be different. The voltage is applied to the entire system.

Writing Ohm's Law for two resistors each with the same voltage:

$$V = I_1R_1 \tag{11}$$

$$V = I_2R_2 \tag{12}$$

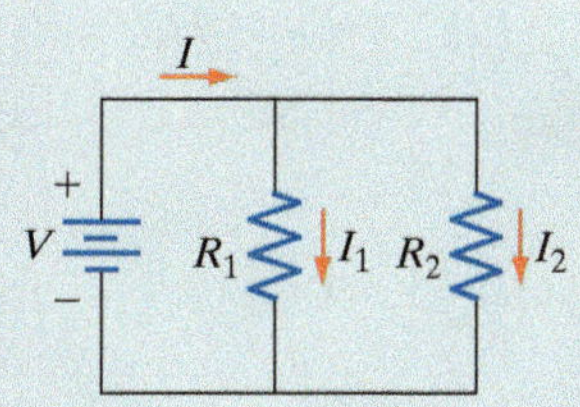

Figure 6-12

Resistors in parallel have the same voltage across both of the resistors.
A system of two resistors in parallel will always have less resistance than either resistor individually.

Solving for I_1 in terms of I_2 since the current is the same:

$$I_1 = \frac{R_2}{R_1} I_2 \tag{13}$$

Writing Ohm's Law as applied to the overall system:

$$V = IR_{\text{eff}} \tag{14}$$

The total current (I) is the sum of the current flowing through each resistor:

$$I = I_1 + I_2 \tag{15}$$

Eliminating voltage (V) from Equation (14) using Equation (12):

$$I_2 R_2 = IR_{\text{eff}} \tag{16}$$

Eliminating current (I) using Equation (15):

$$I_2 R_2 = (I_1 + I_2) R_{\text{eff}} \tag{17}$$

Substituting for I_1 using Equation (13):

$$I_2 R_2 = \left(\frac{R_2}{R_1} I_2 + I_2\right) R_{\text{eff}} \tag{18}$$

Dividing Equation (18) by I_2:

$$R_2 = R_{\text{eff}} \left(\frac{R_2}{R_1} + 1\right) \tag{19}$$

Dividing Equation (19) by R_2 gives:

$$1 = R_{\text{eff}} \left(\frac{1}{R_1} + \frac{1}{R_2}\right) \tag{20}$$

Thus,

$$\boldsymbol{R_{\text{eff}} = \frac{1}{\frac{1}{R_1} + \frac{1}{R_2}} = \left(\frac{1}{R_1} + \frac{1}{R_2}\right)^{-1}} \tag{21}$$

These equations for two resistors connected in parallel and series generalize to any number of resistors. For N resistors in parallel, the effective resistance is

$$R_{\text{eff}} = \left(\frac{1}{R_1} + \frac{1}{R_2} + \cdots + \frac{1}{R_{N-1}} + \frac{1}{R_N}\right)^{-1} \tag{22}$$

This form, along with spring equation (w) shown earlier, is sometimes referred to as the "reciprocal of the sum of the reciprocals."

For N resistors in series, the effective resistance is

$$R_{\text{eff}} = R_1 + R_2 + \cdots + R_{N-1} + R_N \tag{23}$$

WARNING!

Some of you may have seen a "simpler" form of the equation for two springs in series or two resistors in parallel.

For two springs in series: $k_{eff} = \dfrac{k_1 k_2}{k_1 + k_2}$

For two resistors in parallel: $R_{eff} = \dfrac{R_1 R_2}{R_1 + R_2}$

These forms are sometimes referred to as "the product over the sum."

THESE FORMS *DO NOT GENERALIZE* TO MORE THAN TWO ELEMENTS.

If you have three or more elements, then you must use the "reciprocal of the sum of the reciprocals" form given earlier.

EXAMPLE 6-9

Find the current (I) in the circuit shown. Resistor 1 (with resistance R_1) and Resistor 2 (with resistance R_2) are connected in series, and the combination is then connected to Resistor 3 (with resistance R_3) in parallel. Use the following values:

$V = 12$ V
$R_1 = 7.5$ kΩ
$R_2 = 2.5$ kΩ
$R_3 = 40$ kΩ

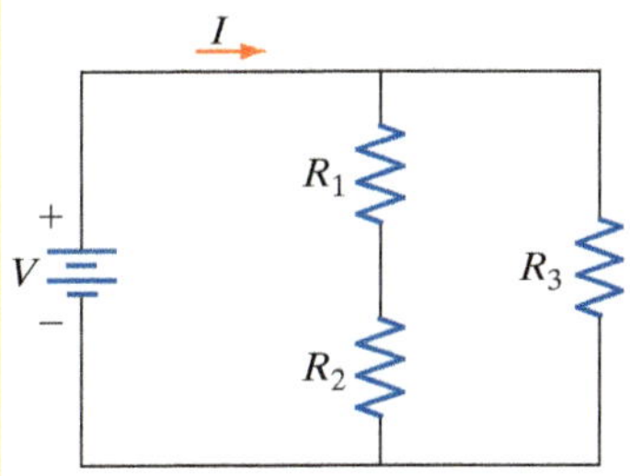

First, we recognize that R_1 and R_2 are in series, so we reduce R_1 and R_2 to a single effective resistor by using Equation (10).

$$R_{eff12} = R_1 + R_2 = 7.5\text{ k}\Omega + 2.5\text{ k}\Omega = 10\text{ k}\Omega$$

Next, we can redraw the circuit so R_{eff12} and R_3 are in parallel. We solve for the effective resistance by using Equation (21).

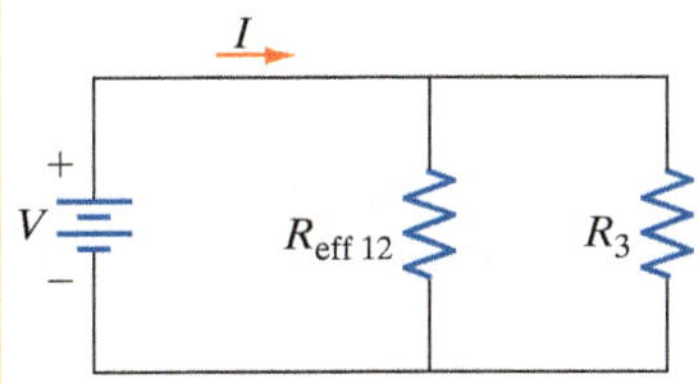

$$R_{eff} = \left(\frac{1}{R_{eff12}} + \frac{1}{R_3}\right)^{-1} = \left(\frac{1}{10\text{ k}\Omega} + \frac{1}{40\text{ k}\Omega}\right)^{-1} = 8\text{ k}\Omega$$

We can now solve the problem with Ohm's Law:

$$V = IR_{eff}$$

$$I = \frac{V}{R_{eff}} = \frac{12\text{ V}}{8\text{ k}\Omega} = 0.0015\text{ A} = 1.5\text{ mA}$$

6.4.10 WHEN ARE COMPONENTS CONNECTED IN SERIES, PARALLEL, OR NEITHER?

Note that in each diagram, the lines with one end loose indicate where the circuit or spring configuration is connected to other things.

Series

When one end (but not both) of each of two components is connected together with NOTHING ELSE CONNECTED AT THAT POINT, they are in series.

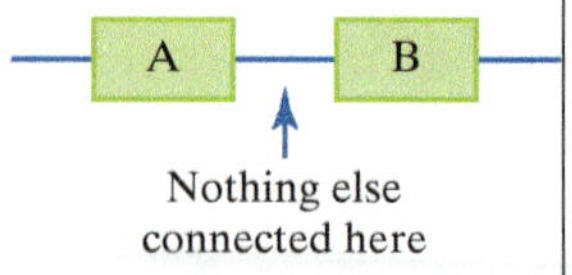

Note that they do not necessarily have to be in a straight line as shown. Electrical components can be physically mounted in any position relative to one another, and as long as a wire connects one end of each together (with nothing else connected there), they would be in series. Two springs can be connected by a string, so that the string makes a right angle direction change over a pulley, and the two springs would be in series.

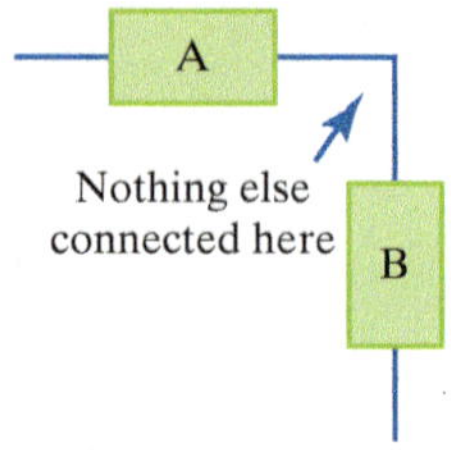

Parallel

When each end of one component is connected to each of the two ends of another component, they are in parallel.

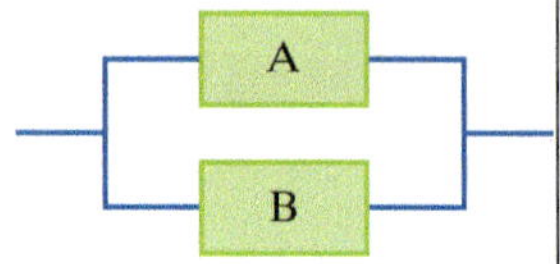

Similar to the series connection, the components do not have to be physically mounted parallel to each other or side by side, as long both ends are connected directly together with no intervening components. This is simple to do with electrical components since wire can be easily connected between any two points. Can you determine a method to physically connect two springs in parallel so that one is vertical and the other horizontal?

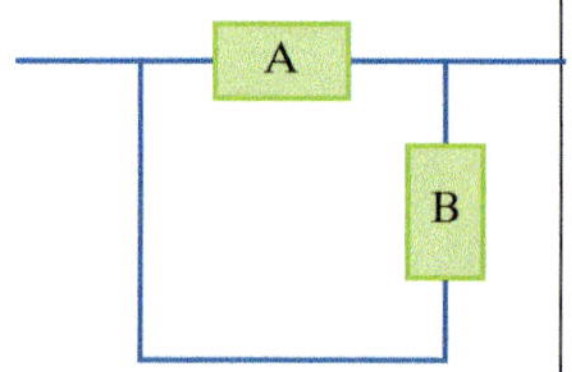

Sample Combinations

In the figure on the right, B and C are in series. A is neither in series nor parallel with B or C since the lines extending to the left and right indicate connection to other stuff. A is, however, in parallel with the series combination of B and C.

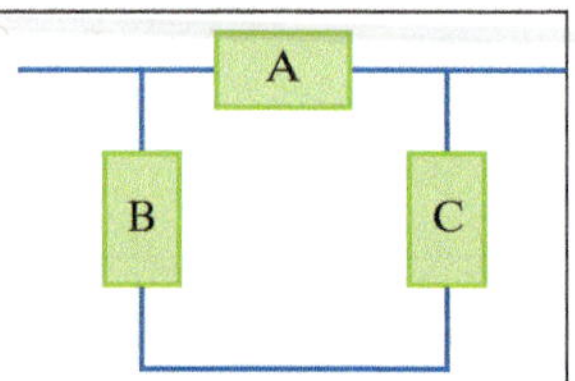

In the figure on the right, no components are in series or parallel with anything since the lines extending to the left and right indicate connection to other stuff. Note the extra line at lower right.

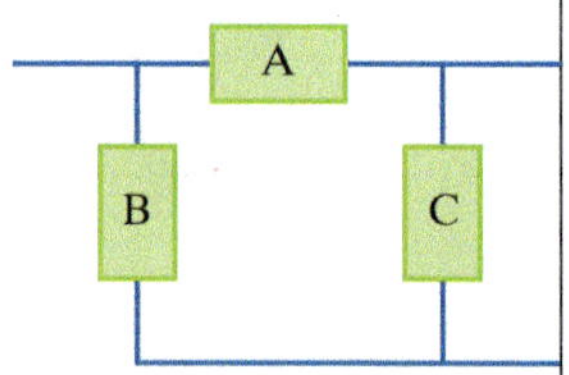

In the figure on the right, A and B are in parallel. C is neither in series nor parallel with A or B. C is, however, in series with the parallel combination of A and B.

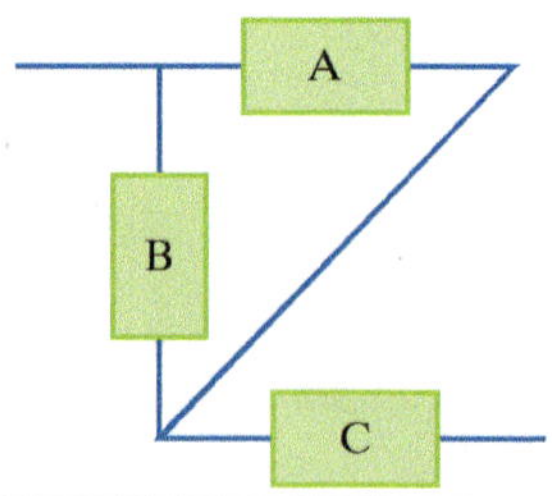

6.5 ENERGY SOURCES

Energy sources may be described as renewable or nonrenewable. Nonrenewable energy sources, such as oil, cannot be replaced. There is a fixed supply that will eventually be used up.

Most of the energy we use today comes from nonrenewable, or limited, energy sources. It takes millions of years for coal, oil, and natural gas to be created. As we use them up, more will not become available. At our present rate of use, it is estimated that we have only a hundred-year supply of oil and natural gas remaining.

Figure 6-13 Materials that accumulate on forest floors may eventually form coal or gas.

6.5.1 FOSSIL FUELS

fossil fuel

Oil, natural gas, and coal are all called fossil fuels. A **fossil fuel** is an energy-rich substance formed from the remains of plants and animals that lived millions of years ago. After the plants and animals died, they did not rot away completely because they were covered by fallen trees, leaves, and mud. Additional layers of soil material, along with movements of Earth's surface, further buried the plant and animal remains. The great pressure over millions of years plus heat changed them into oil, gas, or coal.

Fossil fuels are made primarily of hydrogen and carbon atoms joined together as molecules in a high-energy, chemical bond. During combustion, when fossil fuels are burned, the molecules break down into simpler molecules. As they break down, they release much of the energy contained in the high-energy bonds. This is the energy that we get from fossil fuels. When carbon burns, it produces carbon dioxide. When hydrogen burns, it produces water (hydrogen oxide).

The more energy that is released, the better the fuel. The bonds in gasoline, which is made from oil, release a great deal of energy when the gasoline burns. That is why gasoline is used so widely as a fuel.

Coal

Coal was one of the first fossil fuels to be used widely during the Industrial Revolution. Coal can be mined from the surface of the land by strip mining, or from deep below Earth's surface. In strip mining, very large but shallow holes are dug into the ground and the coal is removed. The hole must be filled in and the land restored by planting trees and grass when mining is finished. To reach deep coal, miners have to tunnel into the ground, dig the coal out, and transport it back to the surface. Coal mining can be dangerous, but modern equipment and safety standards have helped make it safer.

Figure 6-14 Coal can be mined from Earth's surface or from deep underground.

Coal is the most abundant nonrenewable energy resource, but there are drawbacks to its use. Coal must be moved in large quantities by train, barge, or truck to its destination. Often there are traces of sulfur in the coal, which creates sulfur dioxide when burned. When this pollutant combines with water in the air, it forms sulfuric acid. This acid can form acid rain, which can cause serious damage to forests and aquatic life. Engineers and ecologists are working to solve these problems.

Oil

Often called "black gold," oil is the energy resource on which we depend most. Oil is used much more than coal because it is more easily taken from the ground, stored, and transported. It also has much more energy per pound than coal and can be refined into many other useful fuels. Some of these fuels are home heating oil, gasoline, and jet fuel.

Oil is in such demand that oil wells are drilled virtually anywhere oil is discovered. Wells are dug even in areas with harsh climates or hazardous conditions, such as the northern coast of Alaska and the deep ocean floor.

Natural Gas

Natural gas is used for home heating and cooking as well as in industry. Natural gas burns cleaner than other fossil fuels. It has a greater amount of hydrogen than carbon, so the products of combustion contain more harmless water vapor (burned hydrogen) than for other fossil fuels. Natural gas is transported by pipelines from its source in the ground to the places where it is used.

Natural gas is used as fuel in several industries. In the glass industry, huge gas-fired furnaces melt glass in large quantities. Gas is also used to make fertilizers, which has helped increase food production tremendously.

Figure 6-15 Oil is in such demand that it is drilled in all types of locations.

6.5.2 NUCLEAR FISSION

Uranium is another important nonrenewable energy resource. Unlike oil, gas, and coal, uranium is not a fossil fuel. It is not burned to obtain energy. Uranium is a nuclear, or atomic, fuel.

nuclear fission

Matter can be changed into energy by **nuclear fission,** the splitting of an atom's nucleus (Figure 6-16). Recall that an atom's nucleus is made up of protons and neutrons. In nuclear fission, an unstable atom, such as uranium-235, is hit with an external neutron. The atom splits and forms two smaller nuclei and extra neutrons, which give off a great deal of energy in the form of heat and light. The new neutrons hit more atoms around them. They make these atoms split, giving off more energy and more neutrons, creating a chain reaction.

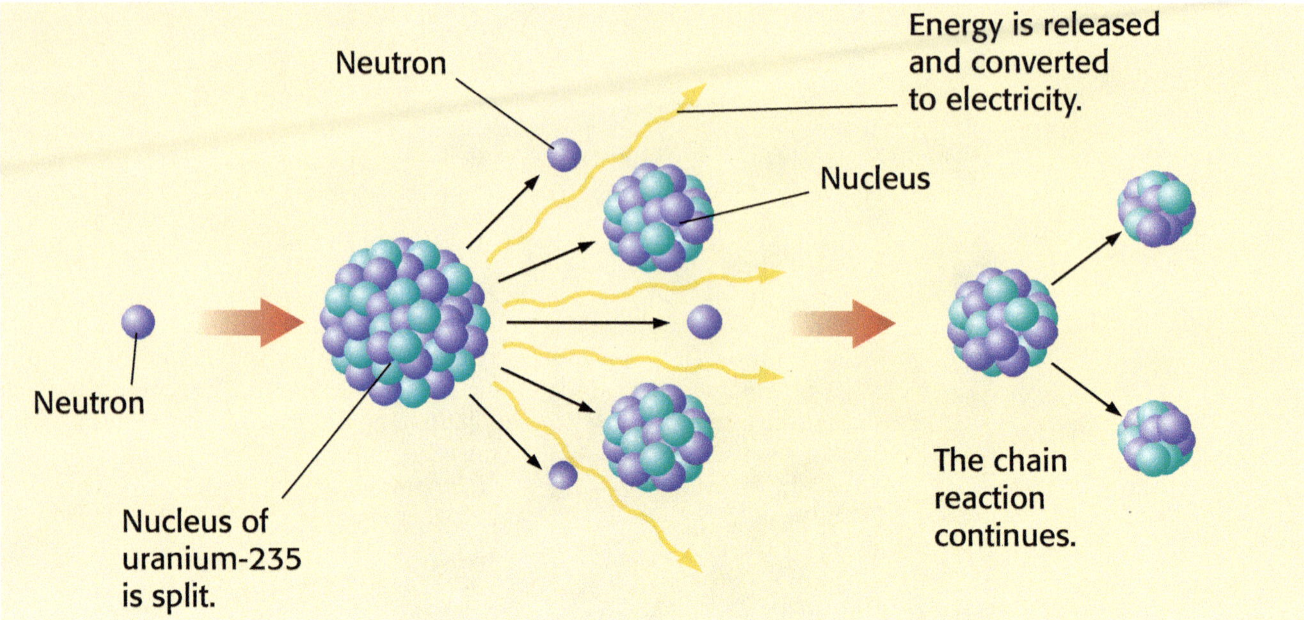

Figure 6-16 Nuclear energy is released when a neutron hits a large atom, such as uranium.

Very large amounts of energy are produced during nuclear fission. In a nuclear power plant, the amount of energy released is controlled by controlling the number of neutrons. Without this type of control, an explosion could occur. In a nuclear power plant, the nuclear energy is converted into electricity.

radioactive waste

A small amount of atomic fuel can produce a huge amount of energy. However, there are risks with nuclear power that are not present with fossil fuels. A nuclear power plant produces excess unusable material called **radioactive waste,** which contains unstable atoms that give off radiation. Radiation can cause burns and illness. At very high levels, it can even kill people. It also remains radioactive for thousands of years. Safety precautions must be taken to make sure that people are not exposed to too much radiation.

6.5.3 RENEWABLE ENERGY RESOURCES

Renewable energy resources are those that are so abundant or that replenish themselves so quickly that they will never run out. For example, solar energy is essentially unlimited because, although the sun will burn out in a few billion years, we could never deplete the amount of energy released by the sun in our lifetime.

It is a good idea to use renewable energy sources such as the sun. However, it can be difficult to find ways to harness these resources. Renewable energy resources have low energy values, so it takes much more energy to get the energy value needed. Research and development of renewable energy sources are ongoing.

Human and Animal Muscle Power

The first source of energy that people used was muscle power. While using your muscle power allows some work to be done, it is rarely good enough for something demanding a lot of power consistently. Animal power is used more often: Teams of horses can pull wagons, or teams of oxen can pull a farm plow, for example. In some parts of the world, human and animal power are still important sources of energy.

Solar Energy

solar energy

We depend on **solar energy**—the energy from the sun—to support life on Earth through photosynthesis. The sun's energy is also purposefully used in many ways. Solar power plants, for example, use many special mirrors called parabolic reflectors to focus sunlight onto a single spot to generate very high temperatures. The heat causes water to boil and create steam, which can be used to generate electricity.

The sun's heat is also used directly for heating homes and other buildings. For example, some solar collectors are placed on the roof of a house to heat water for showers, washing, and cooking. Some buildings also use fans and pumps to distribute the collected heat throughout the building. Passive solar designs for houses use the sun for partial heating. Windows, walls, and doors are carefully placed so the sun can be used for heating during the winter.

photovoltaic cell

A **photovoltaic cell** is a solar cell that can turn sunlight into electricity (Figure 6-17). Photovoltaic cells can be used to charge batteries that run radio relay stations, radio telephones, and other electrical devices. Photovoltaic cells also are used on most satellites and space vehicles. About 10 percent of the sun's energy is converted into electricity in a photovoltaic cell.

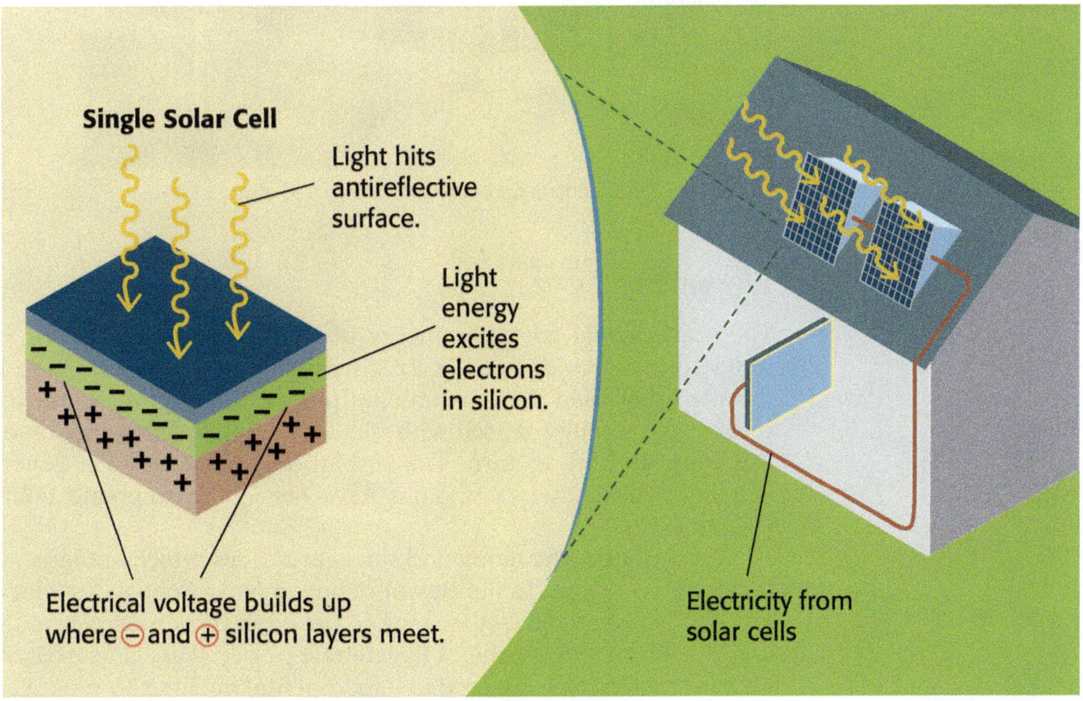

Figure 6-17 A solar cell converts light energy into electrical energy.

Wind Energy

Wind has been used as an energy source for hundreds of years. Machines that harness the wind have been used to pump water, grind grain, and move ships across the water. Today, wind is used to generate electricity (Figure 6-18). In some locations where it is windy, farmers lease their land for wind farms. The farmers earn income and can still grow crops on the land. However, wind-powered generators cannot make electricity when the wind isn't blowing or if it is blowing below a certain speed. The speed of the wind and the diameter of the blades determine the amount of energy produced. Due to this constraint, wind farms are useful only in places where the wind blows briskly most of the time.

Figure 6-18 A wind farm converts wind energy into electrical energy.

Water Energy

The gravitational potential energy of water has been used for centuries. Long ago, water wheels provided power for grinding grain into flour. In the 1800s, a turbine was invented that used the gravitational potential energy of water to produce power. A **turbine** is a circular device with blades. Falling water, wind, or other power sources cause the turbine to turn. The turbine is connected to a generator that produces electricity. In many places, dams have been built to create falling water for power (Figure 6-19).

turbine

Gravity from the moon and sun causes tides, which are large movements of water in oceans and rivers. In the Bay of Fundy in Nova Scotia, the tides can cause the water level to rise nearly 60 feet in some of the bay's tributaries. When the water flows back, it is directed through turbines to generate power. Wave energy, another possible source of water power, is being used in experimental facilities to produce electricity.

Figure 6-19 In a dam, the potential energy of water is converted to electrical energy.

Geothermal Energy

Figure 6-20 The energy of hot steam from underground is called geothermal energy.

geothermal energy

At Earth's center is a hot, molten core. Volcanic activity within the last 3 million years has brought molten rock, called magma, to within 3 miles of Earth's surface. **Geothermal energy** is thermal energy stored below Earth's surface (Figure 6-20). When magma is close to the surface, it can heat underground water to the boiling point, creating steam. It is possible to use this steam to power turbines, which in turn drive electric generators. The challenges to building geothermal plants are the relatively unstable land areas, in geological terms, on which they must be located and water that is very corrosive because of the dissolved minerals and salts. Also, the number of places with magma close to Earth's surface is limited.

Nuclear Fusion

nuclear fusion

Nuclear fusion is another atomic method of changing matter into energy. In **nuclear fusion,** the nuclei of two atoms are forced together to form a new nucleus, and a large amount of energy is released. However, a tremendous amount of energy is needed to force the nuclei together. Under the right conditions, more energy is given off than is used.

A successful fusion power plant has not yet been built, but fusion research continues. It is hoped that someday fusion will prove to be an efficient way to generate electricity. Fusion power plants will use two forms of hydrogen—deuterium and tritium—as fuel. Sources for these fuels are readily available. The waste products of fusion are not radioactive, and the process would not cause pollution. Many people hope that fusion plants will provide most of our electricity in the future.

Biomass

biomass

Biomass is accumulated vegetable and animal wastes. Biomass can serve as a major source of renewable energy. Three basic biomass processes are used to produce energy.

The first process is direct combustion of waste products. Burning wood is a biomass process. A cord of wood (a stack 4 feet by 4 feet by 8 feet) can provide the same amount of heat as about 200 gallons of oil. Wood is renewable, but forests must be managed carefully. Once trees are cut in an area, new trees should be planted to replace those cut down. In parts of the world where there is not much wood, dried animal wastes are gathered and burned to provide heat.

The second way to process biomass is gasification. Methane gas is produced as the biomass rots. It is collected and stored to be used as a fuel. Methane gas can be converted into electricity or used to power a combustion engine.

The third biomass process is fermentation, which uses microorganisms to turn biomass such as corn or other grain into alcohol and carbon dioxide gas. The alcohol can be stored and used as a fuel, or it can be added to other fuels to make them last longer. An example of this is gasohol, which is a mixture of gasoline and alcohol, or ethanol made from corn (Figure 6-21).

Figure 6-21 Ethanol, a type of car fuel, in part uses energy from corn.

PROJECT 6: BASIC ELECTRICITY

Problem Situation:

In order to better understand basic electricity and electrical components, we want to learn how to put some together and to measure electrical quantities like voltage, current, and resistance.

Your Challenge:

multimeter

Using a **multimeter,** measure some basic electrical variables. Once we can measure these, we can verify one of the most fundamental laws of electricity, Ohm's Law. You will then use Ohm's Law to predict values of resistance.

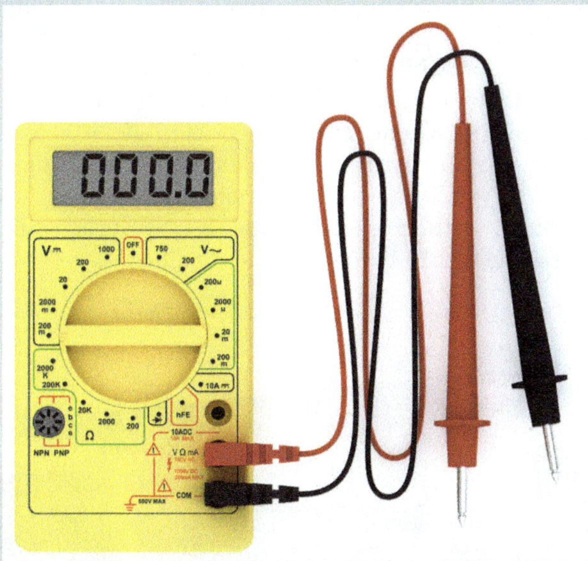

Figure 6-22 Multimeter.

Materials you will need:

Laptop or Computer with Microsoft Excel

- Digital Multimeter
- Breadboard
- 9-V battery
- 9-V battery clip
- Set of jumper wires

(Note: a variety of resistors should be used: These are recommended, but substitutions may be made.)

- Resistor: 100 ohm, Carbon Film, 1/2W, 5%
- Resistor: 150 ohm, Carbon Film, 1/2W, 5%
- Resistor: 220 ohm, Carbon Film, 1/2W, 5%
- Resistor: 330 ohm, Carbon Film, 1/2W, 5%
- Resistor: 470 ohm, Carbon Film, 1/2W, 5%

Resistor: 560 ohm, Carbon Film, 1/2W, 5%
Resistor: 820 ohm, Carbon Film, 1/2W, 5%
Resistor: 910 ohm, Carbon Film, 1/2W, 5%
Resistor: 1000 ohm, Carbon Film, 1/2W, 5%
Resistor: 10,000 ohm (10kΩ), Carbon Film, 1/2W, 5%

1. Clarify the Design Specifications and Constraints

In this scenario, we are not necessarily developing a design for an electrical system. Instead, we are attempting to validate a fundamental law of electricity. Ohm's Law is a mathematical model that relates resistance, current, and voltage through a simple linear relationship. Unlike other equations in engineering, the relationship will be easily spotted.

Since we are working in the real world, a 100 ohm resistor is not necessarily a 100 ohm resistor. That sounds deceiving. However, the truth is that manufacturing a perfect resistor, while ideal, is not possible. The 5% associated with each of the resistors on the materials list is a testament to our conjecture about the existence of a perfect resistor. Whenever a resistor is made, it is assigned a certain *tolerance* as deemed appropriate by the manufacturer. This tolerance number tells the customer the degree to which the true resistance is expected to deviate from the intended value. In our case, the 5% tells us that the measured resistance R could be 5% above the listed value or 5% below the listed value.

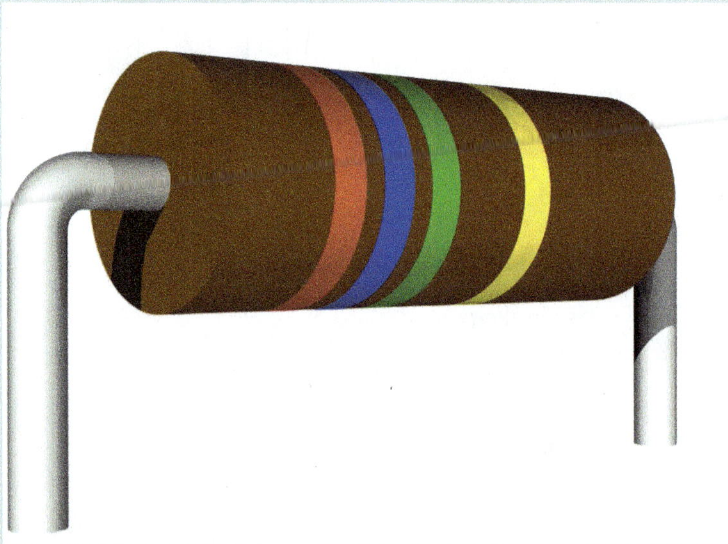

Figure 6-23 A 1.6MΩ resistor.

To find the actual resistance, we can use a device called a multimeter. Because it is able to take a variety of very precise and accurate electrical measurements, the multimeter can tell us the degree to which the resistance deviates from the value listed on the component.

How can we tell the resistance value from the colored stripes on the resistors?

The resistor color code can be used. We put the resistor with the gold or silver band on the right (think of using it as a period at the end of a sentence). We then read the colors of the three stripes from left to right, writing down the first two numbers. The third stripe gives us the multiple of ten.

In other words, write the first two numbers, then the number of zeros for the number of the third stripe.

Resistor color code

Band color	Value
Black	0
Brown	1
Red	2
Orange	3
Yellow	4
Green	5
Blue	6
Violet	7
Grey	8
White	9
Gold	0.1
Silver	0.01

Tolerance color code

Band color	±%
Brown	1
Red	2
Gold	5
Silver	10
None	20

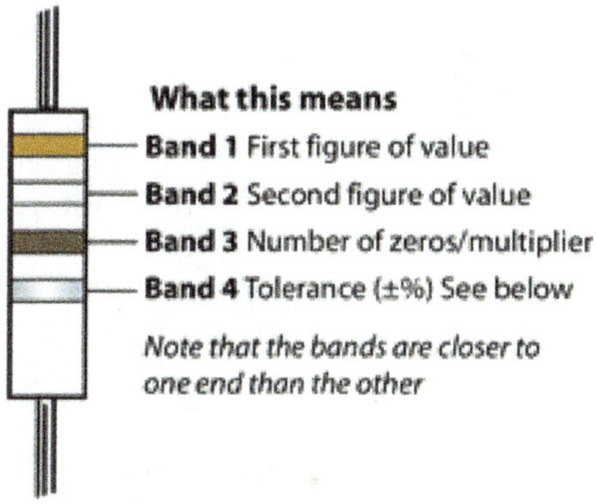

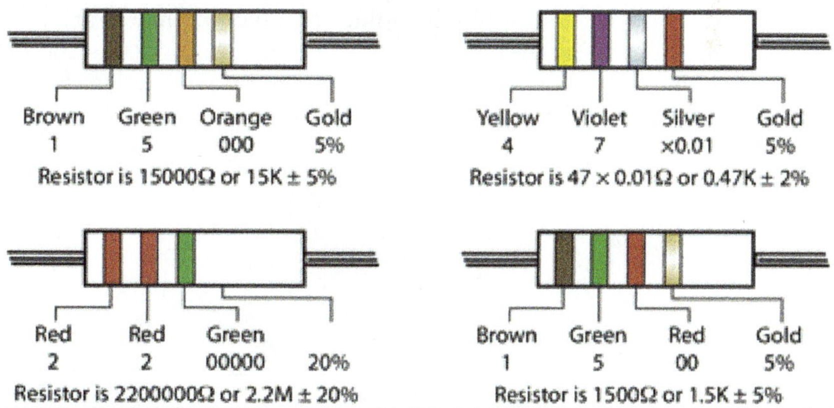

EXAMPLE 6-9

What is the value for a resistor with a red, green, and orange stripe (followed by a silver stripe)?

Red = 2
Green = 5
Orange = 3

2 5 000 = 25,000

This is a 25,000Ω resistor, +/- 10%.

In your engineering notebook, answer the following questions:

1. What is the main constraint limiting our ability to make precise predictions involving an electrical circuit-based model?
2. How will the constraint affect our investigation?
3. Create a table either by hand or in Excel and paste it into a blank space; leave room for more work later. In the table, find the maximum and minimum values of each of the resistors considering the listed tolerances in the materials list.

2. Research and Investigate

Knowledge and Skill Builder 1: Creating a Simple Breadboard Circuit

Breadboards are a platform allowing us to prototype circuits. In completed circuits, each of the components is soldered to whatever board is being used. Breadboards do not require soldering, so components are easily added and removed, making it an excellent tool for building draft circuits and testing components.

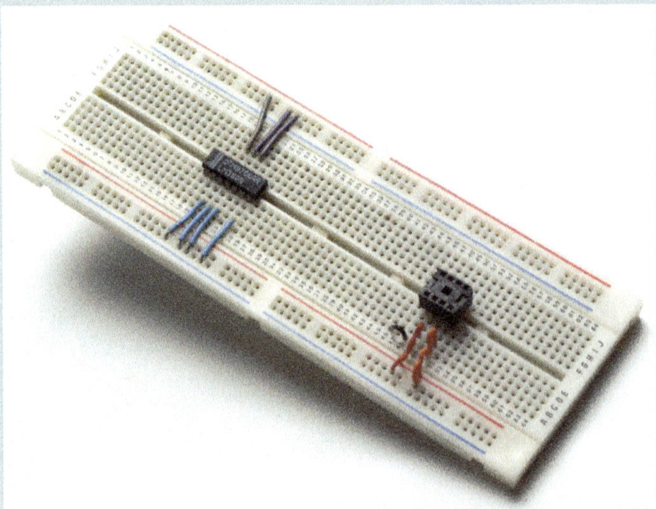

Figure 6-24 A breadboard with components.

Each of the holes in the breadboard can accept the legs of our electrical components. In order to set up a circuit, we need to follow the conductive layout. Underneath the surface of the breadboard are connections between the numerous holes. The outermost strips are reserved for our power source, the battery. To ensure that the circuit components are connected, make sure that no discontinuities exist in the outline of your circuit. In other words, if you insert a component into one hole and another component into a different hole in the same row, the components are connected.

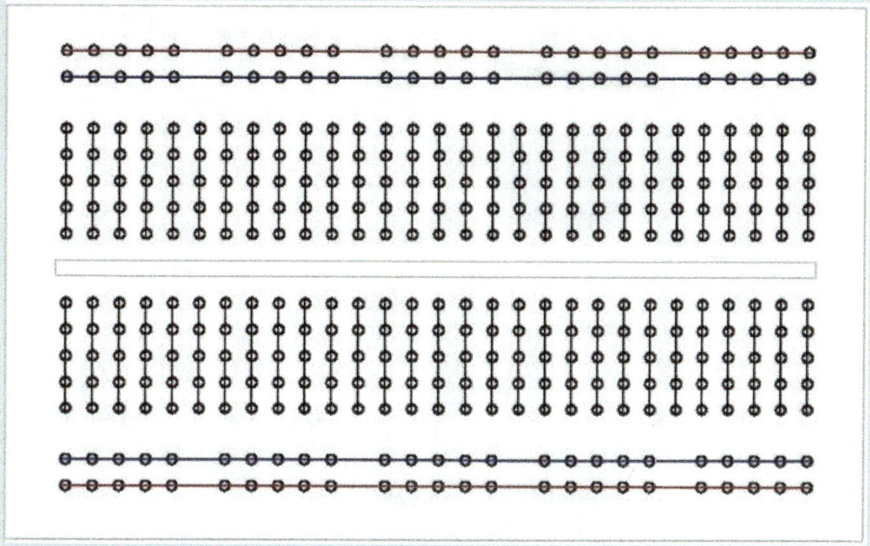

Figure 6-25 A breadboard with connections shown.

(Note that the lines in the image show the connections; they are not drawn on most breadboards.)

To connect the battery to the resistor in the breadboard, we could simply use the battery clips and attach them to the resistor on either end. The positive terminal of the battery would be attached to one leg of the resistor while the negative terminal of the battery would be clipped to the other leg. Alternatively, the exposed conductive part of the leads from the battery can be inserted into the power strips on either side of the breadboard if there are no clips. Be sure you are following the conductive pathways.

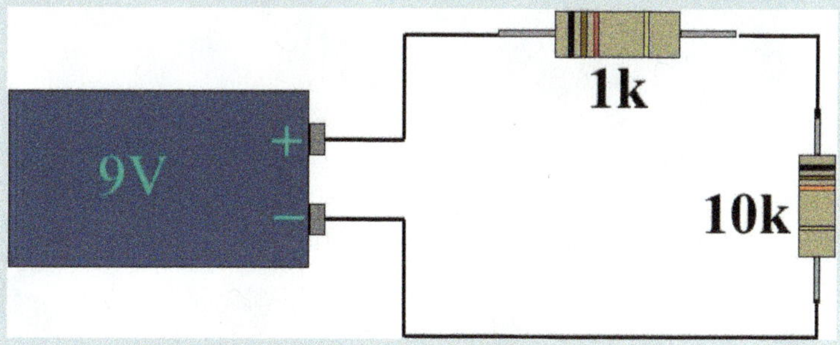

Figure 6-26 Battery powering a 1kΩ and 10kΩ resistor.

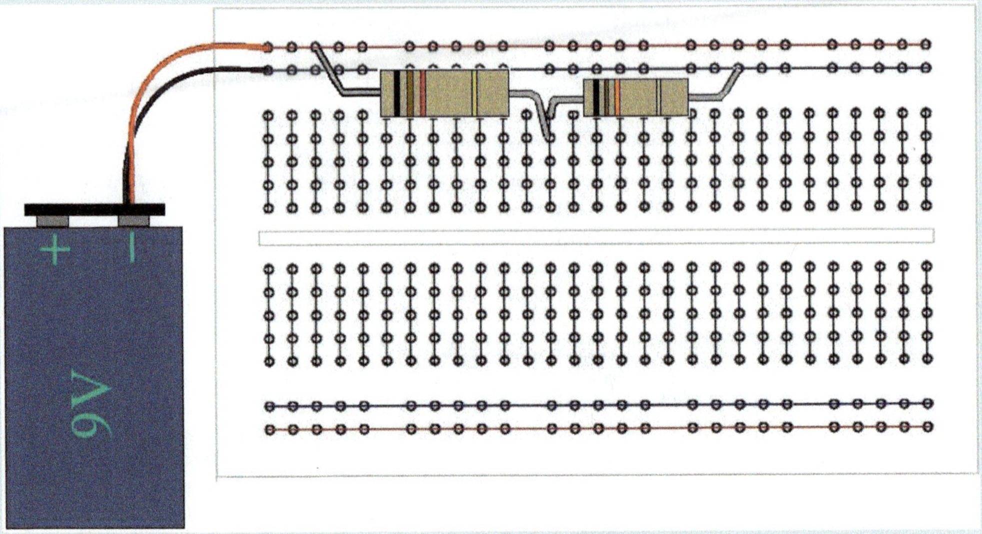

Figure 6-27 Battery with 1kΩ and 10kΩ resistor, attached physically in a breadboard.

Perform the following step:

1. Using the breadboard, set up a circuit with two resistors and the battery.

Knowledge and Skill Builder 2: **Using a Multimeter to Measure Resistance, Current, and Voltage**

To facilitate taking measurements, we have a few tools at our disposal. If we want to find the actual resistance, we can use a multimeter. Because it is able to take a variety of electrical measurements, the multimeter can tell us the degree to which the resistance deviates from the value listed on the component. The multimeter will likely have a dial that can be adjusted to change what it is trying to measure. For resistance, locate the setting with the symbol for ohms, the capital Greek letter sigma (Ω). The multimeter will likely have settings of 200Ω, 2kΩ, 20kΩ, 200kΩ, and 20MΩ. We know these settings correspond to simple SI units and increases in magnitude with resistance. If we pick a resistor out of a hat, place the multimeter on the middle setting, 20k, and attempt to measure the resistance, there are three possible outcomes.

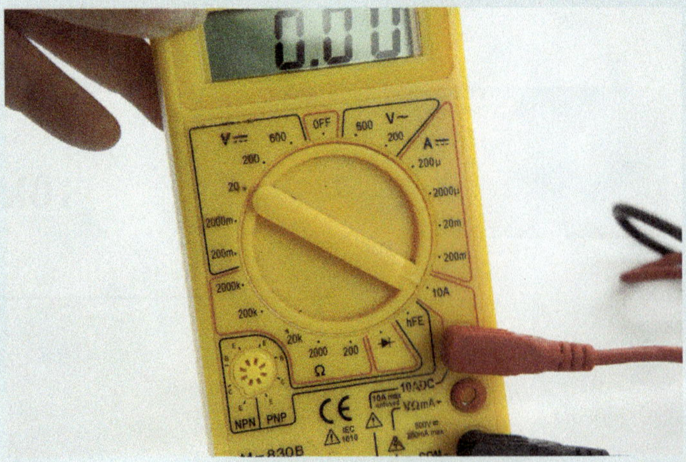

Figure 6-28 Multimeter settings.

1. The screen could read 0.45, which means the measured resistance is 0.45kΩ.
2. If the screen reads 1 or OL, we will need to move the dial to a higher mode, as that means the multimeter is overloaded.
3. If a 0 or 0.00 appears, we will need to move the dial to a lower setting.

The closer we are to the appropriate setting, the more precise our measurement will be. For instance, changing the setting from 20kΩ to 200 Ω may yield a value of 0.462, or 0.462kΩ. However, if we drop even lower to the 200 Ω set, then our estimated resistance will be higher than the setting. This will cause scenario 2 to occur; our meter will be overloaded. That outcome makes sense considering we are attempting to measure a resistor that has a higher resistance than the maximum of 200 Ω allowed due to the setting. Likewise, if we continued to turn the dial in the other direction toward the 20MΩ setting, the value of the resistor in *mega* ohms (10^6) would be too small relative to the setting. Therefore, we would expect a 0 or 0.00 to be displayed on the screen.

To physically take a measurement, take the leads of probes from the multimeter and place them firmly on either end on the resistor's legs. Once a number appears on the screen, it may fluctuate slightly. Remember, resistance is not perfect, so use the fluctuating numbers as an estimate.

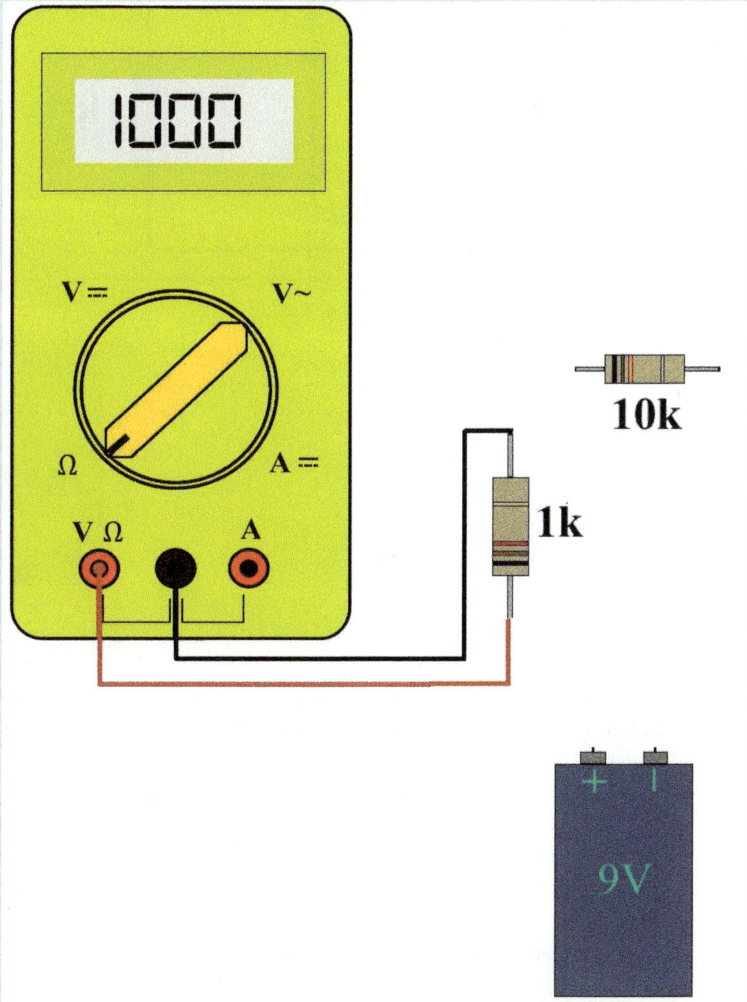

Figure 6-29 Measuring resistance with a multimeter.

(Note: The resistor should never be in a circuit or attached to power to measure resistance!)

Measuring the current is slightly different. We will need to move the dial to the "A" collection of settings, where A stands for amps – the unit for current. The same outcomes still apply, except for the fact that current will be fairly small in comparison to resistance. In practical situations, we will obtain more measurements in milliamps than amps.

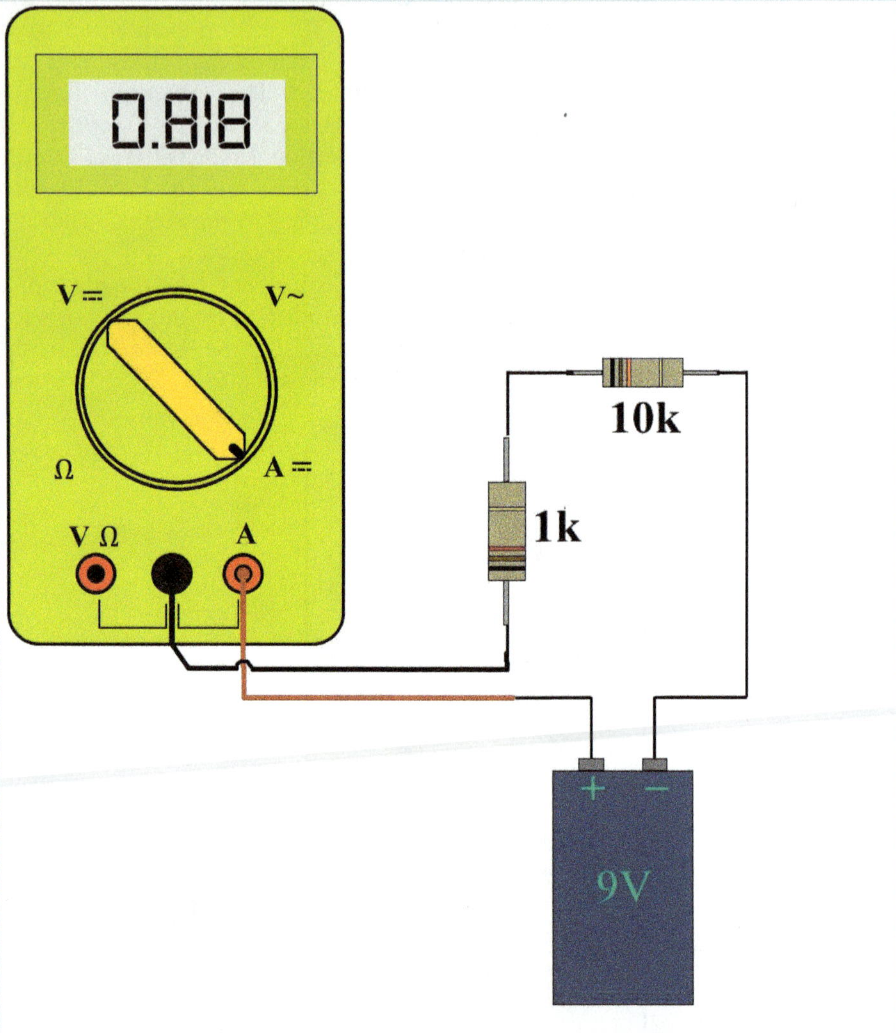

Figure 6-30 Measuring current with a multimeter.

(Note: The circuit is "broken" so the current can flow through the meter itself. The setting must be on A for Amperes, or Amps.)

Finally, we need to adjust the dial one last time toward the "V" collection of settings if we want to measure the voltage. Like current and resistance, we will have the same set of rules. Although we know the battery is 9 volts, we can easily check this claim. Take the leads of the multimeter and place the + lead on the + end of the battery and the – lead on the – end of the battery. It should be no surprise that the multimeter reads 9 V on the screen. What if we switch the placement of our leads? Place the – lead on the + end of the battery and the + lead on the – end of the battery. Now we have – 9 V. What happened?! This phenomena is due to the sign convention for voltage. We define positive voltage as "downstream" from + to – while the negative voltage is defined as fighting "upstream" from – to +.

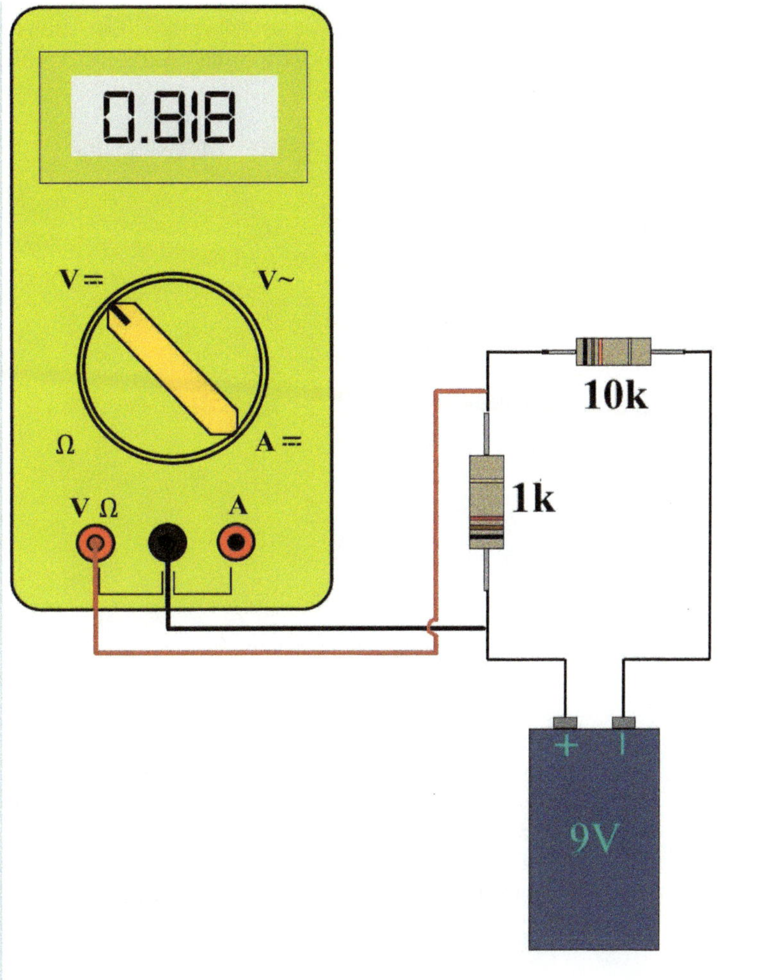

Figure 6-31 Measuring voltage with a multimeter.

(Note: The leads touch either side of the resistor while the power is attached. Remember to set the meter to D.C. volts and not A.C. volts).

Perform the following steps and answer the following questions in your engineering notebook:

1. Create another column in the table and label it "Measured Resistance."
2. Use the multimeter to measure each resistor. Record the value in the new column in the appropriate row.
3. Did you end up with any resistors that violated the tolerance interval? Create a new column and title it "Error." Calculate the error relative to the labeled resistance for each resistor. (How much does the measured value differ from the theoretical value?)
4. With the resistor in the circuit and being powered by the battery, use the multimeter to measure the current through the resistor and record the value in the appropriate row. Exchange the resistor in the breadboard for the other measured resistors and repeat.

Knowledge and Skill Builder 3: **Using Excel to Model Data**

We can use computational and graphical tools like Excel to make our lives easier and allow us to explore the collected data. In this activity, we will need to make a scatterplot and create a trendline in order to see the relationship between the physical quantities. A scatterplot needs two variables, an independent variable X and a dependent variable Y. The independent variable will be the quantity we have control over, while the dependent variable will be the result of us changing the independent variable. For example, we have a table of X values and Y values from some experiment.

X (Independent Variable)	*Y* (Dependent Variable)
0	2.2
10	9.8
20	22
30	29.1
40	38.1
50	53
60	60
70	75.3
80	78.1
90	88
100	99

If we want to find the *trend* of the data, we are searching for a mathematical model of that data. Excel makes this process painless. We begin by creating the scatterplot. Use your mouse to highlight the entire table in Excel. Click on the "Insert" tab and you will see a button labeled "Recommended Charts." Near this button, you will see a handful of chart types. Click the button circled in red. A dropdown menu will appear; click the first option in the top left-hand corner.

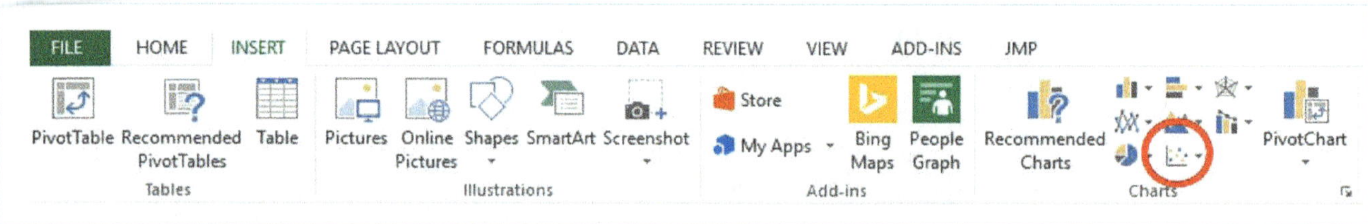

Figure 6-32 Producing a chart using Excel.

Once you click the scatterplot button, your chart will appear. If you would like to edit your X axis and Y axis or add a title, click anywhere on the chart. A new table called "Chart Tools" will appear on your tool bar. Click on "Design," then "Add Chart Element." From here, click on "Axis Titles" and "Chart Title" respectively to add and edit. Note that those features are purely cosmetic; in order for us to add a trendline, we will go to "Trendline" on the same menu. We have a few options: Linear, Exponential, Linear Forecast, and Moving Average. For now, click on "Linear." A line passing through the data is instantly added to the chart. To know the equation of the line, click on it and check the box "Display Equation on chart."

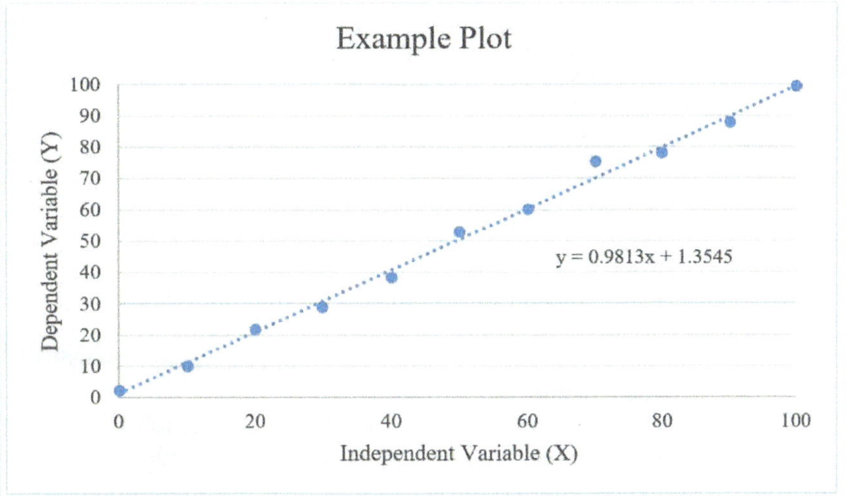

Figure 6-33 Sample plot.

In this example, the trendline equation is $y = 0.9813x + 1.3545$. The trendline can be used to predict values across our data set; for example, the trendline estimates the value of y for $x = 55$ as

$$y = 0.9813(55) + 1.3545 = 55.3$$

It is not necessary for the trendline to hit all of the points in the data set; we are not playing a game of "connect the dots." Rather, we are attempting to determine a pattern the data set is following.

A rule of thumb measure for how well our trendline fits our data is provided by the coefficient of determination (otherwise known as the R^2 value). To figure out the R^2 value in this case, we can click on the trendline again. This will bring up the menu we used to add the equation to the chart. Under the "Display Equation on chart" list, click on "Display R-squared value on chart." Once we click on the correct button, we see that our R^2 value is 0.995. What does that mean? Our R^2 value is bounded between 0 and 1, where 0 means that our fit does not model the data well at all and 1 means the trendline is a perfect fit.

Although we can think of the R^2 value as a measure of how well our trendline fits the data set, we can artificially inflate it by changing from "Linear" to "Polynomial" and increasing the order. Once we have a polynomial of a high enough degree, we will have a fairly high coefficient of determination since polynomials can be stretched and shrunk accordingly to fit any kind of data. To reiterate, do not try to connect all of the dots; you will inflate the R^2 value to the point of it becoming meaningless. If you find yourself using Polynomial Order 3 or more, take a step back and try simpler models. Chances are that you are overcompensating.

In addition, do not feel as though "Linear" is the only feasible option. Try out different fits and critically examine your results.

Complete the following steps in Excel and record any responses in your engineering notebook:

1. Isolate the columns "Current" and "Measured Resistance." Which column is the independent variable and which is the dependent variable?
2. Create a scatterplot of the independent variable versus the dependent variable.
3. Choose a trendline. What is the type of fit you chose and what is the coefficient of determination for the fit. Write down the equation.

4. Look closely at your equation. What is the coefficient (not R^2) of the fit close to? What does this coefficient represent: voltage, current, or resistance?
5. Replace the coefficient from 4, Y, and X with the name of the appropriate electrical quantity.
6. Use algebra to rewrite the equation from 5 into the more familiar form.

3. Generate Alternative Designs

The procedure we went through in the previous step is certainly not the only way to validate Ohm's Law. To take measurements, we fixed one quantity – voltage - by using a simple 9 volt battery, changed the resistances physically, and then measured the current.

In your engineering notebook, describe two other ways we could have derived Ohm's Law. Include any appropriate diagrams, figures, or sketches. The approaches can be similar, but be sure to describe how the methods differ.

4. Choose and Justify the Optimal Solution

Pick your preferred solution. In your engineering notebook, respond to the following prompts:

1. Explain how your solution meets the requirements and constraints.
2. What trade-offs, if any, did you make in selecting your solution?

5. Develop a Prototype

In any technological activity, you will use seven resources: people, capital, time, information, energy, materials, and tools and machines.

In you engineering notebook, indicate which resources were most important in this activity and how you made trade-offs among them.

6. Test and Evaluate

With any model, we need to check and make sure its predictions are plausible.

Perform the following steps and record relevant data and responses in your engineering notebook:

1. Calculate the predicted resistance if the current is 0.05 A, 0.08 A, and 1 mA.
2. How do these values compare to the previous measured values?

7. Redesign the Solution

Mathematical models can always be subject to evaluation and reformulation like any other engineering design problem.

In your engineering notebook, respond to the following prompts:

1. If you had to redo certain parts of the investigation, what problems did you encounter?

2. Did you modify your original solution?
3. If you had to redo this investigation, what would you do differently?

We can go about applying Ohm's Law as long as we have two of the three quantities.

4. Have your teacher provide you with a random resistor. Replace the resistor in the circuit with the new resistor you were provided. Disconnect the battery from the circuit and measure the voltage across the battery. Write down the measured voltage.
5. Reconnect the battery and measure the current through the resistor.
6. Use Ohm's Law to predict the value of the supplied resistor. Write down the measured value.
7. Verify your calculation with your teacher and calculate the error from the theoretical value of the resistor. Record the error. Is the measured resistance within the tolerance?

8. Apply Your Results

We saw how changing resistance values can change values of voltage and current. Answer the following questions in your engineering notebook or in your digital portfolio:

1. Choose an emerging technology in electrical engineering. (You may find one on your own or select solar power, wireless communication/cell phones, or robotics.)
2. How could we use this knowledge to change an aspect of this technology? In other words, consider an important aspect of the technology: By changing a resistance, therefore changing a value of current or voltage, how can we potentially improve the technology?
3. How can we use this knowledge to make a change in the control of a manufacturing process?

9. Communicate Your Achievements

Write a three-page report using the data collected during this investigation and responses in your engineering notebook. Your report must have the following sections:

Introduction: What is this report about?
Purpose: Why did you do this investigation?
Procedure: How did you go about conducting the investigation?
Results: What did you discover? Include relevant tables, charts, and equations.
Discussion: Comment on your results. What were the limitations of your investigation?
Conclusion: Summarize the report and provide the reader with a sense of finality.

Chapter 6 ASSESSMENT QUESTIONS

1. A girl pushes on a bicycle with 40 newtons of force, moving it 30 meters. How much work has she done?

2. Explain the difference between potential energy and kinetic energy.

3. On a cold winter day, why does a thick coat make you feel warmer? What is the source of the heat?

4. A motor with a power of 100 watts is connected to a flywheel. How long, in units of hours, must the motor operate to transfer 300,000 joules to the flywheel?

5. A motor with an input power of 100 watts is connected to a flywheel. How long, in units of hours, must the motor operate to transfer 300,000 joules to the flywheel, assuming the process is 80% efficient?

6. The current through a 12 kilo-ohms [kΩ] resistor is 25 microamps [μA]. What is the voltage across the resistor?

7. A 1000-ohm [Ω] resistor has 120 volts [V] across it. What is the minimum wattage rating of the resistor?

8. A current of 0.1mA flows through a resistor. What is the value of the resistor if:

(a) the voltage across the resistor is 9V?
(b) the voltage across the resistor is 3.1V?
(c) the voltage across the resistor is 0.1V?

9. How much current is flowing through a 1,000Ω resistor if:

(a) the voltage across the resistor is 9V?
(b) the voltage across the resistor is 3.1V?
(c) the voltage across the resistor is 0.1V?

10. A resistor is dissipating 125 milliwatts [mW] of power. The voltage across the resistor is 12 volts [V].

(a) What is the value of the resistance? Express your answer in units of ohms, using an appropriate prefix.
(b) What is the current through the resistor? Express your answer in units of amperes, using an appropriate prefix.

11. A 3-volt [V] battery delivers a constant current of 100 milliamperes [mA] to the bulb in a flashlight for 20 minutes.

(a) What is the total charge in units of coulombs that passes through the flashlight bulb? Assume conventional current, thus the charge will be positive.
(b) What is the total energy delivered to the flashlight bulb in units of joules?

12. A constant current charges a 20-nanofarad [nF] capacitor to 5 volts [V] in 10 microseconds [μs].

(a) Determine the current. Express your answer in units of amperes, using an appropriate metric prefix.
(b) What is the total energy stored in the capacitor? Express your answer in units of joules, using an appropriate metric prefix.

13. You need to purchase a motor to supply 400 joules in 10 seconds. All of the motors you can choose from are 80% efficient. What is the minimum wattage on the motor you need to choose?

14. Consider the following strange, but true, unit:
1 donkeypower = 0.33 horsepower
A certain motor is rated to supply an input power of 2,500 calories per minute at an efficiency of 90%. Determine the amount of output power available in units of donkeypower.

15. The power available from a wind turbine is calculated by the following equation:

$$P = \frac{1}{2}A\rho v^3$$

where P = power [watts], A = sweep area (circular) of the blades [square meters], ρ = air density [kilograms per cubic meter], and v = velocity [meters per second]. The world's largest sweep area wind turbine generator in Spain has a blade diameter of 420 feet. The specific gravity of air is 0.00123. Assuming a velocity of 30 miles per hour and the power produced is 5 megawatts, determine the efficiency of this turbine.

16. A constant voltage of 5 volts [V] is applied across a 250-millihenry [mH] inductor until the current through the inductor is 200 microamperes [μA].

(a) For how many seconds was the voltage applied to the inductor? Express your answer using an appropriate metric prefix.
(b) What is the total energy stored in the inductor? Express your answer in units of joules, using an appropriate metric prefix.

17. A 10,000-microfarad [μF] capacitor is charged to 25 volts [V]. If the capacitor is completely discharged through an iron rod 0.2 meters long and 0.25 centimeter in diameter, resulting in 90% of the stored energy being transferred to the rod as heat, how much does the temperature of the rod increase? Give your answer in kelvins.
Data you may need:
Specific gravity of iron: $SG = 7.874$
Specific heat of iron: $C_P = 0.450\ \mathrm{J/(g\ K)}$

18. What are the color bands around a resistor with the following resistance values?

(a) 1,000Ω
(b) 333Ω
(c) 1MΩ
(d) 47Ω

19. What is the resistance value of each of the resistors shown?

(a)

(b)

(c)

(d)

20. What is thermal energy?
21. What is power? How is it different from work?
22. Describe three different kinds of engines.
23. A boy pushes on a car with 100 newtons of force but does not move it. How much work has he done?
24. List all the energy resources used by your family, telling how they are used and whether each resource is renewable. Consider heating and cooling your home, running appliances and machines inside and outside your home, and your means of transportation.
25. Do research to find out how diesel fuel is different from gasoline and why it is used for diesel engines instead of gasoline.
26. Will the energy generated by falling water in a hydroelectric plant be more, less, or the same as the gravitational energy stored in the water that falls (assuming there are no losses in the generator)? Explain your answer.
27. In a team, investigate how a power plant generates electricity.
    (a) Prepare a shared document (such as a Google doc). It should be a three-page illustrated report, including pictures and/or diagrams.
    (b) Prepare a two-minute presentation for the class. Produce three PowerPoint slides to use as visual aids.
28. Determine the equivalent resistance of four resistors connected as shown.

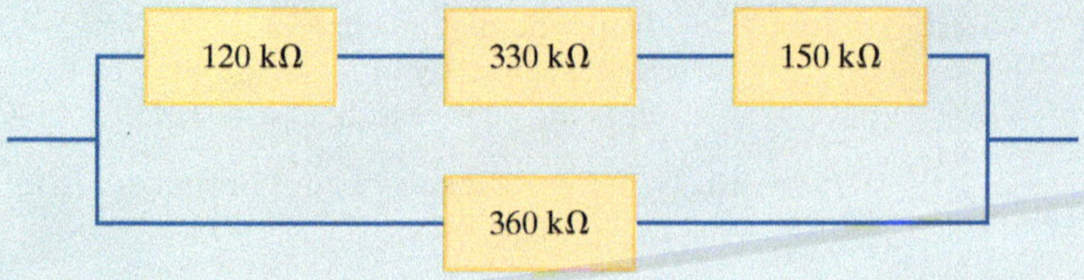

Determine the equivalent resistance of four resistors connected as shown.

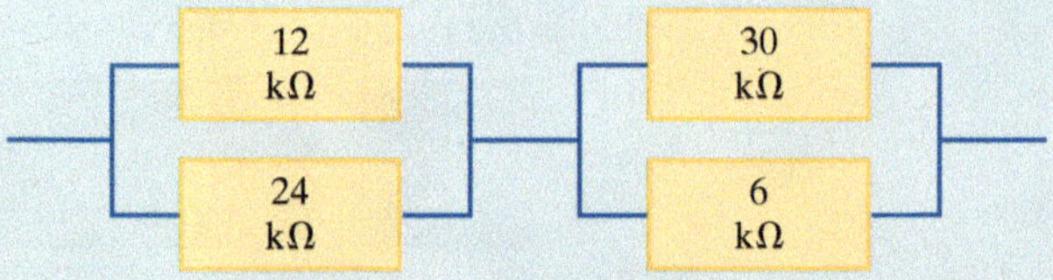

29. If we put a 10kΩ, 2.2kΩ, and 100Ω resistor in series, what is the equivalent resistance? Draw a picture of this resistor configuration.
30. If we put a 10kΩ, 2.2kΩ, and 100Ω resistor in parallel, what is the equivalent resistance? Draw a picture of this resistor configuration.
31. Research: What materials are ceramic disk capacitors and tantalum capacitors made from? Describe the differences in these materials as they relate to capacitance.
32. Research: Using the Internet, find an advertisement for a position in engineering that mentions circuits, circuit elements, voltage, resistance, or current. What are the job responsibilities for this position? Can you find elements of the ad that show that this is a professional position?

CHAPTER 7 ELECTROMECHANICAL DESIGN

Without power, much of life as we know it grinds to a halt. There are many ways to generate power and seemingly endless ways to consume power. If we hope to generate power more efficiently in the future, we will probably need electromechanical solutions.

7.1 POWER

We have seen power before, but it is a concept worth revisiting.

power

In science, **power** is the rate at which work is done. More specifically, it is the amount of work done during a given period of time. This can be illustrated by the following equation:

$$\text{Power (P)} = \frac{\text{Work (W)}}{\text{Time (t)}}$$

The person jogging in Figure 7-1 has a certain amount of power. We can estimate that a person running for 6 minutes would use 315,000 joules. We can determine the amount of power needed for this run using the equation above.

$$\begin{aligned} P &= \frac{315{,}000\text{ J}}{(6\text{ minutes})(60\text{ seconds/minute})} \\ &= \frac{315{,}000\text{ J}}{360\text{ seconds}} \\ &= 875\text{ watts (W)} \end{aligned}$$

Figure 7-1 Power is the rate at which work is done.

One watt of power is equal to 1 joule per second. If the person had a greater amount of power, less time would be needed to jog the same distance. Similarly, a car with a powerful engine can accelerate to a given speed faster than the same car with a less powerful engine. The same amount of work is necessary to move the car to that speed, but the power available determines how quickly the work will be done.

There are other units that measure power, such as horsepower. The term *horsepower* was originally derived from the amount of energy a horse needed to pull a load of coal. One unit of horsepower is equal to 746 watts.

The power to do work comes from energy. A machine that uses energy to do work is called a power

Figure 7-2 All types of machines use energy to do work. The engine of this train uses electrical energy to move.

system. An engine is a power system that converts various sources of energy into mechanical energy to do work (Figure 7-2).

Engines can do many different jobs. They supply the electricity to light homes. They power cars and planes. They carry rockets into space. There are many different kinds of engines, which use many different kinds of fuels. Two types of engines are internal combustion engines and external combustion engines.

7.2 INTERNAL COMBUSTION ENGINES

internal combustion engine

In an **internal combustion engine,** the combustion process occurs inside the engine. When the fuel is ignited, chemical energy is converted into thermal energy and then into mechanical energy. The two most common types of internal combustion engines are gasoline and diesel.

Gasoline Engines

gasoline engine

The first successful internal combustion engine fueled by gasoline was invented in Germany by Nicolas Otto in 1876. It was smaller and more efficient than the steam engines of the time. In a **gasoline engine,** a spark ignites the gasoline, which is a highly flammable liquid derived from oil.

Cars use a gasoline internal combustion engine. In this type of engine, gasoline is burned to produce heat, and then this thermal energy is converted into mechanical energy. Because there are four steps in this process, it is called a four-stroke cycle.

- During **Step 1,** the intake stroke, the piston moves downward, creating a vacuum and pulling a mixture of gasoline and air into the cylinder through an open intake valve.

- During **Step 2,** the compression stroke, the piston moves up and compresses the air-fuel mixture. When the piston reaches the top, the spark plug fires to begin step 3.
- During **Step 3,** the power stroke, the firing spark plug ignites the air-fuel mixture. The heat from the rapid burning raises the temperature and pressure of the gas, and the high-pressure gas pushes on the piston. The gas expands, pushing the piston down.
- During **Step 4,** the exhaust stroke, the piston moves back up. The exhaust valve opens and the burned gases are forced out. Then, the cycle begins again.

The piston is connected to the crankshaft, which transforms the vertical piston motion to a rotary motion. The crankshaft is connected, via gearing and axles, to the car's wheels.

The gasoline engine has come into widespread use. It is used in nearly all cars and many trucks. Small gasoline engines are used in lawn mowers and go-carts. They are also used in portable generators, leaf blowers, snowblowers, and outboard engines in boats.

Diesel Engines

diesel engine

A **diesel engine** is similar to a gasoline engine, except that it burns fuel oil, which is heavier and oilier than gasoline (Figure 7-3). The fuel is injected into the cylinder at the end of the compression step. Because only air is compressed, the compression is much greater. This greater compression means that the temperature of the air is hotter than in a gasoline engine. As a result, the fuel is ignited as it is injected.

Diesel engines tend to be larger than most car engines. Although they are used in some cars, they are used mostly on heavy equipment such as tractors, earth movers, trucks, and buses, as well as on locomotives and ships. They are known for their low maintenance and long life, but they do not have the quick response of gasoline engines.

Gas Turbine Engines

gas turbine engines

Gas turbine engines use pressurized natural gas, kerosene, or jet fuel as their energy source. Gas turbine engines are used on jet airplanes and in rockets.

jet engine

Gas turbine engines are based on a principle that can be observed in a simple garden hose. Take a garden hose and turn on the water. As you hold the hose, there is very little force on your hand from the water leaving the hose. Now, put a nozzle on the end of the hose and turn on the water. You should feel much more force from the nozzle. This is the reactive force created by an increase in velocity across the nozzle.

Attaching a nozzle to a garden hose to increase the water's velocity is similar to what happens in a jet engine (Figure 7-4). A **jet engine** is simply a gas turbine engine that is attached to the wing of an airplane.

First, air is compressed in the compressor, raising its pressure. In step 2, energy is added to the air by burning fuel in the combustion chamber, and the temperature and pressure of the air increase. In step 3, the high-temperature, high-pressure gas expands in the turbine, producing power. The turbine is used to drive the compressor, and the exhaust gases from the turbine go through a nozzle, increasing in velocity. This increase in velocity causes a force on the engine that propels the airplane forward.

Figure 7-3 Diesel engines are used in most large trucks. Diesel engines are heavier than gasoline engines, but they are also more durable.

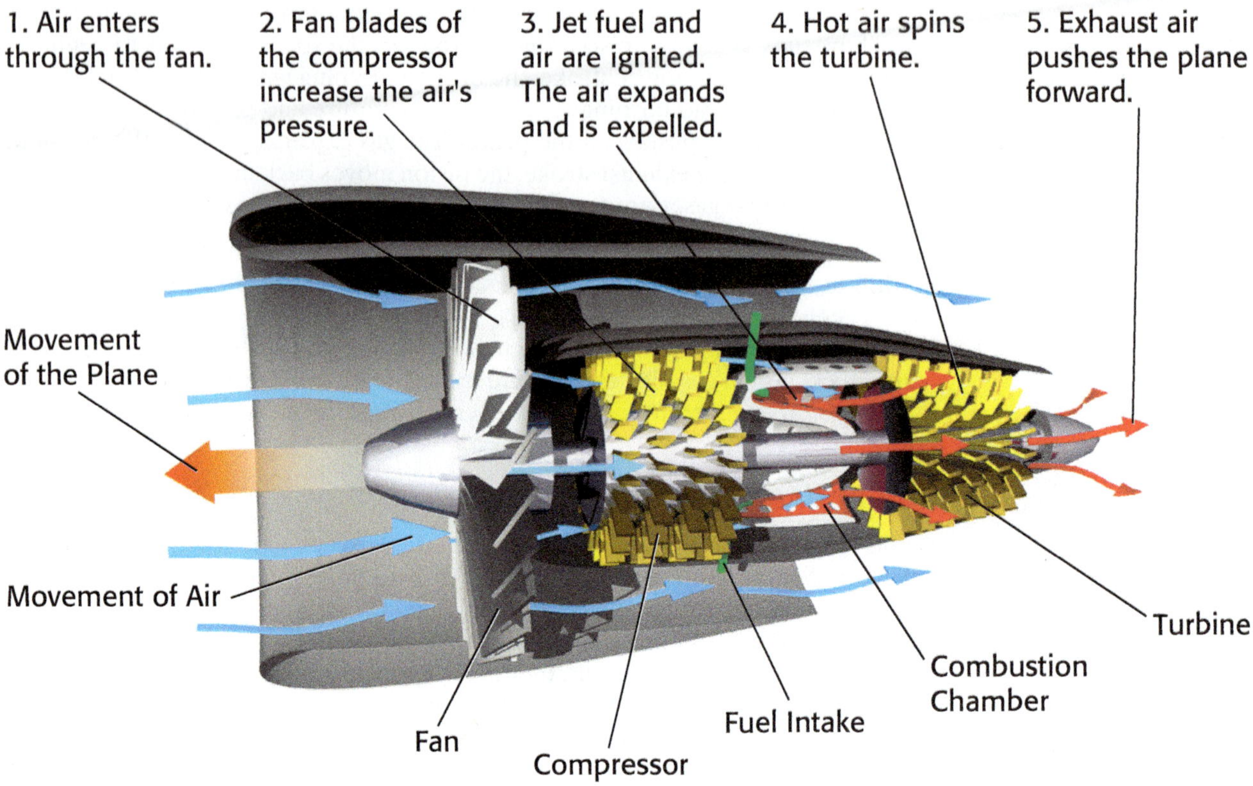

Figure 7-4 A jet engine propels the plane forward by expelling air at very high speeds.

Rocket engines work in the same way as jet engines except that they carry their own oxygen supply This oxygen is used to burn the fuel, which is typically a form of liquid oxygen.

7.3 EXTERNAL COMBUSTION

external combustion

In **external combustion,** fuel is burned outside of an engine and provides heat to another liquid or gas. As with any type of combustion, fuel is ignited and chemical energy is converted into thermal energy. In external combustion, however, the thermal energy heats another liquid, which is converted into mechanical energy to produce the power.

Many power plants use steam as a source of power for external combustion (Figure 7-5). Fuel (gas, coal, or oil) is burned, and the heat causes water to boil under pressure in a tank, producing steam. The high-pressure, high-temperature steam leaves the tank and enters a turbine, causing the turbine to rotate. The spinning turbine is connected to a generator, producing electric power that is distributed along power lines. The steam leaves the turbine at a low pressure and temperature and condenses back to liquid water. A pump is used to increase the pressure of the water so that it returns to the tank to be reheated.

Other energy sources besides combustion can be used to heat the water to produce steam. For instance, solar energy can be reflected with special mirrors, and nuclear energy can heat the water under pressure in a nuclear reactor.

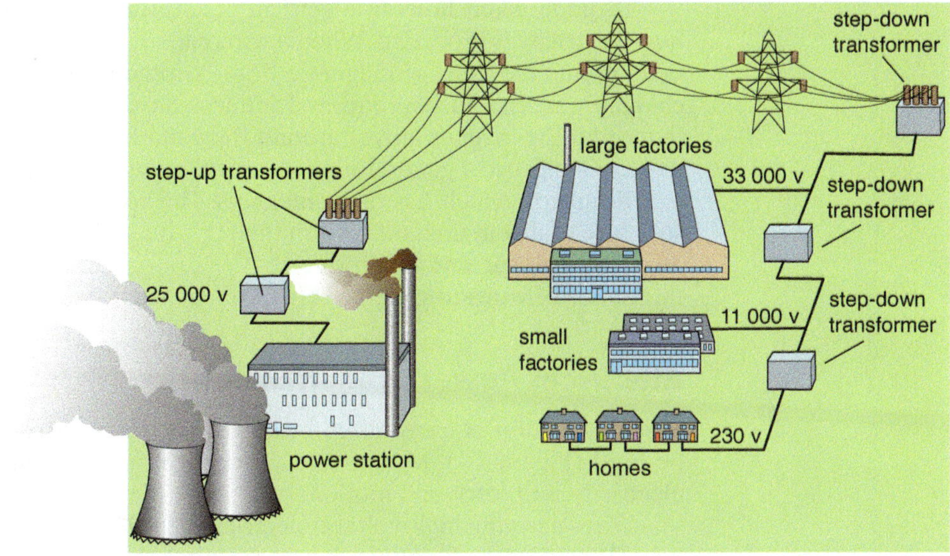

Figure 7-5 Power plants are used to convert different forms of energy into electricity.

Generating Electricity

electrical generator

Although steam is the most common source of mechanical energy for turning turbines, falling water from a dam is also used. In a dam, the energy from moving or falling water is used to turn underground turbines. An **electrical generator** converts the mechanical energy into electrical energy (Figure 7-6).

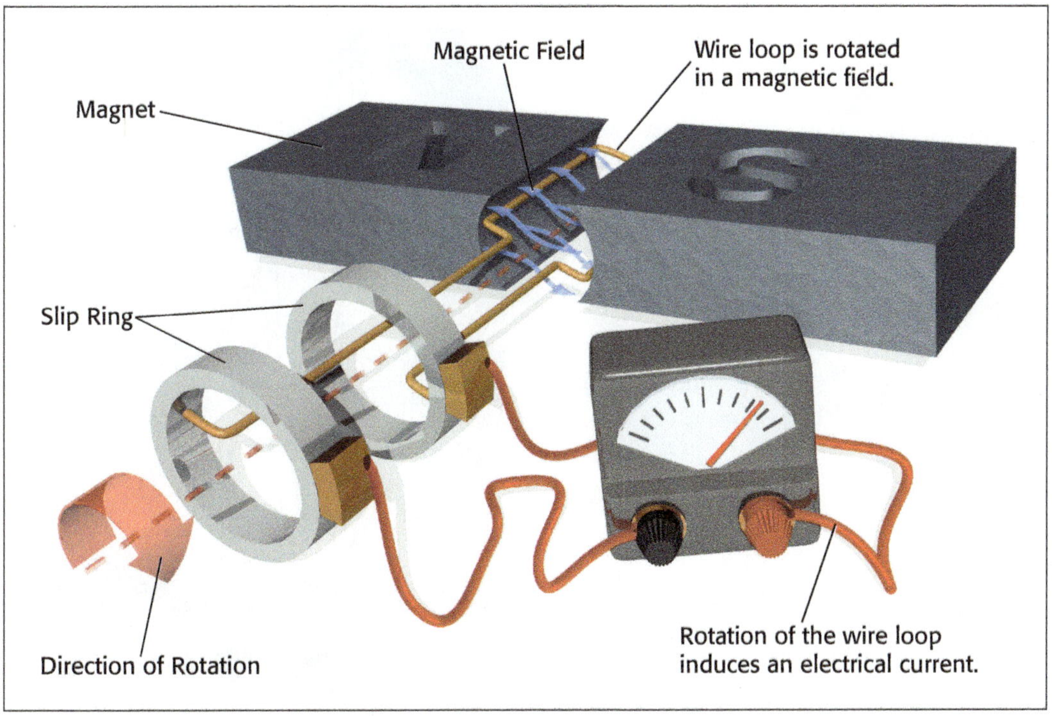

Figure 7-6 Electricity can be generated by rotating a wire in a magnetic field.

To understand how an electrical generator works, you first need to understand how electricity is produced. When a wire, called a conductor, moves through a magnetic field, it cuts the lines of magnetic flow. A force develops in the wire that pushes the charged particles to one end and induces an electrical current. Current can also be induced if the magnet moves around the conductor.

In an electrical generator, mechanical energy turns a turbine that spins a cylinder called a rotor, which has large magnets. The rotor spins within a stator, a stationary structure that contains coils of conducting wire. As the magnets spin within these wires, an electrical current is produced. Generators in power plants use hundreds to thousands of wires to generate high-voltage current.

Electrical Power Transmission

Electrical power is transmitted by wires that carry current to the places where it will be used (Figure 7-7). Motors and other electrical equipment are used to change the electricity into force and motion.

Wires carrying high-voltage electricity lose less energy to heat than do low-voltage lines. Therefore, high voltage is sent on wires that carry electricity over long distances. Transformers are devices that can increase or decrease the force of electrical current. They are used to change the high voltage to lower voltage near the place where the electricity will be used.

Power transmission lines from several power generating stations, or power companies, are often joined in a grid. The high-voltage wires from one power station are joined to the wires from other power stations through a series of switches. A grid is useful when demand is so high that a station or company cannot meet electricity needs. The station or company can then tap into electricity from other stations or companies to supply needed power.

Figure 7-7 Electricity is carried along high-voltage lines to reduce the loss of energy through heat.

EFFECTIVE TEAMING: DEPENDABILITY

You have completed some projects and design challenges, and have probably worked as part of a team during and outside of class. We were introduced to professionalism earlier, an important issue when entering business and industry. Can we demonstrate a level of professionalism while in school?

dependability

Working with a team (or even as an individual) requires **dependability**. A dependable person is one who can be relied upon. This can mean within a team, where teammates can depend on an individual to fulfill the role that person has within the team and to deliver that which he or she is responsible for. As an individual, you can be dependable by doing what you are responsible to do.

reliability

Professionalism also requires **reliability**. Reliability means that you can be relied upon for accuracy and honesty, each and every time. These terms are similar and typically go together.

responsibility

Responsibility implies that you can take ownership of your role and deliverables within a project. Again, the term seems similar to dependability and reliability. If we consider these together, one should be responsible for part of a project (take ownership of something), should be dependable to deliver what he or she is expected to deliver, and be reliable such that the deliverable will be accurate and correct.

punctuality

One more trait toward defining professionalism is **punctuality**—meaning you (and/or your deliverables) should be on time (if not early).

Time Management

time management

Time management involves more than punctuality: Managing a project involves defining a schedule, determining the order in which tasks must be completed (after all, you can't build a wind turbine before designing or obtaining parts), and making sure each member of the team follows the schedule. Consider any project thus far: How did you determine the amount of time to take on any task? How did you ensure that these tasks helped build toward the goal?

For the next project, pay attention to the order in which tasks must be completed, and keep a record of the schedule as you go.

DESIGN ACTIVITY 7

WINDMILL POWER GENERATION

Problem Situation

Windmills are becoming a popular source of electrical power. They convert mechanical energy of spinning blades to electrical energy. Imagine that you are working for a company that is interested in providing windmills as a power source for homeowners in remote locations. You have been asked to develop the initial design of a windmill that generates a great deal of electrical power. The voltage output from the generator will be an indication of its power production. The greater the voltage, the greater the power produced.

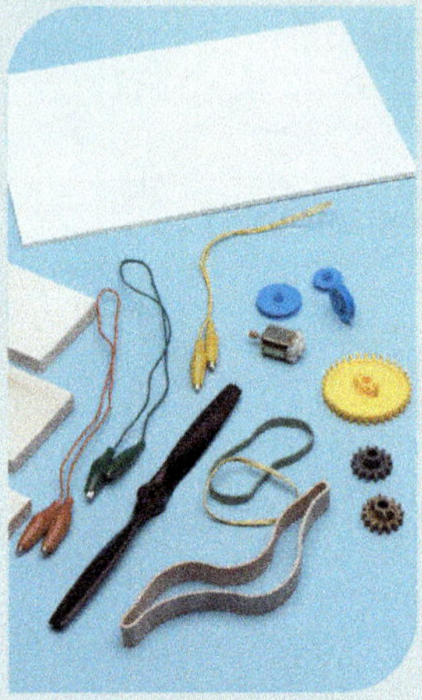

MATERIALS

You will need:

- 3-volt DC motor
- box fan
- foam board
- gears or pulleys with rubber bands
- glue gun
- high-speed rotary tool
- multimeter
- plastic propeller blade
- wire with alligator clips
- wooden sticks, 60 cm long

Your Challenge

You and your teammates are to design and construct a model windmill. In your engineering notebook, describe the challenge in sufficient detail such that it would be understood by a student at a different school.

1. Clarify the Design Specifications and Constraints

To solve the problem, your design must meet the following specifications and constraints:

- The model windmill will generate the most electrical power possible.
- The model windmill will be tested using a box fan provided by the instructor.

Describe the criteria and constraints in your engineering notebook. If the requirements listed here generate questions or you have ideas that will eventually guide your design, include them in your notebook.

2. Research and Investigate

To better complete the design challenge, you need to first gather information to help you build a knowledge base.

Knowledge and Skill Builder 1: **Windmills Around the World**

Investigate how windmills are used. For instance, some are used for pumping water, others for generating electricity. What did you find out?

Enter this information in your engineering notebook or digital portfolio.

Further investigate windmills that are used for generating electricity. What are some of the characteristics of these windmills in terms of speed?

Enter this information in your engineering notebook or digital portfolio.

Knowledge and Skill Builder 2: **Generator Testing**

The small DC motor acts as a generator, producing electricity by rotating the shaft. Attach the propeller blade to the generator. You want to test the generator by having the wind from the fan rotate the blade. You will need to attach the generator to a raised surface so that the blade can rotate freely and the air can move past the blade. Attach wire with alligator clips to the motor and to a multimeter set on DC voltage. Record all questions and answers in your engineering notebook or digital portfolio.

Turn on the fan. What voltage does the motor produce?

____

Remove the propeller blade. With your instructor's assistance, attach a high-speed rotary tool to the generator. Look at the rpm of the tool. Turn it on with the low-speed setting and record the voltage.

____

Move the tool's switch to the high-speed setting and record the voltage.

____

It will be difficult to obtain these high speeds in your design, but now you know the maximum voltage your generator can produce.

Knowledge and Skill Builder 3: **Gearing Up**

You will need to increase the speed of the generator to increase the voltage output, and thus increase power output. Gears or pulleys can be used to speed up rotation or slow it down.

Construct a machine that contains a large gear that engages with a smaller gear. You should be able to rotate the gear set by hand. Perform experiments in which you rotate the larger gear one revolution and note how many revolutions the smaller gear rotated. If your instructor provides pulleys with rubber bands instead of gears, perform the same tests.

Reproduce this table in your engineering notebook or digital portfolio.

Large Gear (Pulley) Rotation	Small Gear (Pulley) Rotation(s)
1	
2	
	1
	2
	3
	4

What is the gear ratio for the gears? This is the ratio of the number of turns the small gear makes for one rotation of the large gear.

The gears or pulleys are part of an energy transmission system. There are frictional losses between the gears because the rubber belts used on pulleys diminish their effectiveness. Even though the gear ratio will cause speed to increase, the friction will cause it to decrease. How does this relate to the First and Second Laws of Thermodynamics?

Enter this information in your engineering notebook or digital portfolio.

3. Generate Alternative Designs

In your Guide, describe two possible solutions that your team has created for the problem. Your solutions should be based on the knowledge you have gathered so far. Remember to consider the specifications and constraints. In your description, indicate what you consider to be each solution's strengths and weaknesses. You will want to consider how the generator and gears will be supported on the windmill's tower.

4. Choose and Justify the Optimal Solution

Choose your preferred solution. Explain how your solution meets the specifications and constraints. Why is this the better alternative? What trade-offs, if any, did you make in selecting this alternative?

5. Develop a Prototype

Construct your model windmill. Include a drawing or a photograph of your final design in your notebook or portfolio.

In any technological activity, you will use seven resources: people, capital, time, information, energy, materials, and tools and machines. In your notebook or portfolio, indicate which resources were most important in this activity, and how you made trade-offs among them.

6. Test and Evaluate

How will you test and evaluate your design?

Describe the following in your notebook or portfolio:

Did your design meet the initial specifications and constraints? Indicate the tests you conducted and the experiments you performed to verify this.

Did you include any additional specifications? Describe the testing procedure, and explain how the design meets these specifications.

7. Redesign the Solution

Your redesign should be based on the knowledge and information that you gained during the activity.

If you had the opportunity to redesign your solution, what problems did you encounter? Did you change your original design concept? Why? If you had to redesign your model, what changes would you recommend in your new design? Explain your reasoning. What additional trade-offs, if any, would you have to make?

8. Communicate Your Achievements

Your teacher may present requirements on how you will present your design. One possibility for which you should prepare is the following:

Form a team of three and review the three designs. Discuss as a team which design is the best and why. Be sure to consider your criteria for success and constraints you faced.

Each team should prepare a 3-4 page report on the best design. The report is to include figures, including a digital photograph of the device, a table with data collected, and your conclusion on why this design is the best.

Each team should prepare a 4-5 minute presentation which must include the information specified in the report requirements using PowerPoint or a similar visual aid. In your presentation, describe how this project relates to an emerging and innovative trend of environmentally-friendly power generation in electrical and mechanical engineering. How could this project help in a career in E.E. or M.E.?

Combine these requirements into your digital portfolio.

Chapter 7 ASSESSMENT QUESTIONS

1. Explain the differences between gas turbine engines and power plants from a combustion viewpoint.

2. Why is it necessary to send electric power at high voltage when it is sent over long distances?

3. How are gasoline and diesel engines similar?

4. Draw a diagram describing how a wind turbine can power a home. Include the inner-working of the wind turbine to the electric breaker box inside the home.

5. Research: Describe the difference between a gasoline and diesel engine.

6. Research: How does a gasoline-powered generator generate electrical power? Include a description and a diagram.

7. In a steam power plant, why is the steam pressurized?

8. Describe "the grid" as it applies to the power supplying a state. If we consider the grid as a system, what are the inputs? What are the outputs?

9. Explain how a DC motor can be used to generate electricity.

10. Use a CAD program to draw a high-voltage tower. Create a drawing with minimal detail such that an engineer could use to design an array of towers. (Save this design.) Using your rough drawing, modify it to add more detail such that an engineer could use to demonstrate the tower itself. (Save this design as well.) Be sure to dimension the drawings using ANSI standards. As part of the research required to draw the tower, answer the following questions:

(a) How far apart are the conductors (wires)?
(b) How tall are typical towers?
(c) How far apart are towers on a typical power run?

11. **(a)** Describe how the precision of the multimeter is useful in the windmill project.
(b) Describe how the accuracy of the multimeter is useful in the windmill project.

12. Research:

(a) How does the windmill project relate to an emerging technology in electrical engineering?
(b) How does the windmill project relate to an emerging technology in mechanical engineering?
(c) Use the Internet to find an engineering career in which working on a project like this would be helpful. Why would this experience be helpful in pursuing that job?

13. **(a)** In the windmill design, describe where physical properties of your prototype are important.
(b) Research: What materials are used to construct blades in large-scale wind turbines?
(c) What material properties of these materials make them ideal for use in blades?
(d) What physical properties of these materials make them ideal for use in turbines?
(e) What is the difference between physical and material properties?

14. Research: What is a Gantt chart? How would developing a Gantt chart prove helpful for this project?

15. Reflecting on the windmill project, write a paragraph for each term reflecting on your reliability, dependability, responsibility, punctuality, and time management.

CHAPTER 8
COMPUTER PROGRAMMING

Computers are controlled by software that can be designed in a variety of programming languages. Computer programs are a translation of what you want to accomplish into something the computer can understand, so the term "programming language" is particularly appropriate. Some computer programs are installed permanently or temporarily on computer chips, and others are installed on a variety of other media, such as hard drives or removable media like CD-ROMs.

Computers relentlessly produce a particular result given a particular set of input conditions. It can be frustrating when you make a simple mistake in a computer program—the computer will do exactly what you tell it to do, even if your mistake would be obvious to a person.

The biggest difference between a computer and a person is that you can ask a person open-ended questions—questions like design questions that can have many answers. Computers can only process questions that have a single answer.

This makes the process of programming a computer a bit like trying to ask another person to solve a problem when they are on the other side of a wall and you can communicate only by passing them slips of paper asking questions that can have only one answer and waiting for the person to pass back a slip of paper with the answer on it.

If a computer always produces the same result every time given the same input conditions, then why does my computer crash sometimes when I am doing something that should work?

The computers you use are simultaneously running a large number of complicated computer programs, including the operating system, background programs, and whatever programs you have started intentionally. Sometimes these programs compete for resources, causing a conflict. Other times, programs are complicated enough that the "input conditions," including the configuration of data in memory and on the hard disk, the time on the system clock, and other factors that change all the time while the computer is running create a combination of circumstances that the programmers never anticipated and so did not include programming code to handle, and the system crashes.

SOME ADVANTAGES OF COMPUTERS

Given our description of how computers work, it may sound to some as if computers are too simple to be useful. The value of programming is linked to a few important characteristics of computers.

- ***Calculation speed:*** Although computers can only answer analytical questions, they can answer such questions very quickly—in small fractions of a second. Computer programs can therefore ask the computer a lot of questions in a short time, and thus find the answer to more complicated problems by breaking down the complicated question into a series of simple questions.

- ***Information storage:*** In "Memory: Science Achieves Important New Insights into the Mother of the Muses" (*Newsweek*, September 29, 1986), Sharon Begley estimates that the mind can store an estimated 100 trillion bits of information. The typical computer has a small amount of storage compared to that, but computers are gaining. Where computers have a bigger advantage is that new information can be incorporated in a fraction of the time it takes a human to learn it.
- ***Information recall:*** Computers have nearly 100% recall of information, limited only by media failures. The human brain can be challenged to recall information in exactly the same form as it was stored.

algorithm

Learning to create effective algorithms is a crucial skill for any aspiring engineer. In general, an **algorithm** is a well-defined sequence of instructions that describe a process. Algorithms can be observed in everyday life through oral directions ("Simon says: raise your right hand"), written recipes ("Bake for 15 minutes at 350 degrees Fahrenheit"), graphical assembly instructions, or other graphical cues. As an engineer, writing any algorithm requires a complete understanding of all the necessary actions and decisions that must occur to complete a task.

scope

When writing an algorithm, you must answer a few questions before attempting to design the process. To even begin thinking of a strategy to describe a process, you must have carefully defined the scope of the problem. The **scope** of an algorithm is the overall perspective and result that the algorithm must include in its design.

For example, if we are required to "sum all numbers between 1 and 5," before thinking about an approach to solve the problem, we must first determine if the scope is properly defined. Does the word "between" imply that 1 and 5 are included in the sum? Do "numbers" include only the integer values? What about the irrational numerical values? Clearly, we observe that we cannot properly define the scope of the charge to add all numbers between 1 and 5.

Likewise, imagine you were charged to design a device that transports people from Atlanta, Georgia, to Los Angeles, California. How many people must the device transport? Does the device need to travel on land? Should it travel by air? Should it travel by water? Does the device require any human interaction?

written algorithms
graphical algorithms

This section covers two methods of defining a process: with **written algorithms** and with **graphical algorithms**. Both methods require properly identifying the scope of the problem and all of the necessary input and output of the process.

8.1 SCOPE

One of the most difficult steps in designing an algorithm is properly identifying the entire scope of the solution. Like solving a problem on paper involving unit conversions and equations, it is often necessary to state all of the known and unknown variables in order to determine a smart solution to the problem. If information is left out of the problem, it might be necessary to state an assumption in order to proceed with a solution. After all variables and assumptions about the problem have been identified, it is then possible to create a sequence of actions and decisions to solve the problem.

To clearly understand the scope of the problem, we often find it helpful to formally write out the known and unknown information, as well as state any assumptions necessary to solve the problem. In the following examples, notice that as the problem statements become more and more refined, the number of necessary assumptions decreases and eventually disappears.

EXAMPLE 8-1

For the problem statement, list all knowns, unknowns, and assumptions.

Problem: Sum all numbers between 1 and 10.

Known:

- *The minimum value in the sum will be 1.*
- *The maximum value in the sum will be 10.*

Unknown:

- *The sum of the sequence of numbers.*

Assumptions:

- *We will only include the whole number values (e.g., 1, 2, 3, . . .) in the sum.*
- *The sum will include the starting value of 1 and the ending value of 10.*

EXAMPLE 8-2

For the problem statement, list all knowns, unknowns, and assumptions.

Problem: Sum all numbers between (and including) 1 and 10.

Known:

- *The minimum value in the sum will be 1.*
- *The maximum value in the sum will be 10.*

Unknown:

- *The sum of the sequence of numbers.*

Assumptions:

- *We will only include the whole number values (e.g., 1, 2, 3, . . .) in the sum.*

EXAMPLE 8-3

For the problem statement, list all knowns, unknowns, and assumptions.

Problem: Sum all whole numbers between (and including) 1 and 10.

Known:

- *The minimum value in the sum will be 1.*
- *The maximum value in the sum will be 10.*

Unknown:

- *The sum of the sequence of numbers.*

Assumptions:

- *[None]*

8.2 WRITTEN ALGORITHMS

A written algorithm is a narrative set of instructions required to solve a problem. In everyday life, we encounter written algorithms in the form of oral instructions or written recipes. However, it is extremely common for humans to "fill in the blanks" on a poorly written algorithm. Imagine you are handed a strongly guarded family recipe for tacos. One of the steps in the archaic recipe is to "cook beef on low heat until done." To the veteran cook, it is apparent that this step requires cooking the prepared ground beef on a stove-top in a sauce pan for approximately 10 minutes on a burner setting of 2 to 3. To a first-time cook, the step is poorly defined and could result in potentially inedible taco meat.

Engineers and Written Algorithms

As an engineer, to write effective algorithms you must ensure that every step you include in a written algorithm must not be subject to misinterpretation. It is helpful to write an algorithm as if it were to be read by someone completely unfamiliar with the topic. Each step in the written algorithm should be written such that the stepwise scope is properly defined. The **stepwise scope** is all of the known and unknown information at that point in the procedure. If a step in an algorithm contains an assumption, you must formally declare it before proceeding with the next step. By ensuring that the stepwise scope is well defined, you ensure that your algorithm will not be subject to misinterpretation.

stepwise scope

All written algorithms should be expressed sequentially. The most effective algorithms are written with many ordered steps, wherein each step contains one piece of information or procedure. While the author of an algorithm may consider each step in an algorithm to be "simple," it might not be trivial to an external interpreter. When writing an algorithm, it is helpful to assume that the reader of your algorithm can only perform small, simple tasks. Assume that your algorithm can be interpreted by a computer. A computer can execute small tasks efficiently and quickly, but unlike a human, a computer cannot fill in the blanks with information you intended the reader to assume.

Decision-making can be expressed in a written algorithm. Assume you are designing a process to determine if the value read from a temperature sensor in a vehicle indicates it is unsafe for operation. To express the decision in a written algorithm, phrase your decisions in questions that have a "Yes" or "No" response.

Format of Written Algorithms

The first step in writing any algorithm is defining the scope of the problem. After you define the scope of the problem, create an *ordered* or *bulleted list* of actions and decisions. Imagine taking an English class and writing a research report on the influence of 19th-century writers on modern-day fiction authors. Before writing the paper, you would create an outline to ensure that your topics have connectivity and flow. Just like the outline of an English paper, an algorithm is best expressed as a sequential list rather than as complete paragraphs of information.

If a decision is required in the algorithm, indent the actions to indicate the action is only associated with the particular condition.

EXAMPLE 8-4

Create a written algorithm to express a temperature given in relative units [°F or °C] in the corresponding absolute units [K or °R].

Known:

- *Temperature in relative units (degrees Celsius or degrees Fahrenheit).*

Unknown:

- *Temperature in absolute units (kelvins or degrees Rankine).*

Assumptions:

- *Since the problem does not explicitly state the temperature of interest, assume that the interpreter of the algorithm will input the temperature and units.*

Algorithm:

1. *Input the numeric value of the temperature.*
2. *Input the units of the numeric value of the temperature.*
3. *Ask if the input unit is degrees Fahrenheit.*
   (a) *If yes, calculate the value in degrees Rankine.*
   (b) *If no, calculate the value in kelvins.*
4. *Display the new value and absolute unit.*
5. *End the process.*

EXAMPLE 8-5

Create a written algorithm to calculate the sum of a sequence of whole numbers, given the upper and lower bounds of the sequence.

Known:

- *Upper bound of whole number sequence.*
- *Lower bound of whole number sequence.*

Unknown:

- *Sum of all whole numbers between the upper and lower bound.*

Assumptions:

- *Since the problem does not explicitly state the upper and lower bounds, assume the interpreter will ask for the values.*
- *Include the boundary values in the summation.*

Algorithm:

1. *Input the lower bound of the sequence.*
2. *Input the upper bound of the sequence.*
3. *If the lower bound is larger than the upper bound,*
   (a) *Warn the user that the input is invalid.*
   (b) *End the process.*
4. *If the upper bound is larger than the lower bound,*
   (a) *Create a variable to keep track of the sum (S).*
   (b) *Create a variable to keep track of the location in the sequence (L).*
   (c) *Set the initial value of S to be zero.*
   (d) *Set the initial value of L to be the lower bound.*

(e) *If the value of L is less than or equal to the upper bound,*
   (i) *Add L to the current value of S.*
   (ii) *Add one to the current value of L.*
   (iii) *Return to step 4.e. and ask the question again.*
(f) *If the value of L is greater than the upper bound,*
   (i) *Display the sum of the sequence (S).*

5. *End the process.*

feedback loop

infinite feedback loop

In step 4.e.iii, we required that the interpreter return to an earlier step in the algorithm after changing the values of our variables. This allows us to create a **feedback loop** necessary to calculate the sequence of values. A feedback loop is a return to an earlier location in an algorithm with updated values of variables. It is important to note that if we failed to update the variables, the feedback loop will never terminate. A nonterminating feedback loop is also known as an **infinite feedback loop**.

8.3 GRAPHICAL ALGORITHMS

flowchart

To visualize a process, a graphical representation of algorithms is used instead of a written algorithm. A **flowchart** is a graphical representation of a written algorithm that describes the sequence of actions, decisions, and path of a process. Designing a flowchart forces the author of the algorithm to create small steps that can be quickly evaluated by the interpreter and enforces a sequence of all actions and decisions. Flowcharts are used by many different disciplines of engineering to describe different types of processes, so learning to create and interpret flowcharts is a critical skill for a young engineer. In fact, in the United States, any engineers who discover a new innovative algorithm can submit their concept to a patent office by representing the process in terms of a flowchart.

Three different shapes are used in the creation of flowcharts in this book; a number of other widely used operators are encountered across the world. In this book, we describe all actions with rectangles, all decisions with diamonds, and all connections between shapes with directional arrows.

Rules for Creating a Proper Flowchart

- The flowchart must contain a START rectangle to designate the beginning of a process.
- All actions must be contained within rectangles.
- All decisions must be contained within diamonds.
- All shapes must be connected by a one-way directional arrow.
- The flowchart must contain an END rectangle to designate the end of a process.

actions

Actions

Actions are any executable steps in an algorithm that do not require a decision to be made. Based on this definition, any defined variables, calculations, and input or output commands would all be contained within action rectangles.

All simple actions are contained within a single rectangle on the flowchart. For each rectangle, two arrows are always associated with the shape, with two exceptions. The inward arrow to the rectangle represents the input to the action. It is assumed that any variables defined in the stepwise scope of an action rectangle are accessible and can be used in the action. The outward arrow from the rectangle represents the output of the action. If any new variables or calculations are performed within the rectangle, those values are passed along to the next shape's stepwise scope.

- Exception One: The START rectangle represents the beginning of the flowchart and does not contain an inward arrow. An oval shape is also commonly used to represent the start of an algorithm.
- Exception Two: The END rectangle represents the end of the flowchart and does not contain an outward output arrow. An oval shape is also commonly used to represent the end of an algorithm.

Decisions

decisions

Decisions are any executable steps in an algorithm that require the answer to a question with "Yes" or "No." All decisions in a flowchart must be represented within a diamond shape.

For each diamond on a flowchart, at least three arrows are always associated with the shape. The inward arrow to the diamond represents the input to the decision. It is assumed that any variables defined in the stepwise scope of a decision diamond are accessible and can be used in the decision. The two outward arrows that exit decision diamonds represent the conditional branch based on the outcome of the question asked within the diamond. If the outcome of the decision is true, the flow of the process will follow the "Yes" branch; otherwise, it will follow the "No" branch. Since no new variables are created in a decision diamond, the stepwise scope that enters the decision diamond is passed on to the next shape of each conditional branch.

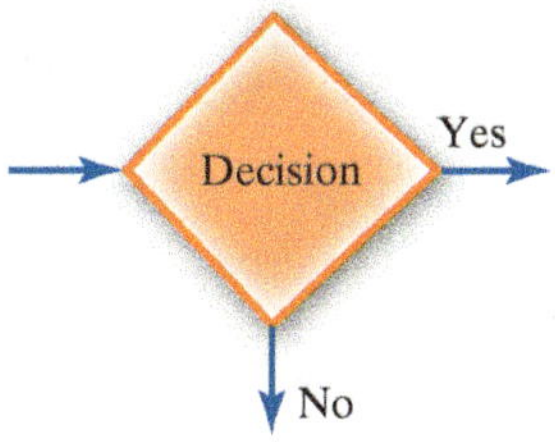

EXAMPLE 8-6

Create a flowchart to express a temperature given in relative units [°F or °C] in the corresponding absolute units [K or °R].

Known:

- *Temperature in relative units (degrees Celsius or degrees Fahrenheit).*

Unknown:

- *Temperature in absolute units (kelvins or degrees Rankine).*

Assumptions:

- *Since the problem does not explicitly state the temperature to be determined, the interpreter of the algorithm will ask for the temperature and units.*

Flowchart:

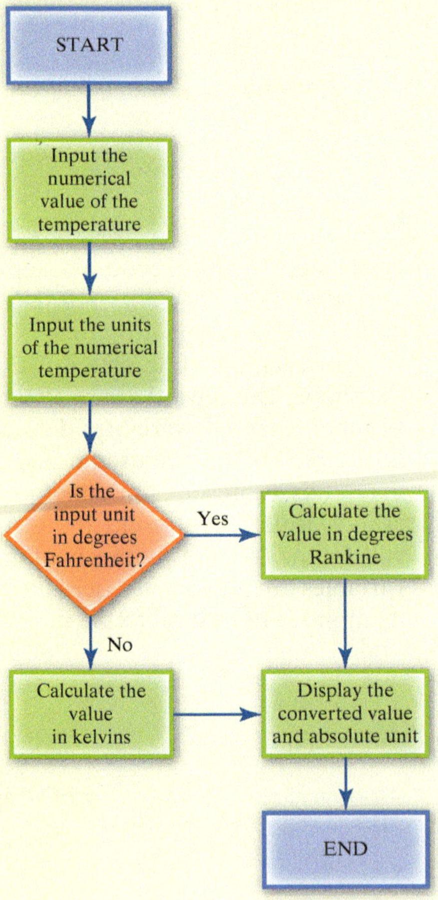

EXAMPLE 8-7

Create a flowchart to calculate the sum of a sequence of whole numbers, given the upper and lower bounds of the sequence.

Known:

- *Upper bound of whole number sequence.*
- *Lower bound of whole number sequence.*

Unknown:

- *Sum of all whole numbers between the upper and lower bound.*

Assumptions:

- *Since the problem does not explicitly state the upper and lower bounds, the interpreter will ask for the values. Include the boundary values in the summation.*

Flowchart:

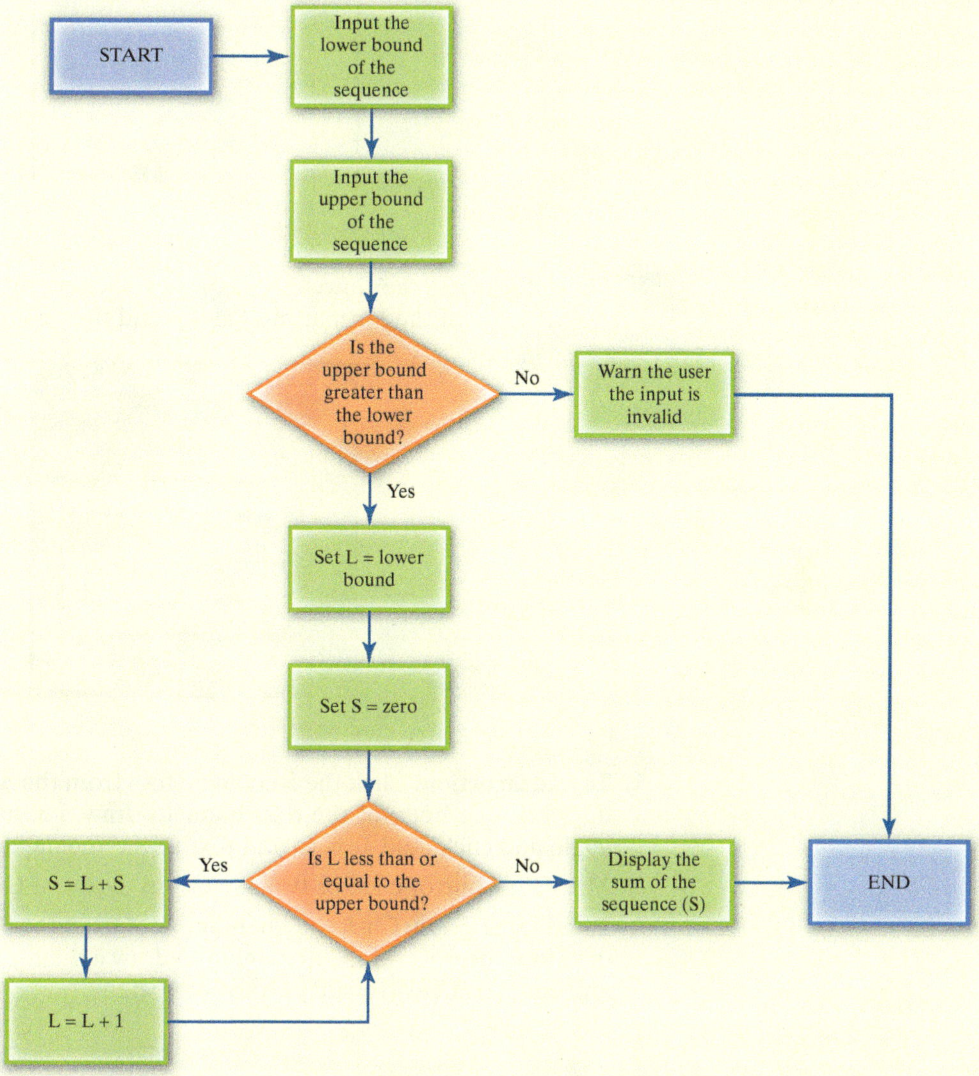

Flowchart Creation in Microsoft Word

Open a new Microsoft Word document. Click the **Insert** ribbon at the top of the Microsoft Word editor window. Click the **Illustrations > Shapes** drop-down menu. You will need to use the rectangle and diamond shapes under the Flowchart section as well as the directional arrows under the Lines section.

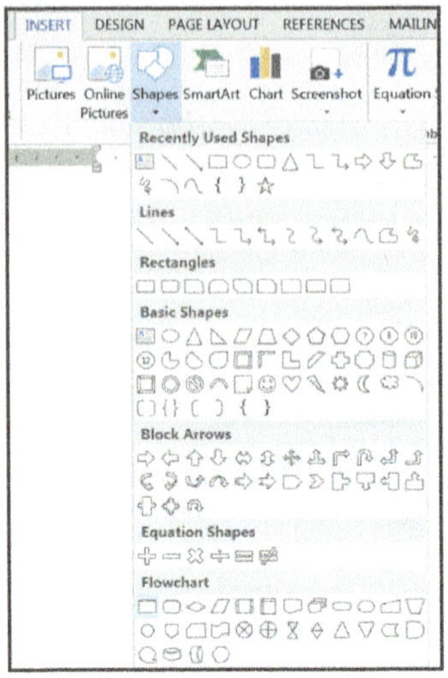

Figure 8-1

⌘ **Mac OS:** Access the flowchart shapes by clicking the Object Palette button near the top of the Formatting Palette and then clicking the Shapes button.

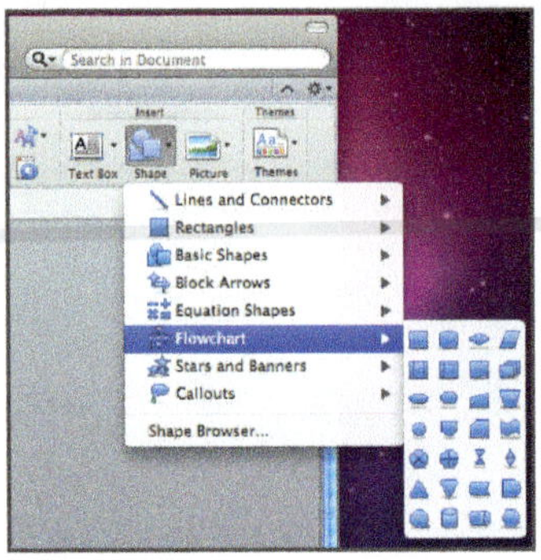

Figure 8-2

↳ **To add an action:** Click the Rectangle tool from the Shapes menu and click-and-drag into the body of the document to draw a rectangle. Right-click the rectangle and click Add Text to add text to the rectangle.

⌘ **Mac OS:** Control-click or two-finger tap to access the **Add Text** menu item.

↳ **To add a decision:** Click the Diamond tool from the Shapes menu and click-and-drag into the body of the document to draw a diamond. Right-click the diamond and click Add Text to add text to the rectangle.

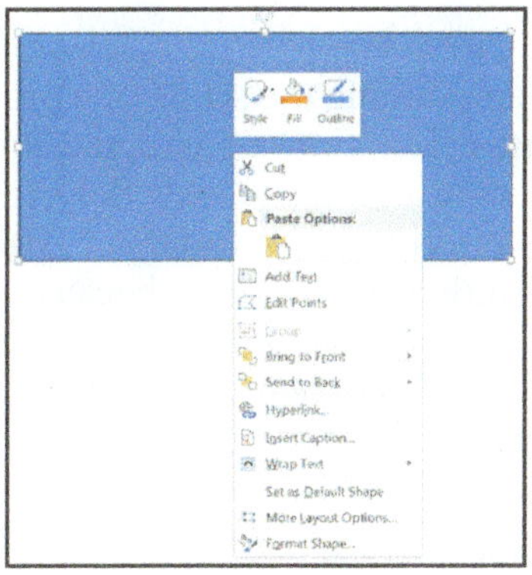

Figure 8-3

⇘ **To insert the YES and NO labels:** Use the Text Box option under the Basic Shapes menu. To remove the border, right-click the text box and click Format Shape. On the Format Shape sidebar, choose **Shape Options > Paint Bucket > Line**, click "No line". Click the "X" to close the sidebar.

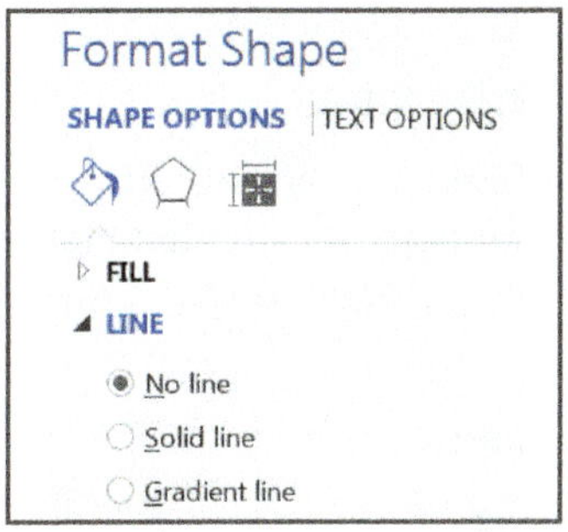

Figure 8-4

⌘ **Mac OS:** In the main menu, click **Insert > Text Box.** The default is probably "no border," but the border can be modified in the formatting palette.

⇘ **To add an arrow:** Click the single direction arrow from the Shapes menu and starting from the source click-and-drag to the destination.

8.4 ALGORITHM BEST PRACTICES

If you have never composed a written or graphical algorithm before, the remaining part of this section details specifics on how to begin planning and writing algorithms from scratch. This section does not intend to be a definitive resource on algorithm development, but it may provide guidance if you are struggling to break down a process into small, achievable steps.

Actions

In every action within an algorithm, there must be a key verb that defines the purpose of that step within an algorithm. The remaining subsections discuss different types of actions and list some of the common verbs associated with that category of action.

Establishing Variables and Constants

After defining the scope of a problem, it might become obvious that there are intermediate calculations or assumed constants that must be contained throughout the process. Along with the explicitly defined known values, these intermediate and constant values are referred to as variables. Algorithmic variables are different from the mathematic definition of a variable because algorithmic variables are treated more like containers to store known values and results of calculations rather than being some unknown entity in a mathematical expression. They are called variables because the stored value can be written, overwritten, and used by other actions or decisions in the algorithm.

Example			**Action**		
We assume the acceleration due to gravity is 9.8 meters per second squared.			**Set** variable *g* to be 9.8.		
Other Verbs					
Set	Define	Assign	Write	Store	Designate
Label	Name	Cast	Insert	Save	Initialize

User Interaction

It is often necessary to write algorithms that can be executed with prompts for input from the person using the algorithm, provide feedback on results, or display any error messages generated in the algorithm.

User Input:

Example			**Action**		
We want the user of the algorithm to provide the amount of water in gallons.			**Input** the amount of water in gallons, save in variable *W*.		
Other Verbs					
Input	Ask	Load	Request	Query	Prompt

User Output:

Example		**Action**		
We want the algorithm to inform the user that the amount of water can't be negative.		**Display** error message to user "Warning: amount of water can't be negative!"		
Other Verbs				
Output	Display	Reveal	Write	Warn

Calculations and Conversions

When algorithms involve calculating a value using an equation, it is helpful to write out the full equation and identify which variables in the algorithm correspond to the variables in the expression. For unit conversions, it is not necessary to write out the conversion factors since those are published standards that are readily available to anyone executing your algorithm. When using conversions, it is best to list them individually so they are easily recognizable to the user. For example, when converting from feet to centimeters, the expression $L = L/3.28 * 100$ is easily recognized as the conversion from feet to meters, and then from meters to centimeters. It is harder to recognize the conversion of $L = L * 30.48$. Furthermore, it is easy to make a calculation error; it is easier to allow the program to calculate for you. When dealing with unit conversions, it is ideal to save the converted value back into the original variable to reduce the number of variables you need to keep track of in your algorithm.

Calculations:

Example	Action
We want to calculate the thermal energy of a substance using the expression $Q = m\ C_P\ \Delta T$, where m is the mass, C_P is the specific heat, and ΔT is the change in temperature.	**Compute** the thermal energy: $Q = m\ C_P\ \Delta T$ All variables should appear in the variable list.

Other Verbs					
Calculate	Adjust	Count	Measure	Add	Multiply
Subtract	Divide	Compute	Increment	Decrement	

Conversions:

Example	Action
We want to convert a variable *t* from minutes to seconds and save the result back in the variable *t*.	**Convert** *t* from minutes to seconds, save in *t*.

Other Verbs				
Convert	Change	Alter	Revise	Switch

Referencing Other Algorithms

functions

When developing a large program, it is sometimes helpful to break that program into several smaller programs, and then reference the smaller programs within the large program; these are called **functions**. As a rule of thumb, each custom function you create should have its own separate algorithm. If you have separate algorithms for a program and the different functions referenced in the code, it makes the algorithms simpler to understand and easier to debug. When calling a function within an algorithm, it is critical to list the variables passed to the function and variables captured by the function. In general, the most common verb used with functions is "call." If you know the name you plan to use for your function, list it; otherwise, this can be set later.

Example	Action
We want to use a function named *Poltocar* that converts coordinates from polar to Cartesian. We will pass in the variable *Z* as the radius and the variable *T* as the angle. We will capture the *x*-coordinate in the variable *X* and the *y*-coordinate in the variable *Y*.	**Call** *Poltocar* In: *Z*, *T* Out: *X*, *Y*

Decisions

All decisions made in algorithms must be constructed as binary decisions. Typical decisions in algorithms involve comparing variables, examining the contents of a variable, or examining the dimensionality of a variable. Any decisions that require some amount of computation in the decision should be split so that the calculations occur in actions before reaching the decision block.

Error Checking

error checking

Including **error checking** in an algorithm allows for the creation of robust solutions to problems that will not lead to incorrect or unstable answers. In general, a check for an error will either terminate the algorithm or lead to some action that will allow the algorithm to continue; the program should notify the user that an error has occurred (see Figure 8-5). If you want your algorithm to re-prompt the user for input to assure that proper data are contained in a variable before proceeding into the remainder of the algorithm, see the section on "Error Prevention." The remainder of this section discusses three different types of errors that may occur in an algorithm, but this is only a starting point. The amount and type of errors that can occur in an algorithm are infinite, so it is up to the designer of the algorithm to decide how and when error checking should occur.

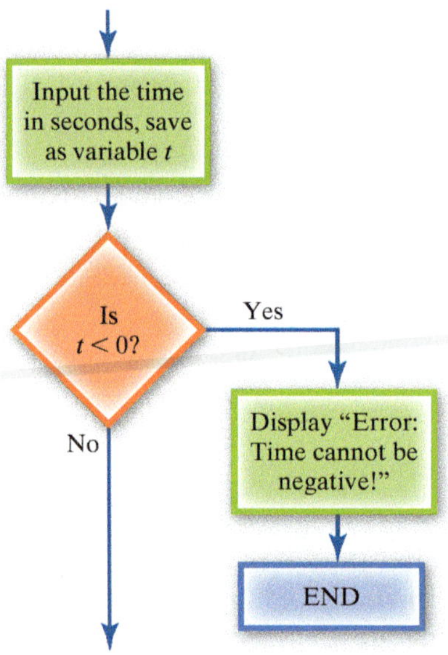

Figure 8-5 An example of error checking embedded in a flowchart.

Division by Zero and Infinite Values

INF

If your algorithm contains a calculation where a combination of one or more variables in the computed expression could lead to a division by zero, it is smart to include a check to see if the result is zero. Some languages like MATLAB will happily compute an expression with a zero divisor and return the result as "**INF**"—a special MATLAB reserved word representing infinity.

Invalid Dimensions of Variables

When an algorithm assumes that one or more variables contain matrices or vectors, any calculations on those variables must follow the same mathematical rules associated with the matrix operation. For example, if an algorithm requires two matrices to be added together, it would be wise to include a check to see if the two matrices have the same number of rows and columns before attempting to add them together. This will prevent algorithms from crashing due to an invalid computation. In addition, this will prevent issues related to accessing elements of a matrix that do not exist.

Invalid Range of Values

Since variables typically represent some measured or computed value, restrictions on those variables that apply in real life may not be directly enforced in your algorithm. For example, if your algorithm prompts a user to input a quantity that cannot be negative (length, volume, time, etc.), it is smart to check if the value in the variable is reasonable. Likewise, if your algorithm should not generate complex values (e.g., $3 + 2i$), your algorithm will need to check to see if the result of a computation would generate a complex value instead of the desired real value.

Error Prevention

Error prevention looks for the same type of errors that are detected in error checking, but error prevention will allow your algorithm to prompt or re-prompt the user to correct the erroneous variables. For example, if your algorithm asks the user to type in a time and the user erroneously types a negative value, your algorithm could detect the incorrect value and go back to the input statement to force the user to type the value again (and again, and again . . .) until the user types a value within the acceptable range (see Figure 8-6). In most programming languages, the structure that enables error prevention is the `while` loop.

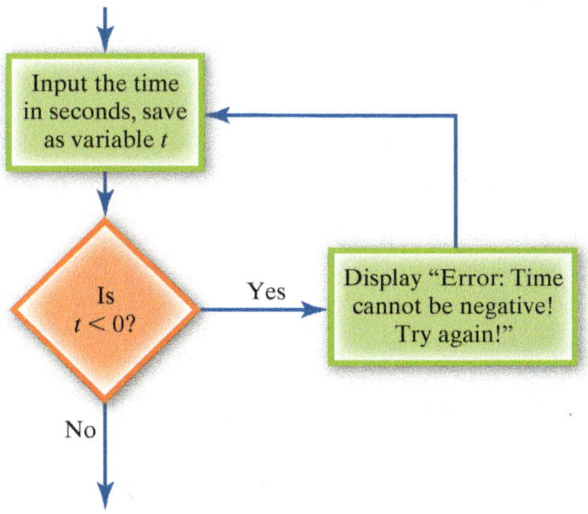

Figure 8-6 An example of error prevention embedded in a flowchart.

Iteration

Some algorithms require repeated calculations that typically involve the use of a sequence of values or some operation on a vector or matrix stored in a variable. Such

counter variable
index variable

algorithms are considered to be iterative because they require a counter variable to keep track of when to terminate. For example, assume we have a vector, **V**, which contains positive and negative values in random order. If we want to create two new vectors, **VN** and **VP**, that contain the negative and positive values of **V** respectively, we will need to iterate through each element of the vector **V**, make a decision about each value, and store it in the corresponding vector. To do this, we would need to create a **counter variable**, sometimes called an **index variable**, that will keep track of the number of times we have repeated a calculation or decision. If we create a counter variable, X, and initialize it to be the number 1, X will actually serve two purposes. In addition to keeping track of the number of times we have repeatedly made decisions and stored new values into **VN** and **VP**, it will also serve as the index variable into the **V** vector so that we can access element V(1), V(2), and so on until we reach the last element in **V**. Figure 8-7 demonstrates this scenario as a flowchart, including the iterated counter variable X.

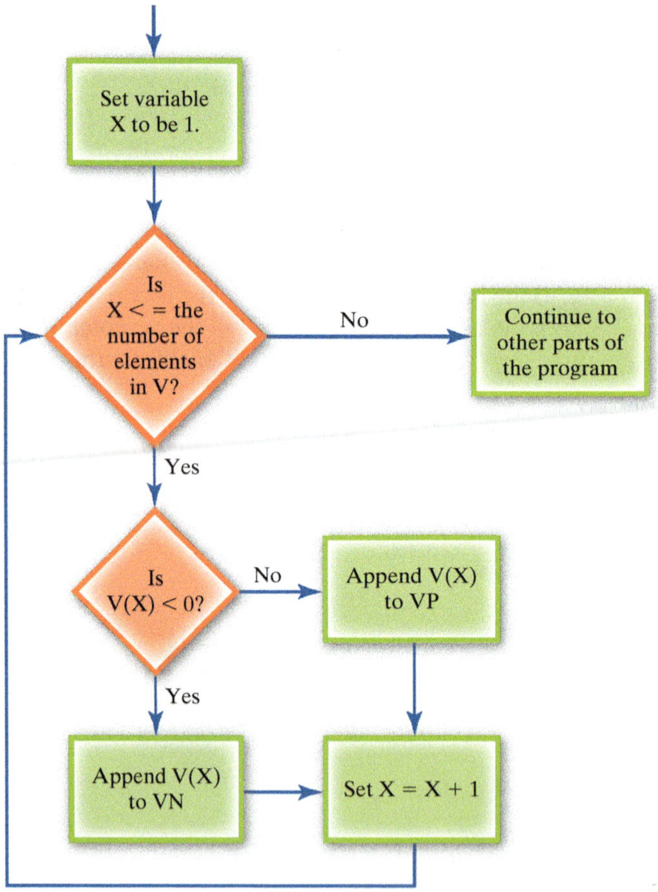

Figure 8-7 An example of iteration embedded in a flowchart.

Algorithms that will require iteration typically have a scenario where you have to repeat some decision or calculation "for each" or "for every" element or value within a sequence or vector. Since there is no "for each" or "for every" building block within an algorithm, this type of structure must be constructed out of the following steps:

- Initialization of a counter variable (e.g., Set X to be 1)
- A decision that involves the value of a counter variable (e.g., if X is less than or equal to the number of elements in $\mathbf{V}$)
- Some action block that increments the counter variable (e.g., Set X equal to the current value of X plus 1)

After the counter increment action block, the algorithm can loop or refer back to the decision made on the increment variable. In most programming languages, the structure that enables iteration in an algorithm is the `for` loop.

Testing Your Algorithm

The last step in writing an algorithm is developing test cases that will reveal whether or not your algorithm behaves as expected. The key to writing test cases is figuring out how many test cases are necessary to confirm whether or not your code works. In general, there should be at least one or two test cases that will demonstrate the proper behavior of the algorithm given good input values. Not only should a test case include a list of all of the inputs used to generate the output, but you should also compute the expected output of the algorithm by hand in order to verify that the algorithm works. In addition, there should be test cases that verify all of the error checking/prevention built in to ensure that the algorithm works properly. Any decisions that lead to different states within your algorithm should have a test case to verify that the logic you designed is arranged properly in order to generate the desired output.

For example, assume we have designed an algorithm that will calculate power given energy and time. In our algorithm, we included two error checks—the first check to see if energy is greater than 50 joules and the second check to see if time is greater than 0 seconds. If either of these conditions is not true, the algorithm will display an error message "Error: Incorrect input value" and terminate. In addition, we added logic to check to see if the energy in the system is greater than 500 joules, and if that is true, added 5 seconds to the time variable; otherwise, if the energy is less than or equal to 500 joules, we left the time variable alone.

A test case for this scenario would look like this:

Input	Output
$E = 0$ J, $T = 30$ seconds	Error: Incorrect input value
$E = 55$ J, $T = -3$ seconds	Error: Incorrect input value
$E = 400$ J, $T = 5$ seconds	$P = 80$ W
$E = 550$ J, $T = 10$ seconds	$P = 36.7$ W

To help you develop good algorithms, you should consider using an algorithm to help you document all of the variables, procedures, and test cases necessary to create correct and verifiable algorithms, such as the one shown below. Listing known/input, unknown/output, assumptions and test cases will help you clearly see the program develop and, if anything goes wrong, will help you efficently troubleshoot.

EXAMPLE 8-8 Create an algorithm to determine the volume of a cylinder, given the radius and height.

Known / Input	Unknown / Output	Assumptions
Radius (r) [cm] Height (H) [cm]	Volume (V) [cm^3] Surface area (SA) [cm^2]	Object is a cylinder $V = \pi r^2 H$ $SA = 2 \pi r H + 2 \pi r^2$

Algorithm: Written

1. Enter values of r [cm] and H [cm]
2. Calculate volume of a cylinder [cm^3] using $V = \pi r^2 H$
3. Calculate the surface area of a cylinder [cm^2] using $SA = 2 \pi r H + 2 \pi r^2$

You may choose a method to show your algorithm – either written or flowchart – you do NOT need to show both methods. Both are shown here for illustration only.

Algorithm: Flowchart

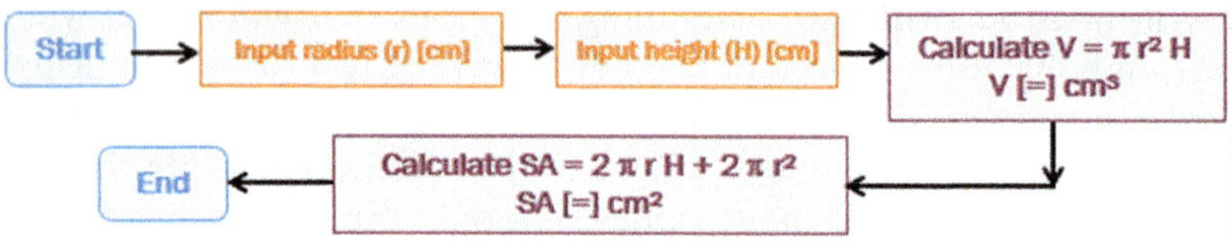

Test Cases

Input:	Output:
r = 3 cm	$V = 113.1\ cm^3$
H = 4 cm	$SA = 131.95\ cm^2$

8.5 VARIABLE BASICS

While there are many different programming languages, a favorite among engineers for computation is MATLAB. Compared to other languages, MATLAB is considerably streamlined. This allows the user to quickly perform the needed calculations and move on or create a substantial mathematical model. Due to these qualities, we will discuss programming primarily in the context of MATLAB. However, a majority of the concepts will transfer to other languages.

A variable in MATLAB is not like a variable in mathematics. In math, a variable is typically an unknown for which you wish to determine a value. In MATLAB, on the other hand, a variable is simply a label for a container (or location) in which information is stored. Variables can be classified according to the type of information they hold. In the workspace, the data type of each variable is represented by an icon shown in the leftmost column of the workspace by the variable name. Table 8-1 below illustrates the different variable classifications in MATLAB.

array

NOTE

An **array** is a list or group of similar variables. If a variable can hold one value of a number, it is considered a scalar.

Table 8-1 Variable Classifications in MATLAB

ICON	Data Type	Description
	Numeric Array	Scalars, vectors, and matrices containing numbers
	Character Array	Text strings. Each character occupies one element.
	Logical Array	0 means false; 1 means true
	Cell Array	Contains mixed data types. Indexed numerically.
	Structure Array	Contains mixed data types. Data organized into named fields.
	Sparse Array	Saves memory by storing only non-zero elements
	Symbolic Array	Allows symbolic mathematics within MATLAB Not covered in this text

Every variable in MATLAB must have a name. This allows us to keep track of large amounts of information and gives a means whereby our code can use the stored information relatively easily. MATLAB names must follow certain rules to avoid unintended side-effects.

Naming Rules

- All names must consist only of letters, numbers, and the underscore character (displayed by holding "Shift" and typing a hyphen).
- Names can *only* begin with alphabetic letters (no numbers or special characters).
- Names cannot be longer than 63 characters.
- Names in MATLAB are case sensitive: `Frog` and `frog` are different.
- Names cannot have the same name as any other identifier (program name, function name, built-in function, built-in constant, reserved word, etc.), because MATLAB's order of execution will prevent the program from finding the correct item.

These rules apply to ANY named item in MATLAB, not just variables.

Always choose names that have meaning within the context of the program, not just random letters. We cannot overemphasize the importance of well-chosen variable names in creating self-documenting code: code that explains itself. For example, if a variable contained the diameter of a bolt, the name `BoltDiam` is immediately recognizable for what it contains; a randomly chosen variable such as `X1` or `Q` has no meaning in the context of the program.

The Assignment Operator

assignment operator

Before beginning our discussion of data types in MATLAB, we need to introduce the concept of the **assignment operator**. The purpose of an assignment operator is to tell the computer to put a value into a variable. Perhaps unfortunately, the most commonly used assignment operator in modern programming languages is =, or the "equal" sign. We say unfortunately, because this notation often confuses novices. In mathematics, a statement such as `A=A+1` is an equation and has exactly two solutions for A: infinity and negative infinity. However, in programming this means "add one to whatever is in variable A, then place the result back in A.

It is important to note in MATLAB the = symbol should be read as "is assigned the value of". Thus, the programming statement `V2=V2+4` is read "V2 is assigned the value of (the original value of) V2 plus four". Note that the original value of V2 is lost in this process.

It is also VERY important to note that the transfer of data in an assignment statement is ALWAYS from right to left.

Consider the programming statement `Ex1=V3−V4^P1`.

When the computer executes this statement, it first recognizes that it is an assignment statement. Therefore it first looks at the expression to the right of the equals and determines its value: Raise the value in `V4` to the value in `P1` and subtract that from the value in `V3`. Next, it looks to the left of the equal sign to find the destination, thus placing the calculated result into the variable `Ex1`.

It is also important to understand that whatever appears to the left of the assignment operator MUST be a variable, or as we will see later, a group of variables. For example, you will get an error if you type `A+1=A` since you cannot have a computation on the left of the equal sign. Also, any variables that appear on the right must already be defined before the statement is executed.

8.6 PROGRAMS

programming

Programming is the process of expressing an algorithm in a language that a computer can interpret. To correctly automate a process on a computer, a programmer must be able to correctly speak the language both grammatically and semantically. In this text, we communicate with a computer by using the MATLAB programming language. In the previous sections, the idea of creating algorithms to represent a process and creating variables in a programming language like MATLAB has been established, so this section intends to combine the two ideas to create programs and functions.

Certain principles apply to programming, regardless of what language is being used. Entire textbooks have been written on this subject, but a few important concepts are addressed here.

Notes About Programming in General

- **Programming style:** Just like technical reports should follow a particular format to ensure they can be understood by someone reading them, all programs should have a few common elements of programming style, particularly the inclusion of comments that help identify the source of the program and what it does. For any program, a proper header should be written describing the scope of the problem, including a problem statement and definition of all input and output variables used in the program.

Properly commenting the source code is also critical for ensuring that someone else can follow your work on the program.

As a rule of thumb, for each rectangle and diamond in a flowchart, there should be a few comments in the source code where that action occurs.

- **Program testing:** When you write a program, testing it is a critical step in making sure that it does what you expect. Lots of things can go wrong.
  - *Syntax errors*: These violate the spelling and grammar rules of the programming language. Compilers (programs that interpret your computer program) generally identify their location and nature.
  - *Runtime errors*: These occur when an inappropriate expression is evaluated during program execution. Such errors may occur only under certain circumstances. Examples of runtime errors are division by zero, the logarithm of zero, or the wrong kind of input, such as a user's entering letters when numbers are expected. Computer programs should anticipate runtime errors and alert users to the error (rather than having the program unexpectedly terminate).
  - *Formula coding errors*: It is not uncommon to enter a formula incorrectly in a form that does not create either a syntax or a runtime error. Evaluating even a few simple calculations by hand and comparing them to the values calculated by the program will usually reveal this kind of error.
  - *Formula derivation errors*: This difficult error to discover in programming occurs when the computer is programmed correctly but the programming is based on a faulty conceptualization. This kind of error is difficult to figure out because programmers generally assume that a program that executes without generating any errors is correctly written. This situation underscores the importance of checking a solution strategy before starting to write a computer program.
- **Keeping track of units:** Some computer programs can be designed to have the user keep track of units as long as they use consistent units. Other computer programs require input data to be in a particular set of units. Regardless of which approach you use, the way in which units are managed must be clear to the user of the program and in the program's comments.

documentation

- **Documentation:** In any programming language, there is usually a special character or pair of characters that tell the program interpreter to ignore everything to the right of the special character to allow the programmer to write human readable notes within the code of the program. These notes are commonly referred to as "comments" within the code. A great deal of controversy surrounds how much documentation is necessary in any program, but this text proposes and follows the following guideline to properly document a program:

  For every numbered item in a written algorithm, you should include that item as a comment in your code.

In addition, at the top of each program, it is extremely helpful to at least write a problem statement and document any variables used within the program. It is helpful to cluster your description of variables into three groups: input variables, output variables, and function variables.

Notes Particular to MATLAB

Since variables in MATLAB are inherently matrices, it is critical that you understand not only some basic matrix concepts, but also the specific notation used in this text (as well as MATLAB) when referring to matrices. Certain details specific to MATLAB are important to note early on.

MATLAB

- **The MATLAB interface:** The MATLAB program window, shown below, has a number of subwindows. The two main windows we are concerned with for now are the Command Window and the Editor Window.

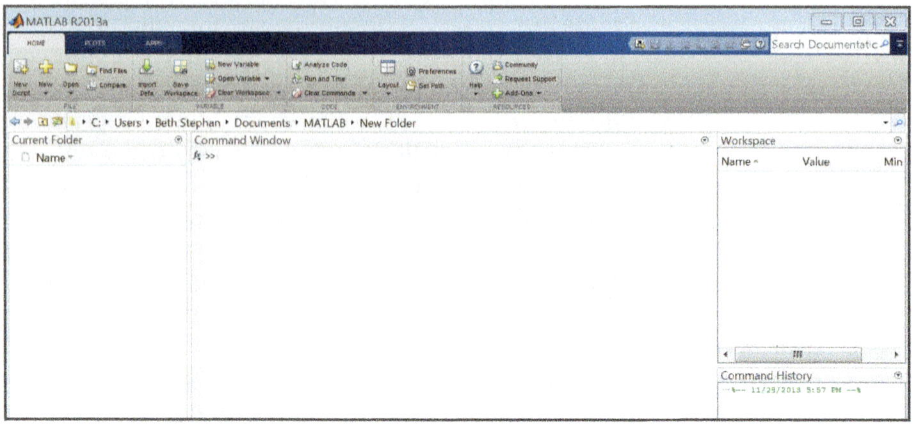

Figure 8-8

- The *Command Window* can be used as a powerful but awkward calculator. The  button at the top-right side of the Command Window menu bar will display a menu for common layout commands, including the ability to close or "undock" the Command Window from the MATLAB program window.

Figure 8-9

To restore the appearance of MATLAB to the default layout, select the **Layout** menu from the **Home** ribbon and click **Default**.

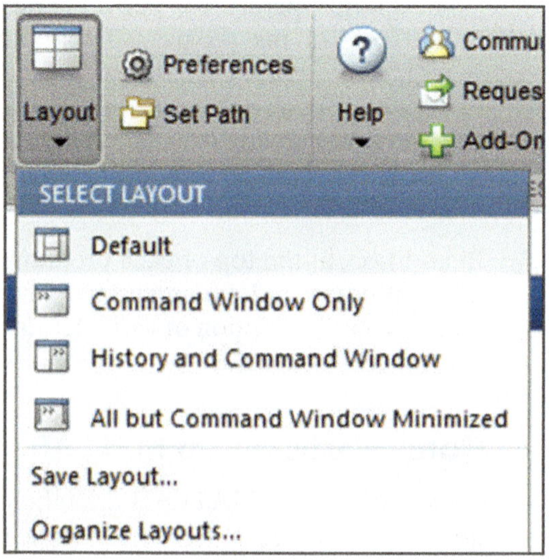

Figure 8-10

To clear any text from the Command Window, enter the special reserved MATLAB expression `clc` in the Command Window.

- Although you can enter MATLAB statements in the Command Window, the *Editor/Debugger window* is the interface of choice for developing MATLAB programs. The editor stores your commands in M-files. This permanent record of your commands allows you to make small changes to either the program or the inputs and to re-execute a complicated set of commands. The editor also assists in the formatting of MATLAB programs, using both indentation and color to distinguish program elements. To launch the editor window, press the **New Script** button on the **Home** ribbon or use CTRL+N. The Editor interface will launch in a new window.

- **Naming programs, functions, and variables in MATLAB:** In MATLAB, program names, function names, and variable names must all follow certain rules to prevent unintended side effects: Refer back to MATLAB naming rules.

order of Search

- **Order of Search:** Order of search is the sequence MATLAB goes through when you type a name in either the Command Window or in a program or function. If you type `Frog` in the Command Window:

1. MATLAB checks if `Frog` is a variable, and displays its value if it is.
2. MATLAB checks if `Frog` is one of MATLAB's built-in functions, and executes that function if it is.
3. MATLAB checks if `Frog` is a program or function in its current directory, and executes the program or function if it is.
4. MATLAB checks if `Frog` is a program anywhere in its path, and executes the program or function if it is. The location where you store your MATLAB programs or functions must be included in MATLAB's path. You can check where MATLAB is looking for programs and functions by typing `path` in the Command Window. Any M-files in the directories listed will be visible to MATLAB. If you try to execute an M-file outside any of the directories listed, MATLAB will ask if you want to switch current directories or add that directory to the list of paths MATLAB can see.

- **what and who:** To display a list of all MATLAB programs in the current directory, type `what` into the Command window. To display all variables currently in MATLAB's workspace, type `who` in the Command Window. Remember that all variables automatically clear after you exit MATLAB.
- **Comments:** For documenting or commenting any code in MATLAB, a special character is used to "comment" out text: the percent symbol (%). The comment character can appear anywhere in a MATLAB program or function, but preferably either at the beginning of a line or at the middle of a line of MATLAB code. Everything after the percent sign to the end of the line is ignored by MATLAB, so you as a programmer can "comment out" any broken lines of code during code testing in addition to providing comments. Program or function file headers will usually have 6–10 lines of comments at the top of every program and more throughout the rest of the program or function.

```
% This is an example of a comment line.
% Comment lines are shown in the Editor Window in green.
```

block comment

A **block comment** is a block of non-executed text between the special block characters %{ and %} that allow the user to type full length paragraphs as a comment without needing to type the % symbol before each sentence. For example:

```
%{
This is an example of using
a block of comments in a code
%}
```

Program Structure and Use

Programs
M-Files

Programs are a set of instructions given to a computer to perform a specific task. In this section, we create MATLAB program files, or **M-Files**, to automate our processes. To create a program in MATLAB, launch the MATLAB editor. By default, the editor will create a new, blank document named Untitled, or Untitled2, or Untitled3, and so on, depending on how many unsaved programs are open in the editor at once. The following example provides a walkthrough for creating a basic program, as well as screenshots of the actions described in the text to help you to follow along while recreating these steps on your own computer running MATLAB.

EXAMPLE 8-9

We want to create a MATLAB program to convert Cartesian coordinates (x, y) to polar coordinates (r, θ). Before we can even attempt to write the program, we need to recall some information about the coordinate systems.

We know the following:

- In the Cartesian coordinate system, the coordinates (x, y) represent how "far away" a particular data point is from the origin (0, 0) by discussing the horizontal and vertical distance with respect to the origin. The x-value represents the horizontal distance and the y-value represents the vertical distance.

- In the polar coordinate system, the coordinates (*r*, *u*) represent the exact distance from the point to the origin (*r*) as well as the angle of elevation (*u*) with respect to the origin (0, 0) in units of radians.
- The conversion from polar coordinates to Cartesian coordinates is

$$x = r\cos\theta$$
$$y = r\sin\theta$$

where θ the angle of elevation in radians.

- The conversion from Cartesian coordinates to polar coordinates is

$$r = \sqrt{x^2 + y^2} \qquad \theta = \tan^{-1}\left(\frac{y}{x}\right)$$

We have the equations necessary to write a program to convert Cartesian coordinates to polar coordinates, and vice versa.

First, we need to ask MATLAB for more information on doing things like calculating a square root and calculating an inverse tangent of a value.

In MATLAB's Command Window, we search for the built-in functions to calculate a square root and an inverse tangent. The `lookfor` *function allows you to search MATLAB help documentation for certain key words. Note that if you type* `lookfor square root` *into the Command Window, MATLAB displays an error message. The* `lookfor` *command is expecting only one keyword following the command "lookfor", so we decide to try our search again, this time including only the word "square."*

Because MATLAB is searching through every built-in function, we need a way to tell MATLAB to stop, because when the function `sqrt` *appears, it seems to be the exact function we need. To terminate a MATLAB command early, we press* **`CTRL + C`** *(i.e., the CTRL key and the "C" key at the same time) on our keyboard, and the prompt will return to the Command Window.* `CTRL + C` *can also be used to stop a program caught in an infinite loop or that otherwise wandered away and got lost.*

To ask for more information on the function `sqrt`*, we type* `help sqrt` *to double-check that the function is going to perform our intended calculation as well as details of how to use it.*

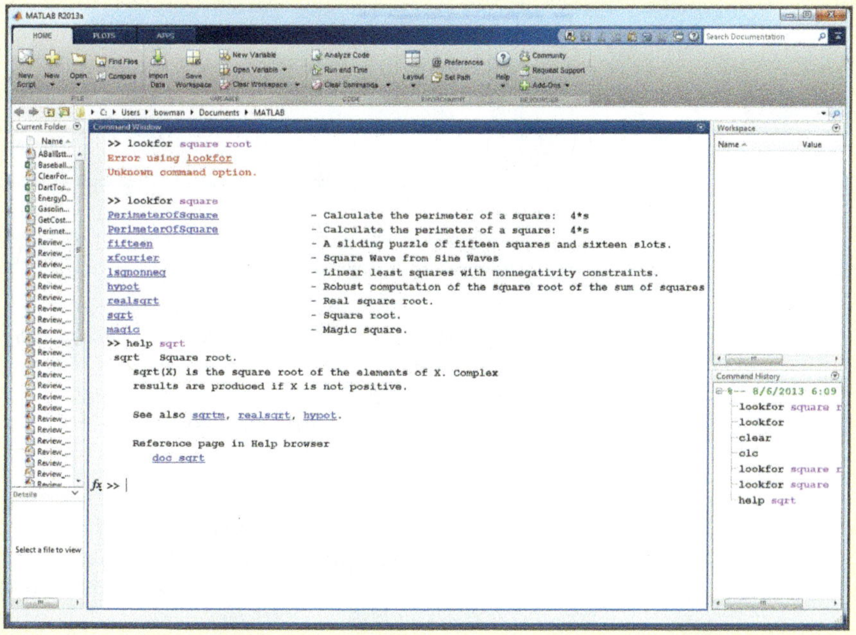

Repeating this process to find the inverse tangent function, we determine that `atan` *is the correct function necessary to calculate the inverse tangent. We note, from the help documentation, that the* `atan` *function will return the inverse tangent in radians.*

To create a program, we click the ***New Script*** *button on MATLAB's Home tab. After creating a new blank M-file, we need to write out our problem statement as code comments. Comments are any non-interpreted text that you would like to include in your code file. All comments start with the percent symbol (%) and everything to the right of the symbol will be ignored on a per-line basis. Comments can follow any executable statements in our code, but for now, we will add the first couple of lines in our MATLAB program by writing out a short problem statement with the % symbol before any text we write.*

Next, we start writing our code. In a MATLAB program, it is good practice to start the program by clearing all of the variables from the workspace (`clear`*) and clearing the Command Window (*`clc`*). That way, we guarantee that if we run our program on a different computer or run it after restarting MATLAB, our program will create all the variables we need to solve the problem and not rely on the workspace having variables predefined before code execution.*

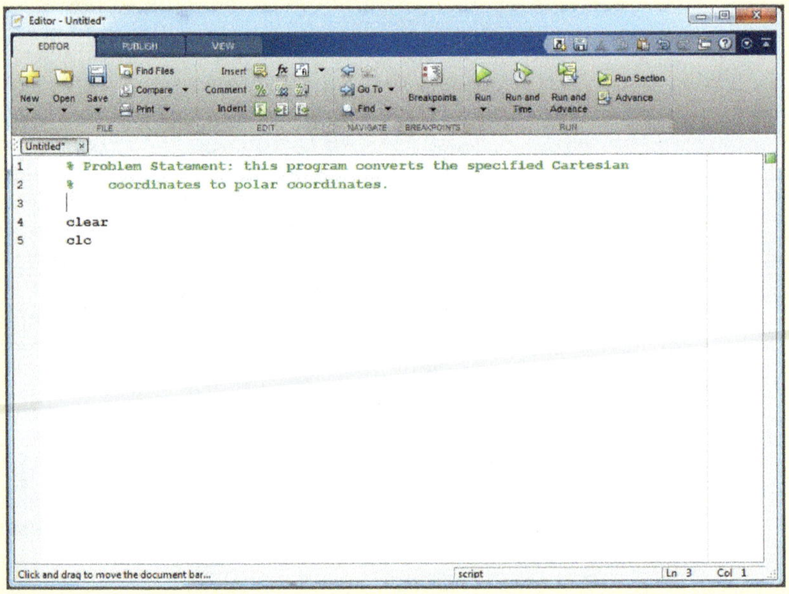

Now we are ready to start our program. Let us look at the step-by-step written algorithm necessary to solve this problem:

1. *Create a variable to contain the x-coordinate as a variable named "x."*
2. *Create a variable to contain the y-coordinate as a variable named "y."*
3. *Calculate the radius by taking the square root of the sum of the squares of x and y and store the result in a new variable called "r."*
4. *Calculate the angle by taking the inverse tangent of the result of dividing y by x and store the result in a new variable called "Theta."*

In the written algorithm, the first thing we have to do is to manually assign the values of x and y. For this example, we assign x to contain the value of 5 and y to contain the value of 10, but it is also possible to allow the user to enter values. Note that in addition to creating the variables x and y, we have also added additional comments in the header of our program documenting what these variables represent.

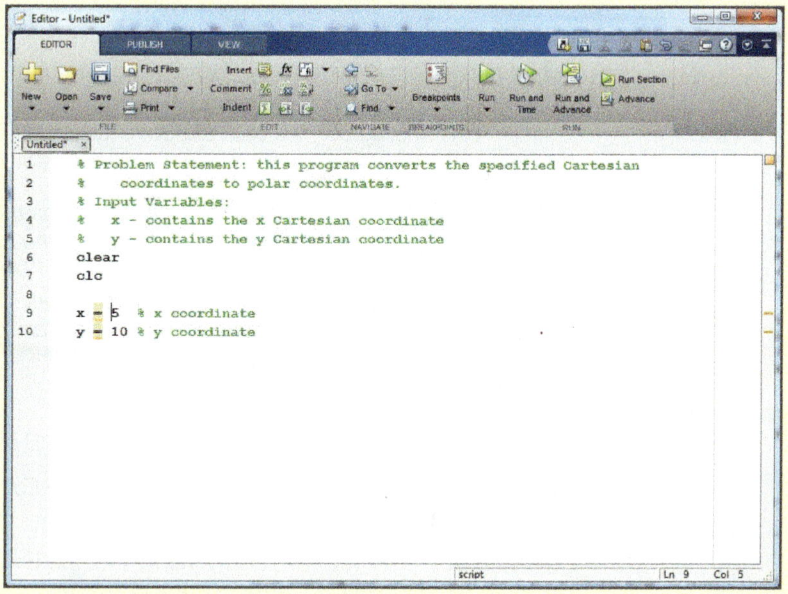

Note that MATLAB highlights the = symbol in the editor. When we move the cursor over the assignment operator (=), a pop-up message tells us that we can suppress the output of the assignment expression by using a semicolon. For now, we omit the semicolon and observe MATLAB's behavior when we leave the code unsuppressed.

Next, we type the equations to calculate the radius and the angle, using the built-in MATLAB functions we discovered with `lookup` *and* `help`. *In addition, we've added extra comments about the new variables representing the radius and angle of the polar coordinate.*

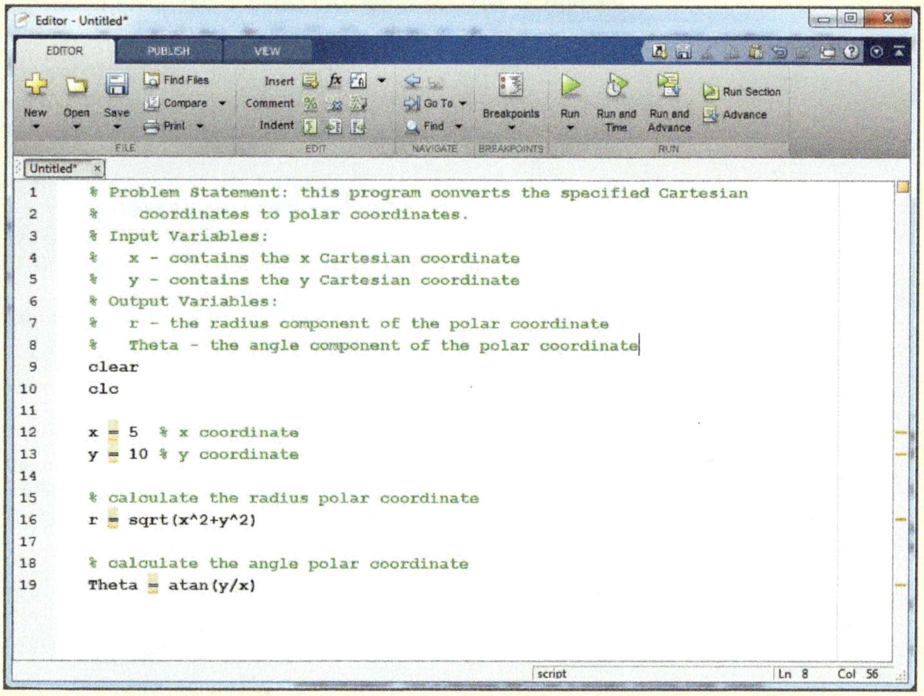

Next, we save the program in MATLAB's path with a file name that follows the program-naming convention. We choose the name "CartesianToPolar" since that name nicely describes the purpose of the program.

To save a program, we click the ***Save*** *icon in the Editor window.*

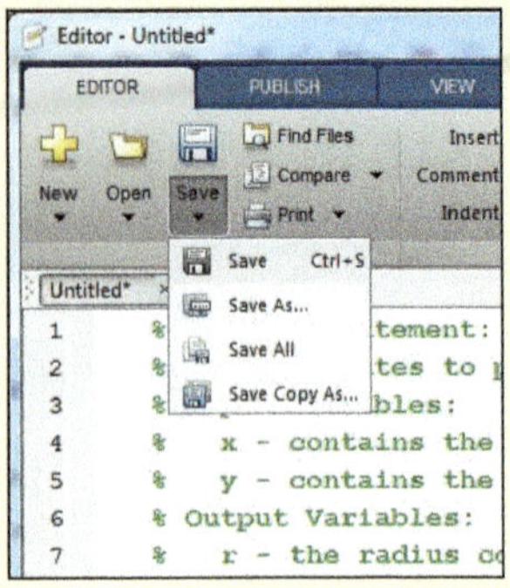

⌘ ***Mac OS:*** *MATLAB is one of the few programs that does not modify the main menu at the top of the screen. Instead, its main menu is at the top of the MATLAB window, just as in the Windows implementation. This is because MATLAB is actually running under an X Windowing System, similar to Linux or UNIX®.*

Running the Program

There are a number of different ways to run a MATLAB program located within MATLAB's path.

Method 1: *In the main MATLAB window (not the Editor), type the name of the program in the Command Window and press* ***Enter.***

Method 2: *In the main MATLAB window (not the Editor), right-click the program you want to run while in the Current Directory box and select* ***Run.***

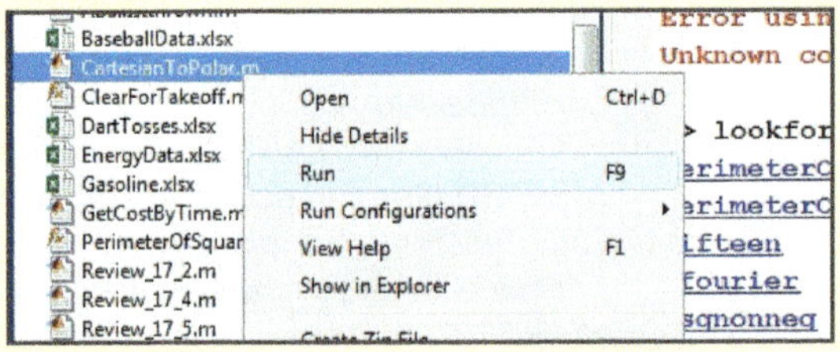

Method 3: *With the program loaded in the Editor window, click the* ***Save and Run*** *button in the toolbar.*

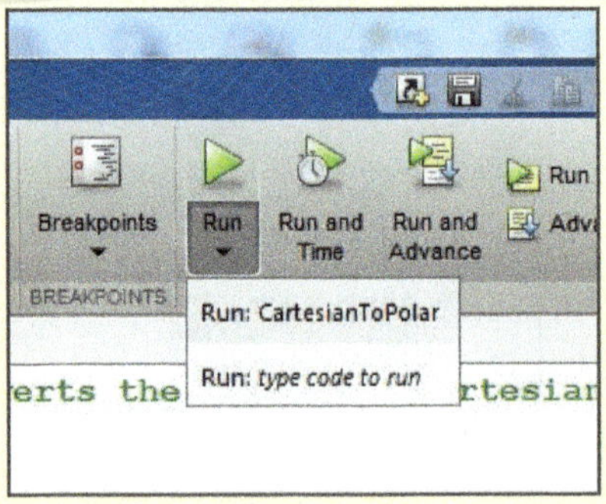

We see that MATLAB spits out the results of every assignment and calculation performed in the program. Let us assume we want to show only the output from the calculations of r and Theta, and not the output of the assignment of x and y. To do this, we add a semicolon (;) at the end of each line of code we do not want showing up in the output of the program.

NOTE

To suppress output, end the code line with a semicolon (;).

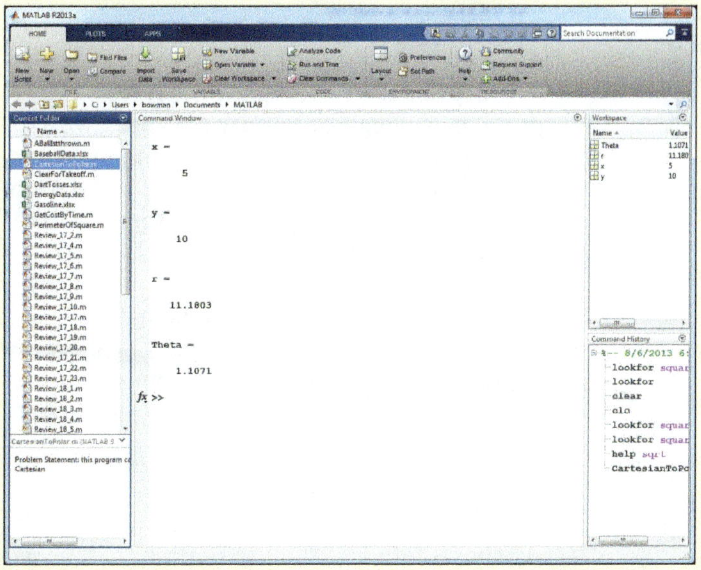

Now when we save and run our modified program, we will see only the output from the calculations of r and Theta.

Programming Features in Our Coverage of MATLAB

MATLAB has many advanced programming features that make sophisticated numerical analysis, including simulation, possible. In this text, you will learn basic programming structures to solve engineering analysis problems.

- **Data types:** There are different data types that can be stored and manipulated by MATLAB. Some of the more common data types used by programs include numeric and string variables, but many applications also use vector and matrix variables for advanced calculations, as well as special types like cell arrays for storing a combination of different types within a single structure.
- **Input and output:** By gathering input, the program can solve a different problem each time the program is executed. Output statements allow you to produce formatted text to communicate with the user. Likewise, being able to read and write Microsoft Excel files using MATLAB code allows your program to perform calculations without needing to type in a bunch of data and permanently save results to special MATLAB data files (called .mat files.)
- **Conditional statements:** These allow a program to take different paths for various user input, intermediate calculations, or final results. Conditional statements allow the programmer to ask questions in order to make decisions.
- **Looping:** Repetition is critical in breaking down complicated problems into many simple calculations.
- **Plotting:** Graphs communicate data relationships in a variety of forms.

EMERGING TRENDS

Engineering is designing for the future and problem solving. As society advances, we start to see emerging trends in programming, robotics, and automation on a small scale, like robotics within homes. Household robots can automatically vacuum while automated smart-homes record energy use to control thermostats automatically. Systems as simple as motion sensor lights can reduce energy use. On a large scale, we see increased use of drones (robotics and automation) and an increased use of analytics in process control.

Analytics

Analytics

Analytics refers to the use of data to determine actions or events – the precise definition of process control. Computer programs are used to control the fine movements required in drone flight; other programs are used to analyze "big data" quickly to make decisions. Imagine examining all of the cell phone conversations of your family and close friends all at once. Now imagine listening to phone conversations from millions of people at the same time - in different languages - talking about work, play, or espionage. As companies grow into multinational corporations, sensors may monitor a production facility in Mexico that manufactures thousands of containers per minute to package items from China; data from Mexico can be fed directly to the factory to slow or speed up production. As millions of gadgets are manufactured, each may be inspected and sent to packaging or repair based on measured data: This is automation in process control.

Engineering and Related Fields – Careers in Emerging Fields

Drones and Robotics: These systems require mechanical engineering for the moving systems and electrical engineering for the control systems and motors. They also require material engineers for the high-strength, low-weight material design or extreme toughness in robotics and computer engineers and computer scientists for the programming and analysis of data from any sensors. (How else will the drone 'know' if it hits a wall?) Related fields: Electrical and mechanical technologists will build and program these devices; project managers are required to ensure the processes are efficient; and, markets are required to sell the devices. *What other fields are related? Can you name more?*

Automation: Home automation, for example, also requires electrical, mechanical, and computer engineers. Beyond this, architects are involved in the design of the home around automation. Industrial engineers and ergonomics

technicians are involved in making the designs easy and practical for all users. Chemists and chemical engineers may be involved in the design of new materials used in new sensors. Manufacturing engineers and technicians, process control specialists, floor supervisors, and inspectors are required in manufacturing the systems and equipment. Each component of each of these systems requires design, and integrating them into a system requires effective engineering.

Many other emerging technologies have career opportunities related to engineering and STEM careers:

Other emerging technologies	Application of	Careers (engineering and related fields)
Driverless cars	Automation Process Control Robotics Advanced Sensors and Control Mathematician	Electrical and Mechanical Engineering Electrical and Mechanical Technology Transportation Specialist Safety Coordinator
"Internet of Things"	Process Control Sensors Energy Efficiency	Electrical and Mechanical Engineering Electrical and Mechanical Technology Computer Science IT Technician HVAC Specialist
Fuel Cells	Chemical Engineering Process Control Sensors Materials Engineer	Chemist Electronics Engineer Power Engineer Technician Electrician
Cybersecurity Analyst	Computer Programming Computer Engineering Big Data Electrical Engineering Mathematician	Computer Programmer Computer and Electrical Engineer Data Scientist Analyst

Science, Technology, Engineering, and Mathematics will continue to build upon these emerging technologies and create new careers and innovations!

Can you identify other emerging fields and the existing careers – and careers that don't yet exist –that could be involved?

Chapter 8 ASSESSMENT QUESTIONS

For 1 to 3, create an algorithm (written and/or flowchart as specified by your teacher) to solve the following problems:

1. Describe the steps necessary to create a jelly sandwich. You may assume that you are starting with a loaf of bread, jar of jelly, a knife, and a plate on the table in front of you.
2. Calculate a temperature provided by the user in units of Fahrenheit in units of kelvin. As a test case, you may assume the user provides the temperature of −129 degrees Fahrenheit, which is the world's lowest recorded temperature.
3. Determine the length of one side of a cube of solid gold, in units of inches. You may assume that the specific gravity of gold is 19.3 and that the user will provide the mass of the cube in units of kilograms. As a test case, you can assume that the user has a 0.4 kilogram cube.
4. The Occupational Safety & Health Administration (OSHA) defines safety regulations on working environments to protect workers from unsafe conditions. The flowchart below shows how OSHA categorizes the safety level of the working temperature given the environment temperature in degrees Fahrenheit. Given the flowchart, for what range of heat index (in degrees Fahrenheit) will the risk level be Lower (Caution), Moderate, High, or Very High to Extreme?

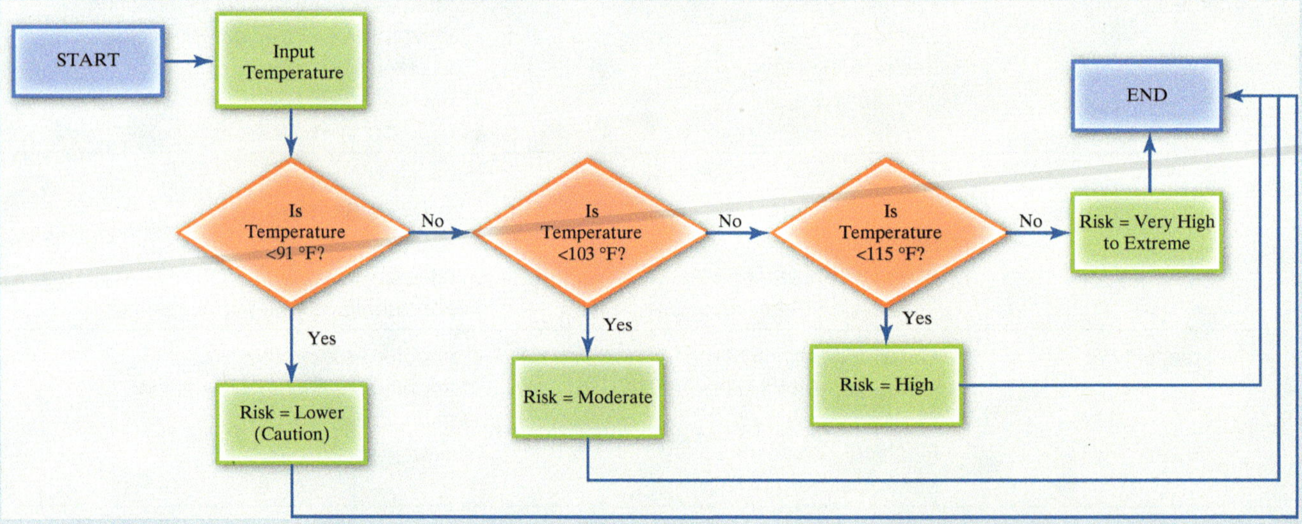

5. Given the flowchart below, for what range of values of pressure (in atmospheres) will the light be each of the colors Yellow, Blue, and Violet?

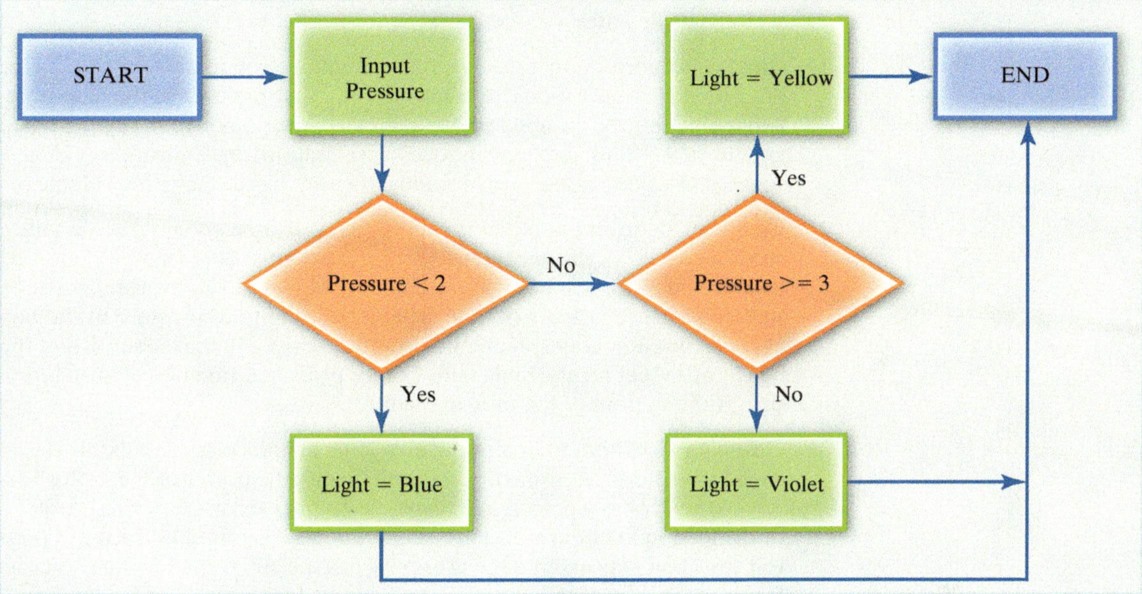

For questions 6 to 10, create an algorithm (written and/or flowchart as specified by your teacher) to solve the following problems:

6. Model the decision of a driver after looking at the color of a traffic light. You may assume that the traffic light contains three bulbs (red, yellow, green) to represent stop, slow down, and go.

7. Describe the steps necessary to cook your favorite meal, including a starting materials list.

8. Describe the steps necessary to walk from your home to school, to a store, or to some other landmark without using any street names. Specify your start and stop locations.

9. Determine the factorial value of an input integer between 1 and 10. Add a statement to check for input range.

10. Ask the user to draw a card from a deck of cards until the user pulls out the queen of hearts.

11. Create a flowchart that represents the following written algorithm.
    - Input the height of person 1 [in units of feet] as P1
    - Input the height of person 2 [in units of feet] as P2
    - If person 1 is taller than person 2
        - Display "Person 1 is taller"
    - Otherwise, if person 2 is taller than person 1
        - Display "Person 2 is taller"
        - Otherwise, display "They are the same height"

12. A reaction vessel is equipped with both temperature and pressure sensors. Create a flowchart that represents the following written algorithm:
    - Read the values of temperature and pressure.
    - If the temperature exceeds the maximum safe temperature (`TMax`), the status should be listed as "unsafe," and a variable `T` should be set to 1. Otherwise, `T` should equal 0.

- If the pressure exceeds the maximum safe pressure (`PMax`), the status should be listed as "unsafe," and a variable `P` should be set to 1. Otherwise, `P` should equal 0.
- If both temperature and pressure are less than their corresponding maxima, the status should be listed as "safe" and both variables `T` and `P` should be set to 0.
- Repeat the entire procedure.

13. Create an algorithm to determine the weight of a rod in units of pound-force on the surface of Callisto, Jupiter's moon. You may assume that the user of the algorithm will provide the volume of the rod in units of cubic meters. Your algorithm may assume that the specific gravity is 4.7 and the gravitational acceleration on Callisto is 1.25 meters per second squared. As a test case, you can assume the user has a rod with a volume of 0.3 cubic meters.

14. Create an algorithm to determine the density of tribromoethylene in units of kilograms per cubic meter in a cylindrical tank. Your algorithm should be written in such a way that the user provides the height measured in units of feet of the tribromoethylene in the tank, the surface pressure in units of atmospheres, and the total pressure at the bottom of the tank, also measured in atmospheres. For your test case, you may assume that the user provides a height of 25 feet for the tank, tank surface pressurization of 3 atmospheres, and a total pressure at the bottom of the tank of 5 atmospheres.

15. Create an algorithm to determine how long, in units of seconds, it will take a motor to raise a load into the air. Assume the user will specify the power of the motor in units of watts, the rated efficiency as a percentage (in whole number form—for example, 50 for 50%), the mass of the load in kilograms, and the height the load is raised in the air in units of meters. As a test case, you may assume the user provides the 100 watts for the power of the motor, 60% for the efficiency of the motor, 100 kilograms for the mass of the load, and 5 meters for the height the load is raised.

16. Create an algorithm to determine how long it will take to boil acetone using a hot plate, assuming acetone has a boiling point of 56 degrees Celsius. You may assume that the user of your algorithm will need to ask for the amount of time (split into the number of minutes and seconds) it takes to boil a user-provided volume of water, measured in milliliters, as well as the temperature of the lab in degrees Fahrenheit, which is also the initial temperature of the water before boiling. In addition, the user will provide the power rating of the hot plate in units of watts and the volume of acetone to be boiled in the experiment in milliliters. Note that your algorithm must determine the efficiency of the hot plate. As a test case, the hot plate takes an average of 8 minutes and 55 seconds to boil 100 milliliters of water in a 75 degree Fahrenheit laboratory with a 283 watt hot plate and it is desired to boil 100 milliliters of acetone.

17. Humans can see electromagnetic radiation when the wavelength is within the spectrum of visible light. Create an algorithm to determine if a given wavelength [nanometer, nm] is one of the six spectral colors listed in the chart below. Your algorithm should ask the user to enter a wavelength, then indicate in which spectral color the given wavelength falls or provide a warning if it is not within the visible spectrum.

Color	Wavelength Interval [nm]
Red	~ 700–635
Orange	~ 635–590
Yellow	~ 590–560
Green	~ 560–490
Blue	~ 490–450
Violet	~ 450–400

18. Create an algorithm to display the letter grade in a course. A course grade is typically reported as a real number with values from 0 to 100. Your algorithm should ask the user to provide a grade in numerical form and display the letter grade corresponding to a numerical grade if the entered value is between 0 and 100 inclusive. If the value is outside this range, ask the user to enter the value again.

Typical range of grades:

A: 90 <= grade
B: 80 <= grade < 90
C: 70 <= grade < 80
D: 60 <= grade < 70
F: grade < 60

19. Construct an algorithm to determine the class rank (freshman, sophomore, junior, or senior) of students enrolled in a school. You may assume the user will enter the total number of students in the school, and the program will ask the user to enter the total number of classes taken by each student. The program will count the number of students in each class rank. The program will produce as output the total number of students in the school and the percentage of students who are in each class rank.

In this algorithm, you may assume that a freshman has taken fewer than 10 classes, a sophomore between 11 and 20, a junior between 21 and 30, and a senior has taken 31 or more classes.

20. Create an algorithm to ask the user to enter a positive value. You may assume the user enters a positive number and you do not need to check if the value is indeed positive. Determine which is greater: the sum of the whole-even numbers between 0 and the entered value or the sum of the whole-odd numbers between 0 and the entered value. At the end of your algorithm, your algorithm should display "The sum of the odd numbers is greater" or "The sum of the even numbers is greater", depending on the outcome of the algorithm.

21. Which of the following are not valid MATLAB variable names? Circle all that apply. For those that are invalid, state why.

(a) `BigNum.docx`
(b) `Two_Nums`
(c) `This is a very long MATLAB variable name`
(d) `2BR02B`
(e) `Is_This_Valid?`
(f) `Var = V + 1;`
(g) `Mult2`
(h) `exp`
(i) `GClefSign`
(j) `AbcDEFGHijKLmnopqrstuvWXyz_Has_Lots_of_Characters`
(k) `DoNotPassGo_DoNotCollect$200`

22. From the following MATLAB Code, find the value of Z.

```
>> X = 5
>> Y = 2
>> X = X + 5
>> Y = Y + 1
>> Z = X + Y
```

23. Find the error in the following MATLAB Code. Explain why the code will fail to complete the process.

```
>> X = 4
>> Y = X
>> X * Y = Z
>> %Calculates the product of X and Y
```

24. Which of the following are not valid program/function filenames? Circle all that apply.

(a) `2b_solved.m`
(b) `calc_circum.m`
(c) `graph-data.m`
(d) `help4me.m`
(e) `MATLAB is fun.m`
(f) `matrix*matrix.m`
(g) `Mult2#s.m`
(h) `pi.m`
(i) `ReadFile.m`
(j) `SuperCaliFragiListicExpiAliDocious.m`

25. Without running these code segments in MATLAB, what will be the value stored in the specified variable after executing each program?

(a) What is stored in **`Dogs`**?

```
clear
clc
Greyhounds = 30;
G_Cost = 250;
Dalmations = 10;
D_Cost = 150;
Dogs = Greyhounds * G_Cost + Dalmations * D_Cost;
```

(b) What is stored in **`W`**?

```
clear
clc
X = 2;
Y = 4;
Z = 2;
W = (Y + Z/X + Z^1/2)
```

(c) What is stored in **`X`**?

```
clear
clc
A = 0;
B = 5;
X = A^B + B^A;
```

CHAPTER 9
AUTOMATION

9.1 MANUFACTURING

People have been using manufacturing technology for thousands of years. Evidence shows that people were making pottery from clay more than 30,000 years ago. The firing process was probably discovered by accident, when people found hardened clay under the ashes of their cooking fires. Once they discovered this process, people began to fire clay items on purpose. This is one of the first examples of **manufacturing**, which is the process of making goods by hand or machine.

manufacturing

Today, all types of products are manufactured in workshops and in factories (Figure 9-1). Some products, such as household appliances and vehicles, are **durable**—they last a long time. Others, such as foods and paper products, are **nondurable**—they are designed to be consumed in a short period of time. Both durable and nondurable products are manufactured in factories or in workshops, using the craft approach.

durable
nondurable

Figure 9-1 Consumer products on display.

9.1.1 THE CRAFT APPROACH

For centuries, people made objects only for themselves or their families. They made their own clothing and tools, and they built the houses in which they lived. Gradually, however, people began to specialize. Shoemakers made shoes not only for family and personal use but also for people in the community. In return, they received goods from farmers and other craftspeople in the area. Different people in the community made candles, cloth, wool, silverware, ceramics, and other items. The craft approach had begun.

craft approach

The **craft approach** is the method of making products one at a time, from start to finish. Although most manufacturing is automated today, craft production continues. Most often, craftspeople work alone at home or in small workshops (Figure 9-2). Often, a master craftsperson hires a young person who wants to learn the trade. These young workers, called apprentices, are trained on the job.

Many people are still willing to pay for one-of-a-kind products. The craft approach allows products to be custom-made to meet the particular needs or desires of the purchaser. People with disabilities, for example, often buy custom-made furniture and other items that are designed to help them carry out daily activities.

9.1.2 THE FACTORY SYSTEM

Until the mid-1700s, all types of goods were made by craftspeople. People worked at home or in small workshops. As America grew, railroads, canals, and highways opened up new markets. More goods were needed, and transportation was needed to move goods to different areas of the country.

Figure 9-2 Many people are willing to spend extra money for handmade goods.

Figure 9-3 With interchangeable parts, entire machines can be assembled in minutes or even seconds.

The Industrial Revolution, which started in the late 1700s, was a period during which many new machines and devices were invented. Eventually, machines such as the steam engine, the power loom, and the sewing machine reduced the time and effort needed to carry out many tasks. Businesspeople took advantage of machinery to improve production. Soon, goods were being made in factories that hired many workers.

Interchangeable Parts

In 1789, Eli Whitney signed a contract with the U.S. Army to make 10,000 rifles in 2 years. Ten thousand was a huge number to produce in those days! Up to that time, rifles had been made one at a time, with each part customized for each rifle. Whitney succeeded in his task by making large batches of each of the different rifle parts. These parts could be put together quickly and easily, thereby increasing production and saving time.

Interchangeable

In Whitney's system, each batch of each part was kept in a separate bin. The parts were standardized—that is, all the parts in a batch were exactly alike. They were also interchangeable. **Interchangeable** parts are manufactured to be exactly alike so that any one can be substituted for any other (Figure 9-3).

Comparing Craft Manufacturing and Mass Production	
Craft Production	**Mass Production**
Workers are highly skilled.	Workers need only limited skills.
Workers make a product themselves, from start to finish.	Workers perform only one part of assembly. They do not see the finished product.
The work is varied and involves different tasks.	The work is routine and very repetitive.
Only one item is produced at a time.	Many items are produced during the production run.
Production is fairly slow.	Items are made relatively quickly.
Each part is handcrafted, so no two are exactly alike.	Parts are machine made and are interchangeable.
The products are usually expensive.	Mass-produced items are relatively inexpensive.
Quality depends mainly on the skill of the craftsperson.	Quality depends mainly on the accuracy of the machines and how well they have been set up by people.

Mass Production and the Assembly Line

mass production

assembly line

Interchangeable parts became one of the most important characteristics of a mass-production system. **Mass production** is the production of goods in large quantities by groups of workers in factories. The production process is divided into steps. Each worker does one step and then passes the item on for the next step. Work is carried out on the assembly line itself and as one task is finished, another begins immediately so time is not wasted in setting things up. An **assembly line** is a system in which each worker performs a specific operation in assembling an item as it is passed along, often on a slowly moving belt, from one workstation to the next.

One of the first moving assembly lines was constructed at the Ford Motor Company in 1913. By applying the same principle that Eli Whitney had used for making rifles to the assembly of a complete car, Henry Ford speeded up automobile production. Auto parts were pushed from one worker to the next, and production time was reduced by about one half. A finished Model-T Ford came off the assembly line every 10 seconds. Today, modern automobiles are manufactured from interchangeable parts.

Impacts of the Factory System

Over the years, the factory system replaced the craft approach in the manufacture of most goods. It also led to the development of other industries. The steel, car, and clothing industries provided millions of new jobs (Figure 9-4).

The factory system allowed more goods to be produced for less cost in a given amount of time. It also created wealth by adding value to the resources that were processed into goods. Luxury items became less expensive. More people could afford to buy the products of mass production. The standard of living improved for people who lived in industrialized nations.

During the early days of mass production, most businesspeople wanted only to improve the process to make a product cheaper and faster. They were not concerned with providing people a number of different choices when it came to the products. They wanted to offer a standard product at a low price. A suggestion was once made to

Figure 9-4 Factory workers.

Henry Ford that he paint his Model-Ts different colors. He replied, "Give it to them in any color, so long as it is black."

This attitude changed with competition. In the 1930s, General Motors started producing a new car model every year. The way cars were advertised and sold was suddenly as big a concern as how they were produced. Marketing and business management became important considerations in manufacturing. Today, companies that can deliver the right product, at the right time, at the right price have an advantage over their competitors.

9.2 RESOURCES

seven technological resources

Manufacturing systems use the **seven technological resources** (people, capital, time, information, energy, materials, and tools and machines) to make a product. As primary resources are processed and assembled into a final product, their value increases. The finished product is worth much more than the individual resources.

People

People are involved in every stage of manufacturing. They design the products. They decide how the products should be made. They choose the materials and the best tools and machines to manufacture the products. They organize the production lines and obtain the necessary capital. They advertise, distribute, and sell the products.

In the past, people provided more of the labor in manufacturing plants than they do today. Today, fewer people are needed to manufacture goods. Computers and more sophisticated machinery have reduced the number of factory workers needed.

Manufacturing has also become more competitive. Industries in the United States have had a hard time competing with foreign companies whose workers are paid much less. To remain in business, U.S. industries have been forced to either improve their production methods or lay off workers.

As a result, more than 300,000 American steelworkers have lost their jobs over the last 25 years. Workers who lose manufacturing jobs often find that without further training, they can get work only in lower-paying service industries.

Quality Circles and Teamwork

quality circles

The roles of people in the manufacturing system have been changing. As workers have become more educated, they have been asked to take part in decisions that affect the company. **Quality circles** are groups of workers and managers who meet to discuss problems and ways of improving production. During these meetings, management can explain problems to workers about costs, profits, and competition. Quality circles allow workers and management to discuss issues openly. The result is a better work environment and improved production.

Quality circles use many attributes of effective teams, such as respect for one another, careful attention to others' opinions, and active participation by all members of the group. Whether a team is made up of employees, students, or athletes, a successful team shares leadership roles and responsibilities so everyone is empowered by the experience.

Entrepreneurship

entrepreneur

Entrepreneurs are a critical type of human resource for manufacturing. An **entrepreneur** is a person who comes up with a good idea and uses that idea to make money. He or she might start a business to fill a specific need, improve the way a product is made or sold, or even come up with an idea for a new product.

Some entrepreneurs are inventors. An inventor comes up with a totally new idea—a safety razor, a laser beam, or a contact lens. Other types of entrepreneurs are innovators in that they improve a product or create a new way to accomplish something. An innovation is an improvement on an invention. Innovations can create new industries or change existing ones. The electric guitar, the diesel engine, and power steering are all types of innovations.

Capital

Capital is an important resource in the development of technology. The reason some technologies evolve and others do not can sometimes be related to the amount of available capital. Companies must have capital to finance their operations. They must buy land, build factories, purchase equipment, pay workers, maintain machines, and advertise their products.

Capital is often obtained by selling shares of company stock to the public. Stockholders, or people who buy shares of stock, become partners in the corporation. If the company makes a profit, the value of its stock may go up.

venture capitalist

Private companies may raise money from investors who contribute capital. A venture, like an adventure, is a trip into the unknown. A **venture capitalist** is a person who supplies money to finance the start-up of a new company. Venture capitalists sometimes take big risks because they expect to make large profits once a new company starts its operations (Figure 9-5).

Figure 9-5 Venture capitalists are interested in the financial success of companies in which they invest.

Time

productivity

In manufacturing, time is money. For a given product and level of quality, the faster it is made, the more profit the company will receive. **Productivity** refers to how quickly and cost-effectively a product is made. A large part of a product's cost is the cost of paying workers. If workers do their jobs faster, productivity generally increases because more products are made for the same labor cost.

Information

Information is critical to all stages of manufacturing. To be successful, companies must find out what consumers will buy and how their tastes are changing. They must also gather information about the costs of materials. This information helps managers choose which materials to buy. Companies use information about production to help them adjust the manufacturing process. They also use university research to learn about new materials and production methods.

Energy

About 40 percent of the energy consumed in the United States is used in manufacturing. Some of the biggest energy users are factories that make metals, chemicals, ceramics, paper, food, and equipment. Most manufacturers use electricity from fossil fuels (coal, oil, and natural gas). Others draw on alternate sources of electricity such as hydroelectric and nuclear energy plants. Factories are often built in places where energy resources are abundant and costs are low. In the late 1800s, natural gas became accepted as the best fuel for melting glass. As a result, the glass industry grew up in West Virginia because the state had plenty of natural gas. Small steel mills that use electric furnaces are built in places where electricity is cheap.

Some industries take advantage of the heat given off during manufacturing. The paper-making industry, for example, reuses the energy to heat water and make steam. The steam is used to turn steam turbines and produce electricity.

Materials

Through manufacturing, raw materials are made into basic industrial materials and then finished products. For example, the steel industry uses coal, limestone, and iron ore to make steel.

The cost of raw materials is important. The cost includes not only the material itself but any shipping or processing that is required (Figure 9-6). Suppose a steel company in Indiana needs iron ore. Should it buy the ore from a mine in Wisconsin, where it is only 20 percent pure, or should it pay the extra shipping charges to buy the ore from Brazil, where it is 65 percent pure? This is the kind of cost-benefit trade-off that manufacturers must consider in choosing materials.

The physical properties of raw materials are often crucial to design. Aluminum is a lightweight metal, highly conductive and with a lower melting temperature. Titanium is stronger and has a higher melting temperature, but is more expensive, has different stress and strain properties, and is less plentiful.These physical properties are important as we design.

Figure 9-6 If goods have to be transported over great distances, their cost is greater.

Rapid Prototyping

A prototype is an original model that can be used as the basis for making more of the item, often using more efficient, mass production methods. Rapid prototyping is a method for creating these models that works from a design created by CAD software. There are a variety of rapid prototyping technologies, such as stereolithography, 3-D printing, and Fused Deposition Modeling. These methods use expensive machines to create an object by laying down multiple layers of material and building up thickness until the object is complete. The various methods use different materials. Three-D printing mixes ink and plastic and works like an ink-jet printer, spraying the ink and plastic mixture on a surface. Stereolithography uses light-sensitive plastic and lasers to melt thin layers of plastic together. Other methods use metal or ceramic powders to create the layers. Rapid prototyping is currently a very expensive technology. The machines and the plastics that they use are costly. The cutting edge of this technology, though, is machines that can make their own parts. Once this is accomplished, the technology should become much more affordable.

Tools and Machines

The tools used in modern factories are advanced and precise. Most of them are automatic. Computer programs and sensing devices provide feedback and guide their operation, especially when used in emerging fields within process control and automation.

The design and use of tools and machines rely significantly on the mechanical properties of the materials, subsystems, and components. Factors such as strength, hardness, and impact resistance play a role in design.

9.3 AUTOMATED MANUFACTURING

automation

A robot can be programmed to pick up a part, move it a certain distance, and drop it into a bin. This simple process, done entirely by a machine, is an example of automation. **Automation** is a system in which machines are controlled automatically. In the example of the robot, a list of instructions is used to direct a machine's activity in a process called program control.

Using Feedback

Feedback control uses feedback to adjust the way a machine is working. Feedback control depends on a sensor, a device that gathers information about its environment. For example, a sensor can detect when a drill has drilled deep enough. At this point, it sends a signal to switch off the drill. A machine operator is no longer needed to turn the drill on and off. In automated factories, these types of sensors also provide quality control. The sensors make sure the machines perform their operations to exact specifications.

Lower Costs

Automation saves money on labor costs. To compete with lower-priced foreign products, more and more factories are being automated. Although this change requires extra capital at the start, the money is usually regained over time. Automated systems require fewer people to run and maintain them, which also helps conserve capital.

The Robot Revolution

You may have heard of the term "robot revolution." Why is the use of robots considered revolutionary? Because robots that exist today can actually take the place of many workers in a manufacturing plant, they are changing the way manufacturing is done. Using just the right amount of pressure, robots can hold and manipulate delicate parts. They have sensors that allow them to "feel" things around them. Other robots can "see" with television eyes to identify differently shaped parts.

Unlike people, robots can work 24 hours a day without tiring. Most of the work they perform relieves people from dangerous, boring, heavy, or unpleasant jobs (Figure 9-7). Robots are used for loading objects onto platforms and conveyors. Because robots are computer-controlled, they can also be reprogrammed. For example, a robot can be programmed to weld a fender on one kind of automobile and then reprogrammed to do the same job on a different model.

The use of robots has allowed manufacturers to make products of higher quality at lower cost. In many factories, robots are taking the place of people, although some workers are still needed to make sure that everything is operating the way it should. People are also needed to install, service, and program the robots. While robots may replace jobs on a manufacturing line, they introduce career opportunities in robotic design and automation for engineers in mechanical, electrical, computer, and materials engineering, among others.

Figure 9-7 Robots and computers are at the center of automated manufacturing.

9.3.1 TYPES OF AUTOMATED MANUFACTURING

Some manufacturing methods lower costs and create a better product. The different types of manufacturing used today include different types of computerized manufacturing as well as flexible manufacturing, agile manufacturing, and just-in-time manufacturing.

Computer-Aided Manufacturing

computer-aided manufacturing

Computer-aided manufacturing (CAM) is a software application that uses computers to control factory machines. CAD/CAM is a technology that joins CAD and CAM applications. It lets a person create a design on a computer screen and then send it directly to a machine tool (Figure 9-8). The machine can make a prototype of the part. It can also make many copies of the final product.

Computer-Integrated Manufacturing

computer-integrated manufacturing

Similar to CAD/CAM technology, **computer-integrated manufacturing (CIM)** uses computers not only for design and manufacturing, but also for business needs. It is used to store information about raw materials and parts, to set times for the purchase and delivery of materials, and to report on finished goods. CIM is also used for billing and accounting. Managers involved in purchasing, shipping, accounting, and manufacturing can get a total picture of all the processes by looking at the computer screen.

Figure 9-8 CAD/CAM technology allows designs to be sent directly to the machine that makes the part.

Flexible Manufacturing

Today, many products are made in batches of a hundred or a thousand, rather than millions. Making products in smaller batches allows companies to cater to customers' special needs. For example, the General Electric Company in New Hampshire makes 2,000 different versions of its basic electric meter.

flexible manufacturing

This type of customized manufacturing is possible with flexible manufacturing. **Flexible manufacturing** is the efficient production of small numbers of products. The same production line is used for each version of a product. To make the different versions, the machines are simply programmed to do different operations.

The John Deere factory in Iowa can make 5,000 different versions of its tractor for farmers with different needs. Each tractor may have different components, or it may be a slightly different size. Computers are used to program the manufacturing process so different parts can be assembled with each production run.

Agile Manufacturing

agile manufacturing

The word *agile* means quick, or possessing ease of movement. **Agile manufacturing** is an organizational strategy in which decisions are made quickly and a greater variety of products is produced quickly and inexpensively. Often, it includes a production system capable of making different products simultaneously.

One key influence on agile manufacturing is the role of the customer. From the beginning of the product-delivery cycle, the agile manufacturing strategy is to understand what customers want, how much they will spend on the product, and how the product can be improved and customized to suit customers better.

How Technology Works...

Getting the Order Right

Making sure that products are uniformly made and of high quality is an important goal for manufacturers. Whether it is a complex product such as a computer or a simple product like this pen, consumers expect it to have no defects. Go online to see how companies manufacture to ensure the fewest defects.

Figure 9-9 Just-in-time manufacturing keeps storage to a minimum. Materials arrive at the factory only when they are needed.

Just-in-Time Manufacturing

just-in-time manufacturing

Just-in-time manufacturing uses careful scheduling to keep materials and products in storage for as little time as possible. In other types of manufacturing, raw materials are most often delivered in large amounts that take up a lot of costly storage space. Storage space can be costly because companies must pay to rent, heat, and light the space (Figure 9-9).

In just-in-time manufacturing, materials and parts are ordered so that they arrive directly at the factory only when they are needed. It also requires that equipment and people are made available when the materials arrive. Once the final product is made, it is shipped immediately to the customer. There is little need for storage space or for workers to stack and store parts in warehouses.

Rapid prototyping and just-in-time manufacturing can be considered emerging trends in manufacturing, with career opportunities in many types of engineering. Many disciplines of engineers are involved in the design of manufacturing systems, while those in related fields such as engineering technologists are often in charge of manufacturing lines, robotics, process control, and automation.

CONNECTING TO STEM

Lasers

Today, lasers are used in products as dissimilar as a CD player and surgical tools. Why have lasers become so important in technology?

A Different Kind of Light

All the light in a laser beam has the same wavelength and travels in the same direction. It is called coherent light. Because coherent light is so uniform, it is powerful and can travel long distances.

Just like a flashlight, a laser produces a beam of light. However, the types of light produced by a laser and a flashlight are vastly different. Ordinary light has many different wavelengths, all moving at various speeds and directions. This is called incoherent light.

Structure of a Laser

A typical laser consists of a gas-filled tube that is closed on both ends. Wrapped around the tube is a light source called a flash lamp. When the laser (and flash lamp) is turned on, the energy from the flash lamp causes tiny atoms to become energized. These energized atoms move around the tube, releasing energy as light.

On the back end of the glass tube is a reflective mirror. Photons bounce off the mirror and pass through the gases again. More atoms become energized and release light. At the other end of the tube is a mirror that is only partially reflective. Only light of a certain wavelength can pass through. This mirror acts to filter any incoherent light that is bouncing around in the tube. The coherent light particles that pass through make up the laser beam.

A Powerful Tool

Because the light in a laser beam is so uniform, the laser is a very powerful tool. In medicine, laser surgery is used to correct vision by reshaping the cornea of the eye. In communications, lasers are used inside optical fibers that transmit information. In supermarkets, lasers scan the bar code that contains a product's price. In manufacturing, lasers are used to find flaws or mistakes in parts. Lasers help manufacturers ensure that their products are of high quality and are made to specifications.

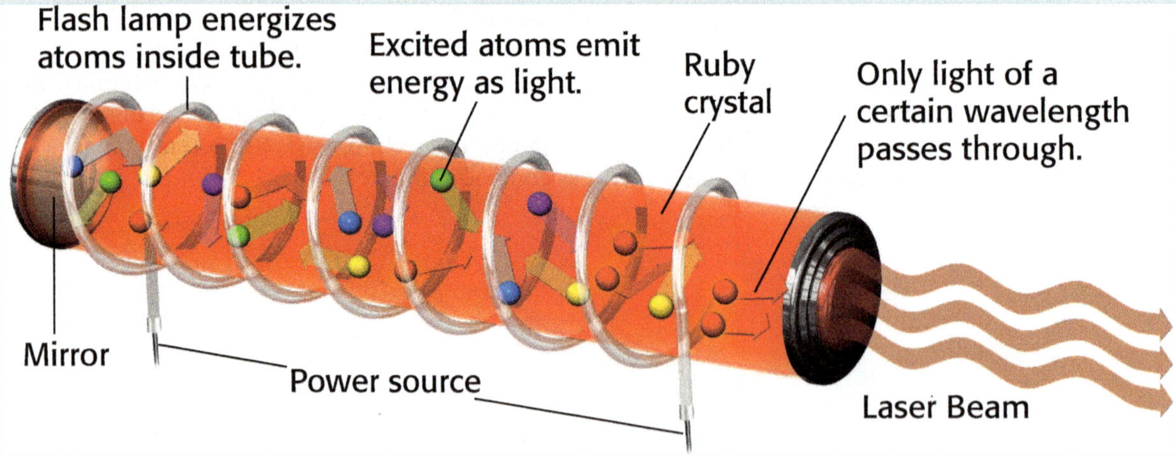

9.4 PROGRAMMING AS A MEANS OF AUTOMATION

programming

Programming, or coding, is the process in which a problem is formulated in a way that a computer can understand and solve. Effective programming is crucial to the success of the computer's use in manufacturing. In the same way that mathematicians view math as an art form, computer scientists find beauty in the ways of programming. The problem could be a simple mathematical exercise, such as finding the closest value of *n!* (pronounced *n* factorial) for any input, or more applied, such as creating a video game.

In order to program a computer to do a meaningful task, we must first speak one of its languages—programming languages, that is. Communication with computers can be easily compared to everyday verbal interactions. There are thousands of spoken languages in the world; while the number of programming languages is not nearly as massive, the sheer number can be intimidating to new coders. A sizable number of the spoken languages can have *dialects*, a form of a language only spoken in a specific region. The same concept happens within the world of programming languages, except the type of language will depend on the application.

9.5 ARDUINO

For the project in this chapter, we will be discussing a particular programming platform called Arduino.

open source

Arduino is a popular **open source** electronic prototyping platform that is based on a microcontroller board and is useful for a variety of applications. Since the platform is designated as open source, the licensing to the software and internal workings of the platform are free. In addition, since the Arduino environment is open to the public, anyone can make improvements to the overall design and functionality and then share it to other users. This means that it is widely used: You might even use it at school and home!

Physically, the Arduino is a programmable device. This is an integrated circuit or a "chip"—physical hardware usually mounted on a circuit board (Figure 9-10).

Figure 9-10 An empty Arduino board.

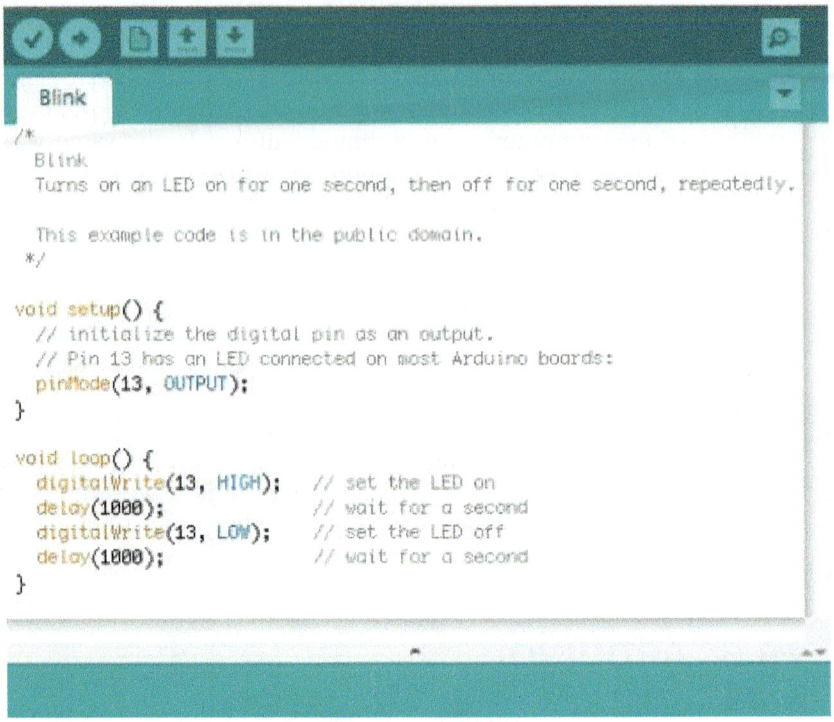

Figure 9-11 Arduino environment.

sketch

The platform works by taking inputs from a series of sensors and switches and regulating various physical outputs like lights and motors. The software associated with Arduino has a text editor for the code that is saved as a **sketch**—Arduino's name for files containing code—with a file extension of .ino. The individual files are called sketches (Figure 9-11), since Arduino uses the concept of a "sketchbook". The sketchbook is like the folder where the files are stored.

In the project for this chapter, we will explore the basic functionalities of Arduino by designing a system that could be considered a part of a manufacturing process. For example, we may need physical hardware that indicates each time a completed part flows by a point in our manufacturing line. We will build and program a device that mimics this by testing an LED and automating the manner in which it turns on and off.

PROJECT PLANNING

We have explored important concepts of professionalism such as dependability, reliability, responsibility, effective teamwork, and punctuality. These qualities build your level of professionalism and can help you get a job in engineering or a related field.

Project Management and Time Management

project management

Project management is related to these professional traits. Fully managing a project involves prioritizing tasks and following schedules to produce efficient results.

Gantt Chart

Gantt chart
dependencies

A **Gantt chart** is a project schedule, listing each task and showing how long it takes, and assigning tasks to show who has responsibility for tasks. Further, **dependencies** are shown: For example, you can't do a *Poster Review* until you are done with *Develop Poster*. However, you can begin to *Build* before completing *Order Parts*. You can see where the start and end of tasks are in the chart below. If we use a program that allows us to enter dependencies (to show that a specific task has to complete before a new task begins), the later tasks will adjust and we can easily see a new end date if we see a delay in that earlier task.

This planning allows you to perform goal-relevant activities in a way that will produce efficient results. Further, knowing tasks that are your responsibility allows you to demonstrate punctuality and dependability!

critical path

If we follow the chart from the first task and push tasks back until we see the shortest time to complete the project, we can see the time of the **critical path**: the minimum amount of time to complete the project.

Programs such as Microsoft Project are used almost exclusively for Gantt charts. Researching on the Internet allows us to find versions suitable for Microsoft Excel.

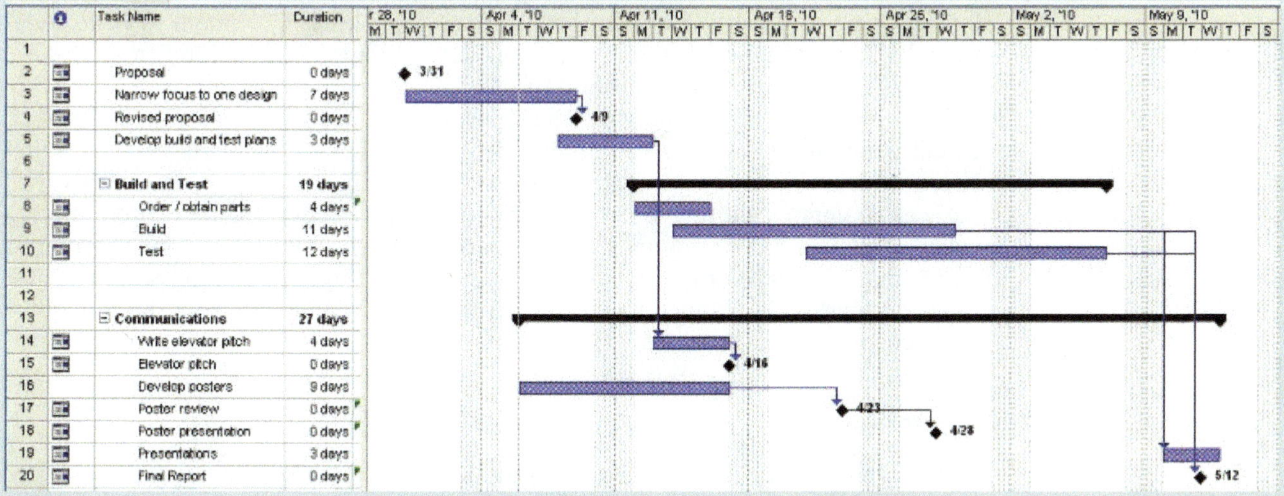

PROJECT 9: USING AN ARDUINO TO AUTOMATE PROCESSES

Problem Situation

In order to better understand how programming can be used to automate processes, we will learn to use an Arduino board to control the behavior of a circuit.

Your Challenge

Using the Arduino board, your team will be responsible for programming the board to turn a light on for 5 seconds and off for 3 seconds repeatedly.

Materials you will need

- Computer or laptop connected to the Internet
- Arduino board kit
  - Arduino Uno, USB to mini-USB cable, breadboard, resistors, LED lights

PROGRAMMING THE ARDUINO REQUIRES THE ARDUINO DRIVERS TO BE ON YOUR COMPUTER.

The software is free of cost and may already be installed on the computer you will use. Please check with your teacher if working on a school computer.

To download, visit Arduino.cc/en/Main/Software

This site also has a lot of sample code you may want to try.

1. Clarify the Design Specifications and Constraints

In this challenge, we need to find a way to have a circuit interact with the Arduino board. Our plan will be to use the Arduino development environment to write code to "talk" to the board. We need to ensure that the blinking criterion is met; that is, the LED lights up for 5 seconds and turns off for 3 seconds.

In order to meet the specified challenge, we will need to learn how to connect circuit components to the Arduino and decide whether to build the circuit on the board itself using the provided pins or by other means. Our next step will involve basic programming within the Arduino environment. The code will be written in a file called a "sketch," which you may recall is part of a larger "sketchbook" that contains all of your other relevant sketches. Actions performed by the Arduino will be facilitated through built-in functions, but it is up to us to figure out how the functions interact with one another.

2. Research and Investigate

Knowledge and Skill Builder 1: **Building the LED Circuit**

There are two methods of creating the LED circuit, one with the Arduino board itself and one using a breadboard. Arduino boards often have an LED already integrated into the board on pin 13, so if you try to run the experiment with no additional hardware, that light will blink. Nevertheless, the circuit can be constructed using only two circuit elements, a 250Ω (or 330Ω) resistor and an LED (Figure 9-12).

1. Connect one leg of the resistor in pin 13 on the Arduino board. Note that it may be helpful to connect a wire from pin 13 to the same row as the resistor leg.
2. Attach the longest leg of your LED to the other leg of the resistor; again, this is easiest to do in a breadboard.
3. Take the other leg of the LED and connect to the ground pin (GND).
4. Done!

Figure 9-12 First Arduino circuit.

Knowledge and Skill Builder 2: **Programming in Arduino**

comment

If you want to add comments in the sketch to explain certain pieces, then you can do so by pressing the forward slash twice, //, and then typing your **comment**. For example,

```
// This is a comment.
```

NOTE

Whatever comes after // is ignored as part of the program, even if it 'makes sense' to you:
// This is a comment, and it is ignored
// delay(7000) – this is also ignored because of the // at the start of the line

In order to speak Arduino, we need to know a few basic commands. Before beginning, we need to include two specific functions to ensure that our sketch functions properly. We add these lines to the *setup* area of the code:

```
void setup() {
 Serial.begin (9600);
 pinMode(13, OUTPUT);
 //whichever pins you'd like to initialize, but in this case, let's use 13...
}
```

The "setup" function is used to initialize any of our variables or pins. Within these beginning lines is the "Serial.begin" function as well, which defines the rate at which data is transferred to the board. For this application, a rate of 9600 bits per second is sufficient.

For the rest of the code, we need to know how to *initialize* the pin containing the LED, essentially telling it to get ready for commands. We can prepare a pin to receive commands using one function, "pinMode." For example, to *initialize* pin 13, we write,

```
pinMode(13, OUTPUT);
```

Without this line of code, the Arduino will not know how to treat its various pins. If we write a command and forget to specify whether the Arduino is supposed to treat the pin as an input or output, digital headaches – errors - are bound to occur.

Since we need to initialize a pin for our project, the pinMode function should go in the "setup" function.

We need one more concept, the idea of a loop. A loop is simply a function in Arduino that allows us to repeat a certain bit of code repeatedly. To introduce this into our program, we say

```
void loop() {
... code / commands you want to loop...
}
```

We know from our knowledge of circuit that we need to provide a voltage drop over the LED to produce a current. Without a current, we have no hope of lighting up the LED. Therefore, we will need to use the Arduino as a power source. The Arduino can provide the necessary voltage for us by using a simple line of code. The function will supply 5 volts to the pin of our choosing. If we wanted to supply a 5 volt drop for pin 10, we would write,

```
digitalWrite(10, HIGH);
```

Why HIGH? Since we want to supply a voltage at a *higher* potential than what is currently in the circuit, we need to tell the Arduino to do so. On the other hand, if we want to cut the supply, we can replace 'HIGH' with 'LOW.' The meaning of LOW is precisely the opposite; this command takes the pin back down to the ground state of zero volts.

To tell the Arduino to "wait" before going to the next command in programming terms, we need to introduce a delay. The delay command uses milliseconds as units, so be sure to keep that fact in mind when declaring the amount of time Arduino needs to hold. For example, if we want Arduino to wait for 7 seconds, we write

```
delay(7000);
```

In more complex sketches and programs in general, the use of comments can help immensely when organizing these functions. Due to the multiple ways in which problems can be solved in computing, you need to be able to communicate your solution to others. For example, suppose you are working for a large software company and are working on a team of programmers to complete a certain project. If you do an initial pass and write a horrendously complicated script without comments, then the next programmer who lays his or her eyes on your work could become confused, frustrated, or both.

Writing in comments can also be a useful method of reinforcing concepts while learning to program. By forcing yourself to explain the line of code, you are engaging in a form of reflective practice, which further cements the concepts in your memory. If you do happen to become "rusty" with programming, returning to an old script with your comments can help refresh your memory.

Practice

1. Open the Arduino development environment and enter the code as shown in Figure 9-13.

```
void setup() {
  // put your setup code here, to run once:
Serial.begin (9600);
pinMode(13, OUTPUT);

}

void loop() {
  // put your main code here, to run repeatedly:
digitalWrite(13, HIGH);
delay(1000);
digitalWrite(13, LOW);
delay (100);

}
```

Figure 9-13 Practice code.

2. Describe what happens and why (based on the code).

Project

Perform the following steps and respond to the prompts in your engineering notebook:

1. Write the script to meet the criteria specified in the problem statement.
2. Why are comments useful when programming? Are there other benefits beyond what was mentioned in the passage?
3. Can you modify your code to make a different pattern for the light?

3. Generate Alternative Designs

Another parallel with mathematics is the flexibility in which problems can be solved. The solution your team used to the make the LED blink is not the *only* method of achieving the same result.

In your engineering notebook, respond to the following prompts:

elegant

1. Describe two other ways we could have made the LED blink using Arduino. Include any appropriate diagrams, figures, or sketches. The approaches could be similar, but be sure to describe how the methods differ. If possible, use the Internet to find alternate solutions.
2. Computer scientists often describe pieces of code as **elegant**. Elegant code uses the simplest possible algorithm and design needed to solve the problem at hand efficiently. Is your code elegant? How can you make it elegant?

4. Choose and Justify the Optimal Solution

Pick your preferred solution. In your engineering notebook, respond to the following prompts:

1. Explain how your solution meets the requirements and constraints.
2. What trade-offs, if any, did you make in selecting your solution?

5. Develop a Prototype

Our prototype in this activity is the draft script you will be writing. Note, however, that in any technological activity, you will use seven resources: people, capital, time, information, energy, materials, and tools and machines.

In your engineering notebook, indicate which resources were most important in this activity and how you made trade-offs among them.

6. Test and Evaluate

Testing the Arduino is as simple as uploading the sketch you created onto the board. To do so, click the "Upload to I/O Board" button (Figure 9-14). If you cannot locate the button, it looks like an arrow pointing to the right at a brick wall. The Arduino environment will alert you if the upload was successful by displaying, "Done uploading." on the status bar.

Answer the following questions in your engineering notebook:

1. Our goal was to illuminate the pin 13 LED for an 8 second periodic interval where it is on for 5 seconds and off for 3 seconds. What happened?
2. If your LED did not follow the specified pattern, why do you think an incorrect cycle or pattern was displayed?

Figure 9-14 Upload and verify.

Expand on your design

We aren't limited to using pin 13 for our Arduino environment (assuming we are using a breadboard to build our design).

1. Change the code to that shown in Figure 9-15. Record the new code in your engineering notebook, along with your observation (what happened?) and explanation (why?).

```
void setup() {
  // put your setup code here, to run once:
Serial.begin (9600);
pinMode(13, OUTPUT);
}

void loop() {
  // put your main code here, to run repeatedly:
digitalWrite(13, HIGH);
delay (1000);
digitalWrite(13, LOW);
delay (1000);
digitalWrite(13, HIGH);
delay (100);
digitalWrite(13, LOW);
delay (100);
digitalWrite(13, HIGH);
delay (100);
digitalWrite(13, LOW);
delay (100);
}
```

Figure 9-15 Modified code.

2. Add more LEDs. Given the sample code and picture provided, and referencing Figures 9-16 through 9-18, do the following:
   (a) Create the sketch and circuit.
   (b) Test the sketch and circuit.
   (c) Record the new code and describe the new circuit in your engineering notebook. What did you observe? How can it be explained?

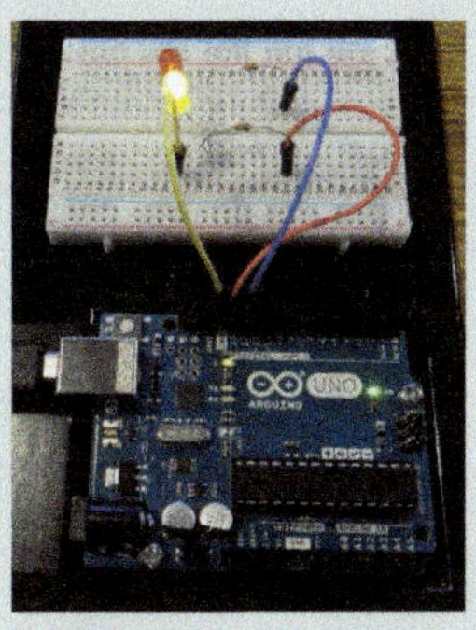

Figure 9-16 Image of Arduino using pins 12 and 13.

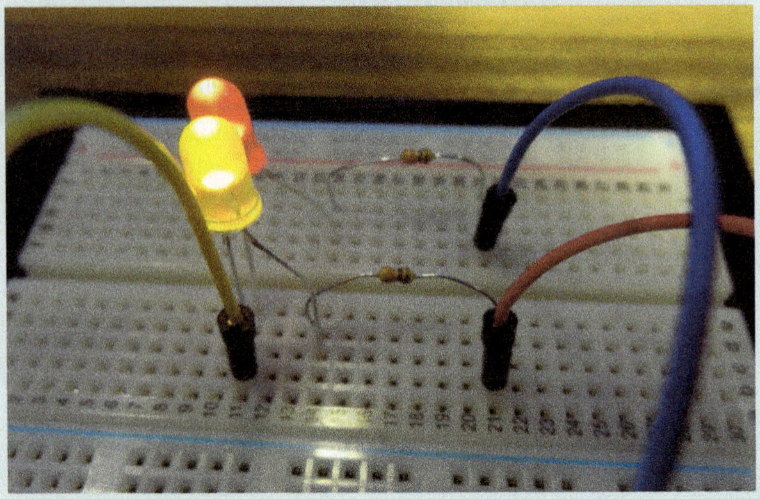

Figure 9-17 Closeup of breadboard.

```
void setup() {
  // put your setup code here, to run once:
Serial.begin (9600);
pinMode(13, OUTPUT);
pinMode(12, OUTPUT);

}

void loop() {
  // put your main code here, to run repeatedly:
  digitalWrite(13, HIGH);

  for (int i=0; i <= 10; i++){
     delay(100);
     digitalWrite(12, HIGH);
     delay(100);
     digitalWrite(12,LOW);
  }

digitalWrite(13, LOW);
delay (1000);

}
```

Figure 9-18 Code for dual-LED circuit.

7. Redesign the Solution

More often than not, the first draft of code will not be correct. While we often allow grammatical errors when speaking without much recoil, making a mistake in the context of talking to a computer could be disastrous. Unlike verbal communication in various languages where one error makes a sentence sound a tad awkward, computers will act like the strictest writing or speech teacher you have ever had. Imagine if your teacher stopped reading your term paper completely upon finding the slightest error. In the same manner, computers will halt whatever process they are doing upon encountering an error, so we undergo a stage of **debugging**.

debugging

Through debugging, we can identify and correct any errors in the code. Note that we could still be receiving an output, but it is different from what we intended. This issue can also be corrected during the debugging stage.

In your engineering notebook, answer the following questions:

1. Is there a way your code can be improved? Explain.
2. Implement the necessary changes in your code and retest. Did your results improve?
3. How can this device be used in a manufacturing environment? Be creative in your response.

8. Communicate Your Achievements

Write a three-page report using the data collected during this investigation and responses in your engineering notebook. Your report must have the following sections:

Introduction: What is this report about?

Purpose: Why did you do this project?

Procedure: How did you go about completing the objective?

Results: What happened? What did you discover? Include relevant tables, charts, and equations.

Discussion: Comment on your results. What were the limitations of this project?

Conclusion: Summarize the report and provide the reader with a sense of finality.

Chapter 9 ASSESSMENT QUESTIONS

1. Give five examples of how manufacturing technology has helped satify people's needs and wants.
2. Explain how craft production differs from mass production.
3. Idendify several ways in which a manufacturer might sell products.
4. Suggest an invention that would help you with your homework.
5. How might you innovate a container for a soft drink?
6. Choose a manufactured product (e.g., a car, chocolate bar, bottle) and explain how the chosen product has changed from its historical beginning to its present state.
7. Many of the products you use every day are manufactured by machines and automated equipment. How would your life change if everything you use had to be made by hand?
8. Do you think that robots should be used on assembly lines instead of people? Explain why or why not.
9. How have computers affected the manufacturing industry and society as a whole? Predict possible changes to the status quo provided that computers continue advancing.
10. Describe how to set pin 11 on an Arduino to deliver output to a device, like an LED.
11. Write a short block of code (or an Arduino patch) to light 3 LEDs in a sequence such that only one is on at any given time, and each different state lasts ½ second. If possible, program and test your design.
12. Research: Why is the fact that an Arduino is "open-source" important?
13. Research: Find sample code for the Arduino on the Internet. Copy the code and describe what it does, step by step. If possible, program and test your design.
14. Describe the difference between agile, just-in-time, and flexible manufacturing.
15. Create a flowchart for the code in Figure 9-13.
16. Create a flowchart for the code in Figure 9-15.
17. Create a flowchart for the code in Figure 9-18.
18. Draw a Gantt Chart to plan moving everything in your bedroom to a different bedroom. Include each large item that needs to be moved, clothing packing/carrying/moving, set up, etc.
19. Research: What material and mechanical properties make aluminum a good choice of material to build a bicycle frame? What properties make it a poor choice?
20. Identify a career path in an emerging field within robotics, such as robotic surgery, or an emerging trend of your choice.

CHAPTER 10
CHEMICAL TECHNOLOGY

SETTING THE SCENE: Using Chemistry Every Day

Many of the things we use every day are created and produced using chemical technology. The water we drink, the food we eat, and the clothes we wear are all modified or made using chemicals. Medicines and personal-care products, such as toothpaste, are also made using chemicals. Many industrial processes also rely on chemical technology, such as

- processing iron ore into steel
- making rubber products from natural rubber
- deriving paper from wood pulp
- making glass and ceramic tiles
- preparing foods and soft drinks
- dyeing textiles

Engineers and scientists study and apply chemical processes in order to improve products and lower their manufacturing costs. This type of development helps raise our standard of living and quality of life.

10.1 UNDERSTANDING CHEMISTRY

Chemistry is the study of the composition and properties of matter. Using chemistry, scientists explore methods of combining various substances to create new substances. Chemists look at the interactions between different substances to find processes for creating new products. The clothes you wear, for example, may be made of human-made fabric or dyed using chemical processes. The hair spray or styling gel you use on your hair was made using chemistry. Many foods you eat have been preserved and made more flavorful using chemicals. Everything we see, smell, touch, and taste is made of chemicals.

Chemical Elements and the Periodic Table

element

periodic table

All matter is made up of atoms. An atom is the smallest unit of any substance that still retains the properties of that substance. An **element** is a pure substance made up of only one type of atom (Figure 10-1).

The **periodic table** lists all the known elements in a way that reflects the special properties of each (Figure 10-2). Each element has its own chemical symbol. For example, look at the information shown for H, the symbol for the element hydrogen. The atomic number, which is 1 for hydrogen, indicates how many protons are contained in that element. Take a look at other symbols in the table. The symbol for the element oxygen is O. The symbol for gold is Au, which comes from the Latin word for gold, *aurum*. Not all abbreviations for the elements in the periodic table are derived from their English spelling.

Figure 10-1 Neon, which is used to light up this sign, is a chemical element.

Chemical Compounds

chemical

compound

A **chemical** is a substance that contains any of the elements found in the periodic table. Atoms of these elements can bond together to form chemical compounds. A **compound** is a substance that is formed by the combination of two or more elements. Water, for example, is a compound formed from the elements hydrogen and oxygen. The symbol for a chemical compound indicates which elements make up that compound. The symbol for water, which is H_2O, indicates that it is made of hydrogen and oxygen.

Molecules and Molecular Models

molecule

Just as the smallest unit of an element is an atom, the smallest unit of a chemical compound is called a **molecule**. Molecules are made of two or more atoms that act as a unit. A molecule of water contains two atoms of hydrogen and one atom of oxygen (Figure 10-3).

The molecular structure of a substance can be thought of as the skeleton of the molecule. It is made up of arrangements of atoms that are unique to that substance. These unique groupings provide clues about how the molecule will react with other molecules. Chemists and chemical engineers use models and pictures to depict molecules.

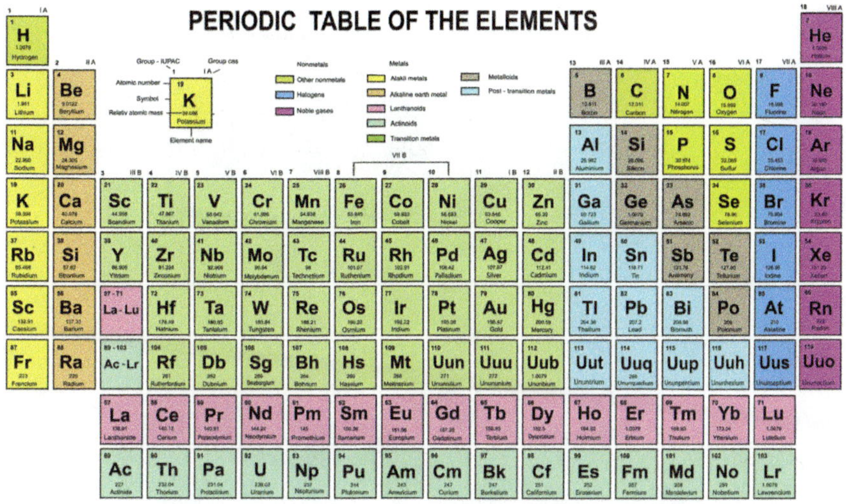

Figure 10-2 The periodic table shows information for every known element.

Figure 10-3 Molecular model of water.

Natural and Synthetic Chemicals

Chemicals can be extracted from a natural plant or animal source, or they can be made (synthesized) in a laboratory or chemical factory. There, chemical elements are combined to form compounds with the desired properties. Because of these different sources, chemicals are sometimes marketed as either natural or synthetic products.

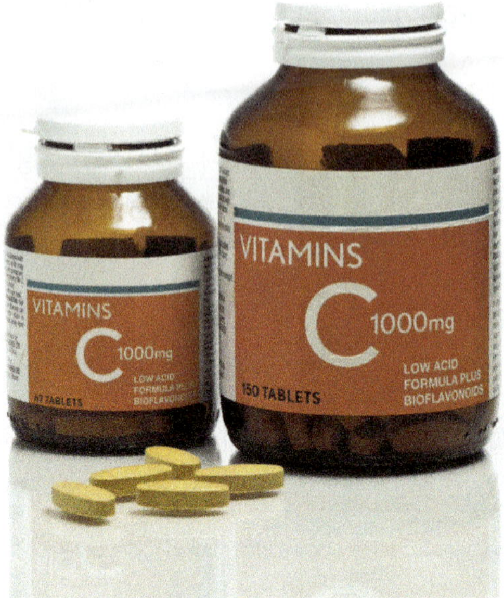

Figure 10-4 Natural products, such as these vitamins, are made of chemicals that have been derived from plants.

natural product

synthetic product

A **natural product** includes substances that appear in nature (such as edible plants and fruits) as well as substances that can be made from natural ingredients (such as wine and soap). A **synthetic product,** on the other hand, is made in a laboratory using specific chemical processes. Examples of synthetic products are chemical fertilizers and most prescription drugs.

Many people think of "artificial" and "synthetic" as being equivalent, but there is a difference. Artificial products, such as cakes and cookies, require human intervention, but they are made from natural ingredients. Synthetic products also require human intervention, but they must also be produced by chemical synthesis in a laboratory.

Synthetic chemistry uses sophisticated knowledge of chemistry to design new compounds. Using computer modeling, chemists can predict molecular properties. For instance, synthetic oils such as Mobil 1™ use a combination of natural and synthetic lubricants to create a very slippery oil that reduces friction between moving parts.

More Similar Than You Might Think

The public often considers natural chemicals to be better than synthetic chemicals. However, a synthetically produced chemical is made up of precisely the same molecules as those found in that same chemical produced in nature. In fact, the goal of chemists is often to figure out how to produce synthetic products that contain the same molecules as in the natural product. Vitamin C (ascorbic acid) is an example of a chemical that can be either extracted from plants or manufactured in a chemical plant. The largest demand for ascorbic acid is in the food industry. However, about one third of the total supply of vitamin C is used in vitamin preparations (Figure 10-4).

10.2 RESOURCES FOR CHEMICAL TECHNOLOGY

Using Chemical Resources

Chemical technology systems produce more than 70,000 commercial products—from paints to plastics to prescription drugs. They do this by using the seven technological resources.

People

Nearly 1 million workers are employed by chemical industries in the United States. These people work in 12,000 manufacturing facilities throughout the 50 states.

Capital

Industrial processing plants are huge industrial complexes costing many millions of dollars (Figure 10-5). One of the biggest expenses is the cost to control pollution and reduce hazardous wastes.

A sizable amount of money is also spent on basic research. New products and improved processes drive the industry. The chemical industry spends more on research and development than any other industry in the United States.

Time

Time is important in chemistry. Each chemical reaction has its own specific rate of reaction. This rate—how quickly or slowly the reaction takes place—is dependent on the temperature and concentration of the materials. Different chemical reactions take varying amounts of time to occur.

Time is also important in production. Time is usually the determining factor for success in the marketplace. Once a market has been identified, a product is developed as quickly as possible to meet the need and, in turn, make a profit for the company.

Figure 10-5 The cost of a large, highly automated chemical plant often runs into many millions of dollars.

Some products are ready for sale within a few months of creation. Other products, such as medicines, may take many years before they are approved for sale because extensive testing is required to see if they are safe.

Information

Information must be gathered about consumer markets for new products and about the sources and costs of various chemicals. Information is also needed to determine methods of developing products and procedures for monitoring and testing the quality of finished products, determining their effectiveness, and quantifying any harmful reactions to humans or the environment.

A great deal of scientific and technical information is needed to understand how chemicals are produced and processed into intermediate chemicals and into specialty products. Much of this information is gathered through basic research. For example, research is done to figure out ways to make natural products artificially. In that way, the availability of these chemicals is ensured and the price of these important chemicals can be kept to a minimum.

Information is also required about the best and safest ways to handle, use, and dispose of chemicals without harming humans and the environment. This information is important not only for chemical workers who may come in contact with harmful chemicals but also for the average person. For example, anyone who changes the oil in a car should know the proper way to dispose of used motor oil, which is considered a toxic material. Motor oil should *not* be dumped into a drain or buried in the yard. It should be returned to a service station that collects used motor oil for recycling.

Some chemical industry employees need to perform analyses of chemical samples. In doing so, they draw on technical information to prepare the samples and use special techniques to separate mixtures into their component chemicals.

Energy

endothermic reaction

Some chemical reactions require energy, whereas others give off energy (Figure 10-6). A reaction that absorbs energy in the form of heat is called an **endothermic reaction.**

Figure 10-6 Especially after exercise, the human body releases heat through breathing.

An example of an endothermic reaction is the decomposition of water: Energy is needed to decompose, or break down, a molecule of water into its elements, hydrogen and oxygen. Most chemical reactions are endothermic.

exothermic reaction

A reaction that releases energy in the form of heat is called an **exothermic reaction.** An example of an exothermic reaction is the formation of water from hydrogen and oxygen gases.

This reverse reaction releases the same amount of energy that is absorbed in decomposing the water molecule. It is important for chemists and engineers to take into account the energy needed in an industrial process because equipment must be designed to handle those requirements. For example, if a reaction is exothermic, conditions need to be right to prevent the reactor and pipes from blowing up.

Materials

Chemicals are produced from many different kinds of raw materials (Figure 10-7). These materials include petroleum and natural gas, metals, minerals, vegetable oils, and animal fats. The chemical industry uses these materials to manufacture more than 70,000 different products. Most products are made in huge quantities, totaling tens of billions of pounds every year. Some products, however, such as those used in specialized medicines, may be produced in quantities of only a few pounds each year.

synthesize

Chemists **synthesize,** or create, molecules for a particular material by combining two or more substances in a chemical reaction. Once the molecules are formed, they are usually mixed with additives or fillers and then processed into other forms that are useable by consumers. Molecules can be transformed into other substances through chemistry.

catalysts

Certain chemical materials called **catalysts** speed up a chemical reaction. One example of a catalyst is a chemical called ptyalin, which is found in human saliva. Ptyalin speeds up the conversion of starch, which is contained in many of the foods you eat, such as pasta and cereals. Starch is converted to sugar rapidly in the presence of ptyalin. Without this catalyst, it would take weeks for your body to digest certain foods.

Figure 10-7 Copper is a raw material that is used in many chemical processes.

Tools and Machines

A great deal of specialized equipment is used to process chemicals (Figure 10-8). This equipment is used to synthesize, analyze, purify, and manufacture products. Reactors are used to contain chemical reactions. Separators and filtering devices are used to purify products.

spectrometer

gas chromatograph

Other types of equipment include instruments for analyzing, or evaluating, chemical samples. A **spectrometer** measures the light energy absorbed or transmitted through a sample. Spectrometers are often used to measure the amount of pigment or dye in a liquid sample. A **gas chromatograph** tests the purity of a substance by heating a product past its boiling point to create gases. The gases are then passed through a tube filled with tiny ceramic or polymer beads. Individual components of the mixture are separated, and a signal indicates the purity of the sample.

Safety equipment, such as exhaust hoods and shields, is used for all types of chemical processing. People working with chemicals wear safety goggles and special clothing for protection.

In the classroom, you should also be careful when using chemicals. It is important to develop good safety practices. For instance, do not inhale chemicals—they might be dangerous. Some chemicals undergo a chemical reaction when they are heated, so what you smell could be a different substance. You should always wear goggles and a safety apron. Also, dispose of chemicals in appropriate containers, not down the sink (unless directed to do so).

Figure 10-8 A gas chromatograph is one of the tools used to process chemicals.

CONNECTING TO STEM

Reducing Pollution with Catalytic Converters

Automobiles burn gasoline, which is a type of hydrocarbon fuel. A hydrocarbon fuel is made up of hydrogen and carbon atoms. In an ideal situation, all these atoms would combine with the oxygen in the air to form water (H_2O) and carbon dioxide (CO_2). However, the combustion process is not ideal, and some harmful emissions are also produced in car engines. These emissions include carbon monoxide (CO) and nitrous oxides (NO and NO_2).

A catalytic converter, which is part of a car's exhaust system, is designed to remove these pollutants. A catalytic converter is a ceramic structure shaped like a honeycomb and coated with the metal catalysts platinum, rhodium, and palladium. The honeycomb structure creates a lot of surface area for the exhaust to contact.

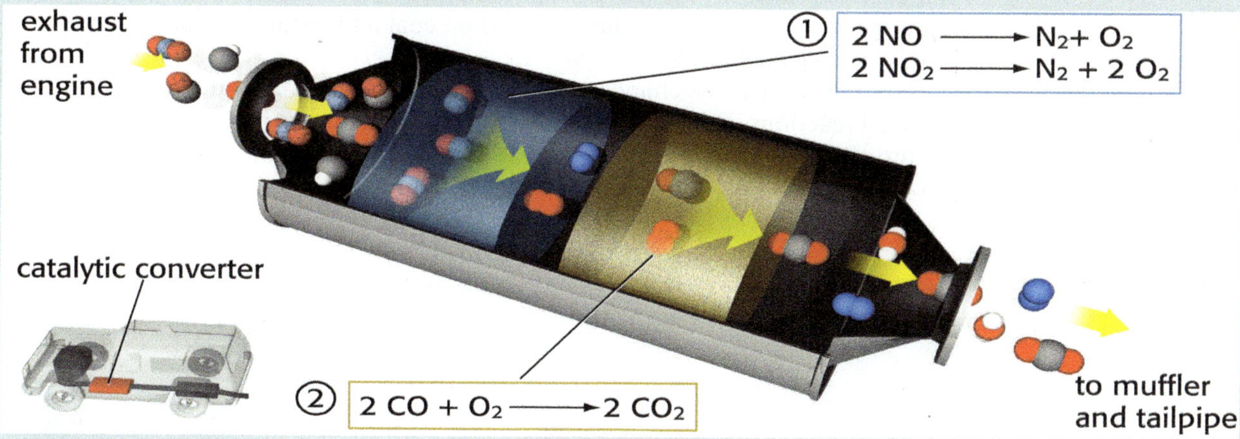

Removing Nitrous Oxides

The first stage of a catalytic converter converts NO_2 emissions. The catalyst holds onto the nitrogen molecule, freeing the oxygen atom. The oxygen atom bonds with another oxygen atom and forms O_2, the oxygen molecule found in air. The nitrogen atom bonds with another nitrogen atom and forms N_2, which is part of the air we breathe.

Removing Carbon Monoxide

The second stage of a catalytic converter converts carbon monoxide into carbon dioxide and water. Hydrocarbons are burned to form water and carbon dioxide. Because oxygen is required for this reaction, an oxygen sensor monitors the amount of oxygen in the exhaust. If oxygen is not sufficient for the catalytic converter to work properly, more air is admitted into the engine.

10.3 USING CHEMICAL TECHNOLOGY

What Is Chemical Technology?

One characteristic of the chemical industry is that it provides chemicals to be used by other industries in manufacturing products. The chemical industry provides resources to virtually all sectors of the economy.

The numerous products supplied by different areas of chemical technology impact our lives every day. Chemical reactions are all around us. Cooking produces a chemical change in food, as when you bake a cake. When our bodies absorb oxygen from the air, another chemical reaction is taking place.

Advances in Chemical Technology

The chemical industry in the United States advanced tremendously during the last century. It moved from an industry based on coal and organic chemicals to one based mainly on natural gas and oil.

Accompanying this change was a better understanding of the role of catalysts in chemical reactions. This led to the development of synthetic products such as fertilizers (Figure 10-9). New plant strains were created that were highly responsive to fertilizers, providing higher crop yields, not just bigger plants.

Figure 10-9 Most fertilizers today are made from chemicals. Fertilizers are used to increase the productivity of farms.

Polymers and Synthetics

The study of polymers, which are large molecules, has become an important branch of chemistry. Just after World War II, research on polymers led to the development of plastics (Figure 10-10). At the same time, the development of synthetic rubber became important because the source of natural rubber in the Far East was cut off during World War II. The rubber industry was transformed from one that depended on a supply of natural raw material to one that draws on the chemical processing of oil for its primary material inputs.

The knowledge gained during World War II also led to transformations in the clothes we wear. Synthetic fibers—nylon, acrylics, and polyesters—were developed and now are commonly used in the textile industry.

Chemically Based Fuels

The expanding automobile industry created a demand for more and improved gasoline and oil. The petrochemical industry again rose to the challenge and developed high-quality liquid fuels. In the process, the industry developed the ability to build large-scale production facilities that were more efficient than earlier, smaller-scale plants. Improved design information allowed this expansion of scale. New processes were defined that led to the development of polymers and synthetics. Today, the petrochemical industry makes up just one of the many different chemical industries (Figure 10-11).

Figure 10-10 Plastics, such as these Legos, are one example of a product made using chemical technology.

Category	Examples of Products
Agriculture and Food	Pest and weed control, fertilizers, animal feed, food flavorings, sweeteners, preservatives, new foods, food packaging
Intermediate and Specialty Chemicals	Acrylic acid (used to produce superabsorbent polymers), Biocides (used to slow the growth of bacteria, molds, fungi, algae), Amines (used to manufacture pharmaceuticals, cosmetics, plastics, and dyes), Catalysts (used to speed up other chemical reactions)
Electronics	Semiconductors, superconductors, integrated circuits, photocopier cartridges, coated wiring
Energy and Fuel	Fossil fuels (natural gas, oil, coal), synthetic fuels, solar cells
Environmental Technologies	Pollution control methods, hazardous waste disposal, water and sewage treatment, recycling processes
Health and Medicine	Vaccines, vitamins, medicines
Housing and Construction	Insulation, roofing materials, concrete, vinyl siding, floor tiles and coverings, countertops
Industrial Coatings and Colorants	Primers, paints, stains, finishes, solvents, dyes, caulking, ceramic glazes, resins, glues
Paper and Printing	Pulp production, papermaking, pens, pencils, inks, packaging
Polymers, Materials, and Textiles	Plastics, fiberglass products, composites, nylon, denim, permanent-press and spandex fabrics
Personal-Care Products	Mouthwash, cosmetics, hair-care products, detergents

Figure 10-11 Chemical technologies are used in many different ways.

cracking

Cracking is a chemical process in which a complex chemical compound is broken (cracked) into simpler compounds. This change is accomplished using high temperatures and pressure (thermal cracking) or by using a catalyst (catalytic cracking). These processes allow chemists and chemical engineers to create processes to convert crude oil, which is very thick, into gasoline and oils used in automobiles and homes.

Pharmaceuticals

pharmaceutical

The pharmaceutical industry is another important chemical industry. **Pharmaceutical** products include medicines, vaccines, and drugs used for healthcare. Again, World War II provided the impetus to link pharmaceutical companies with universities that were advancing the state of the art in biomedical research. For instance, penicillin was discovered in 1928, but its production was done on a small scale. Chemical engineers linked with microbiologists to solve the problems posed by large-scale production of penicillin, creating the first major achievement in biochemical engineering.

Because the development of new medicines had become a national priority, the National Institutes of Health (NIH) and the National Science Foundation (NSF) provided research funding that spurred the industry's growth. The advances in pharmaceuticals resulted in new antibiotics, anti-inflammatories, anti-ulcer drugs, antidepressants, and antihistamines.

Surfactants

surfactant

In ancient times, clothes were cleaned by beating them on rocks in streams. Certain soaps, such as the saponin found in the plant soapwort, were used to facilitate the cleaning process. Soap is a **surfactant,** or *surface-active agent*, and is normally made from plant and animal fats. Soaps are naturally biodegradable (i.e., they break down into environmentally safe chemicals) and inexpensive, and they are processed from renewable resources. However, during World War II, these fats were in short supply and were very expensive. During this time, the first synthetic detergent was made. Detergents are synthetic versions of soaps derived from petrochemicals.

Surfactants have a unique chemical structure (Figure 10-12). Part of the molecule loves to combine with water and hates to combine with fats. The other part hates to combine with water but loves to combine with fats. When these chemicals are in the laundry, part of the molecule bonds with the dirt and part of the molecule bonds with the water so that the dirt is washed away with the water. If air is trapped in water that contains a surfactant, bubbles may be formed. The amount of foam produced by a soap is not necessarily a measure of its cleaning power, although many consumers believe this to be true.

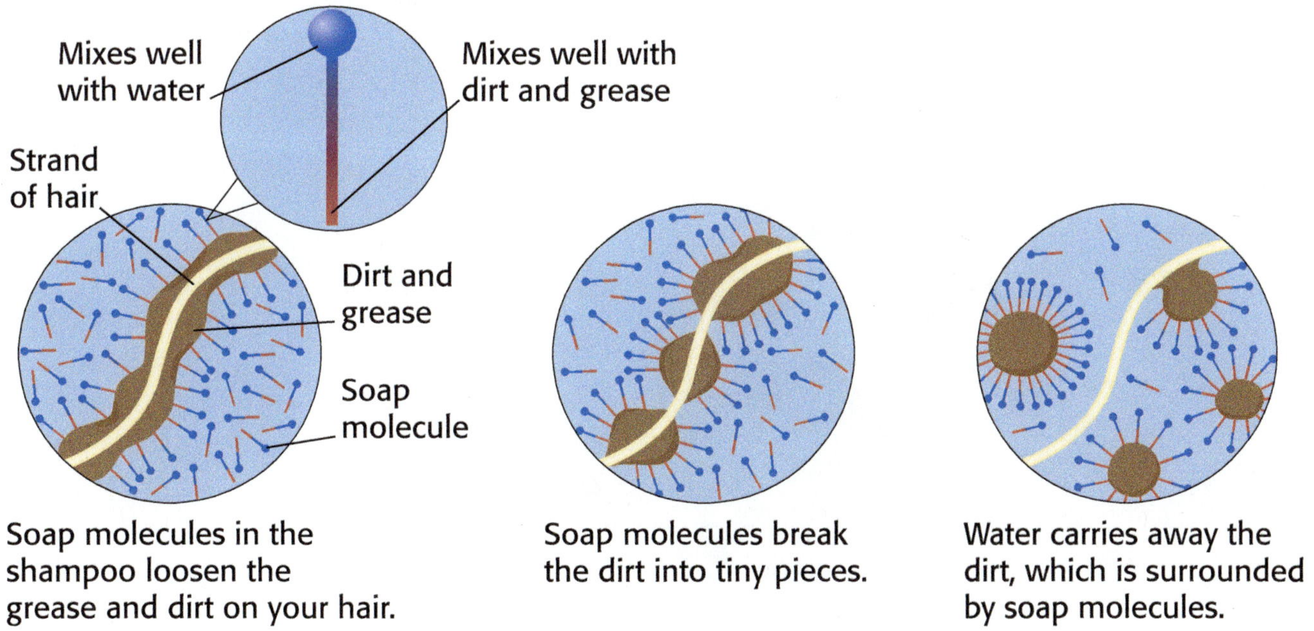

Figure 10-12 Soap has a special structure that allows it to surround and attach to dirt and grease.

Using Chemicals Today

Chemical technology has improved many aspects of our everyday lives. Medicines and vaccines, for example, are allowing people to live longer, healthier lives (Figure 10-13). However, chemicals are used in many different ways—some of which people may not be aware of. To be wise consumers of technology, we need to understand how chemicals are used and what risks they pose.

Commonly Used Chemicals

Many of the chemicals that may pose health risks are used as food preservatives. Sulfites are used as a preservative in many fruits, vegetables, seafoods, and wines. BHA is used in dry foods, such as cereals. Nitrites are used to preserve red meats, poultry, and fish.

The levels of chemicals in a food that can be considered acceptable, or safe for human consumption, are regulated by the Food and Drug Administration (FDA). The FDA determines what levels are appropriate based on the results of many laboratory tests. Often, the effects of a chemical are difficult to assess. BHA, for example, has been shown to cause cancer in laboratory mice when they were given high levels of the chemical. However, in these same tests, BHA also reduced the effects of other carcinogens.

Balancing the Risks with the Benefits

Although the public perceives chemical additives as a major cause of cancer, only 1 to 3 percent of cancer cases can be attributed to human-made chemicals. Additional testing and monitoring are needed to determine the exact effects of chemicals on our long-term health.

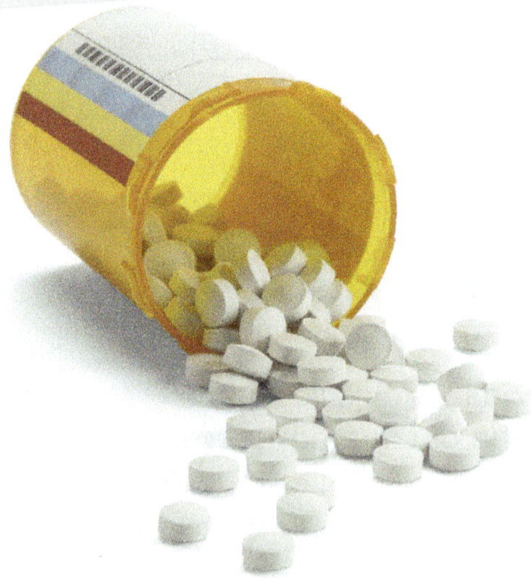

Figure 10-13 Prescription drugs play a key role in the medical industry.

PEOPLE IN TECHNOLOGY: WALLACE CAROTHERS

Today, the clothes we wear are made using all types of chemicals. Gore-Tex is used in raingear and boots, lycra and Spandex are used in athletic wear, and neoprene gel is used as a cushion in everything from wristbands to bicycle seats.

One of the first fabrics to be produced chemically was nylon, which was first made into women's stockings. Nylon was developed and produced due in large part to the efforts of a chemist named Wallace Carothers.

From Harvard to DuPont

Carothers was born in Iowa, where his father was a college professor. Carothers was always interested in science and chemistry. In high school, he built a chemical laboratory in his bedroom. He went on to earn his Master's and Ph.D. degrees, and then became an instructor at Harvard.

In the 1920s, while Carothers was still at Harvard, he was approached by the chemical company DuPont. DuPont wanted to develop a synthetic fiber that could replace silk, which had come into short supply because of the United States' trade problems with Japan. DuPont hired Carothers to pursue basic research on polymers—long, chainlike molecules.

Figure 10-14 Colorful stockings.

Developing and Making Nylon

In 1934, only a few years after he had come to DuPont, Carothers and his team of researchers produced the first long, flexible strands of nylon. This material was both strong and flexible—features that natural silk also possessed.

Once the formula for nylon had been determined, DuPont set about to produce millions of tons of nylon fiber at a relatively low cost. In large quantities, nylon had to be made in a hot, steel-lined vessel called an autoclave.

Nylon Hits the Shelves

The first nylon stockings, known simply as "nylons," were sold on October 27, 1939. They sold out within hours.

Due to his persistent efforts, Carothers brought to the world not just an incredible new material but also the knowledge to develop many more chemical products.

10.4 DEVELOPING CHEMICAL PRODUCTS

Product Development

You may know that furniture and buildings are products that are designed, but chemical products are also carefully designed by scientists and engineers. Designing and developing a chemical product is a complicated, time-consuming procedure. There are several steps in the product development cycle that must be completed before a product can be sold on the open market.

Market Analysis

As with most other technological designs, human needs and desires drive the development of a chemical product. A team of people, from business managers to environmentalists, works together. They determine the need for a new product, its possible impact, and whether it will be profitable.

Product Design and Synthesis

Once the market need is established, researchers design the new product. The appropriate starting materials are identified, and chemical reactions are performed to synthesize the desired product.

The product must have certain properties—it must meet specific requirements or design criteria for the particular application. For example, a dish detergent might have to have a certain viscosity (thickness), color, and appearance (Figure 10-15). These specifications are set directly by the product manufacturer and indirectly by the consumer. For example, consumers probably would not want to buy a cloudy, gray, two-layer dish detergent, so manufacturers set tight specifications to ensure product desirability.

Figure 10-15 The color and texture of dishwashing liquid are the result of careful testing and market research.

Figure 10-16 Scientists perform many tasks in a chemical laboratory.

Characterization and Purification

A team of scientists, engineers, and technicians will characterize a product to determine its physical properties (Figure 10-16). This work is done through a series of tests to find out exactly which molecules are in the product.

purify

Once the exact properties of a product are known, scientists **purify** the product so that the substance is made of only one kind of matter. A product can be purified by filtering, distilling (evaporating and then condensing it), or subliming (heating a solid material into a gas and then condensing it back into a solid).

Formulation and Testing

formulation

Once the key ingredients have been identified, synthesized, characterized, and purified, they are formulated with other chemicals to create a product with the desired properties. A **formulation** is like a recipe, a systematic method of creating a result. The ingredients and their amounts are varied until the desired result is obtained. Formulators (people who do formulation) use computerized methods to create the desired product using a minimum number of trial formulations.

Sample formulations are thoroughly tested before the new product is manufactured. Some of the tests carried out on a formulation try to answer some of the following questions:

- Does the product work equally well in cold or hot temperatures?
- How does the product react with water?
- How does the product react chemically with the container in which it will be stored?
- How does the product look, feel, and smell, or what is its viscosity?

Scaling Up and Processing

Once the desired properties have been achieved, the formulation and its ingredients can be scaled up, or manufactured in large quantities. A pilot batch is usually made as a further test before sending the formulation to the manufacturing plant. A pilot batch is larger than the quantity produced for research, but many times smaller than the actual production run. The pilot batch is tested to make sure the formulation works for larger quantities.

batch process

Chemical formulations can be manufactured in a batch process or in a continuous process. In a **batch process,** all materials are included to complete one operation. A batch process is similar to making a giant pot of spaghetti sauce. All the ingredients are added and cooked in the pot (the chemical reactor) until the sauce (formulation) is done. At the end of the reaction, you have one giant pot of spaghetti sauce (the product).

continuous process

In a **continuous process,** materials are constantly added to the reactor, where they are mixed and heated according to the formulation, continuously producing the desired product. A continuous process is usually more cost-effective than a batch process.

EMERGING AND INNOVATIVE TECHNOLOGY: MATERIAL PROPERTIES

Engineering is a constantly evolving field filled with applications of emerging technology, demonstrating the fast paced innovation typical of this century. In particular, materials and chemical engineering have found a niche for solving interesting problems.

mass

Mass is a fundamental property of materials, so it is imperative to have a standard definition of the scientific unit, kilogram. Most scientific units can be measured anywhere in the world by using natural properties; however, we still based the kilogram on a physical object—a platinum and iridium cylinder. Despite measures to preserve the cylinder, the mass was inexplicably changing throughout the years. The Australian Centre for Precision Optics took a materials engineering approach to solving the problem by creating nearly perfect spheres composed of pure silicon, which boasts high stability to ensure that little to no fluctuation in mass occurs. The mass of the sphere is found by calculating the weight and volume. Then we can know the number of silicon atoms in the sphere and have a more reliable definition of mass.

density
conduct

Another property related to mass is the material's **density**, the amount of mass per unit of volume. We can describe the ability of the material to **conduct** electricity and heat (electrical and thermal). Another useful property is how well a material can transmit and absorb sound.

More mechanical properties concern how the sample will behave when it undergoes some sort of physical transformation. We could provide a measure to the amount of force we exert on a material by stretching it out (tension) or by pressing it down (compression).

tensile test
stress strain curve
strain
stress

When testing for these quantities, one can use a **tensile test** to plot the **stress strain curve** of the material. The stress strain curve plots the **strain** (the ratio of the change in length to the original length) against the **stress** (the applied load pulling the material apart).

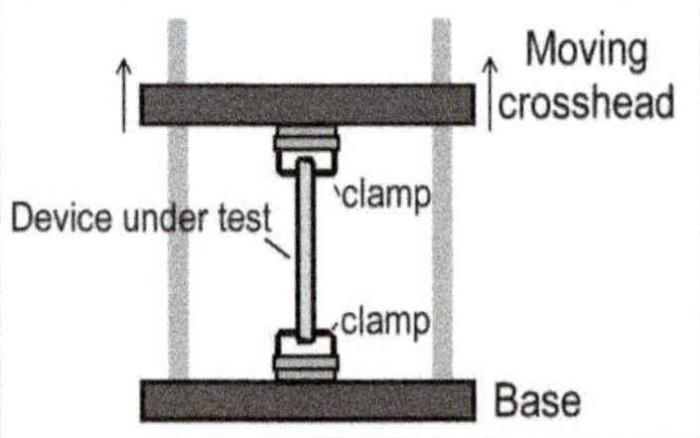

How does this test work? If you have ever pulled on taffy or licorice, you know that you can gently pull and the candy goes back to its original shape. Pull a little harder and it stretches, and there is no going back: It's permanently stretched. Pull harder still and it snaps. This process works for metal as well.

Modulus of Elasticity
Young's Modulus
stiffness
yield stress
Ultimate Tensile Strength
UTS

The stress-strain curve shows how a material responds when a stress (force per unit area) is applied (whether it's compressed or stretched). For lower stress values, the strain is linear (the elastic region): When we remove the stress, the material goes back to its original shape. In this region, we can measure the slope of this line, λ, called the **Modulus of Elasticity** (or **Young's Modulus**), which is also known as **stiffness**. Once we hit the **yield stress**, we enter the plastic region. If stress is applied in the plastic region and then removed, the material will be permanently deformed. The maximum stress we can apply before damage is the **Ultimate Tensile Strength**, or **UTS**.

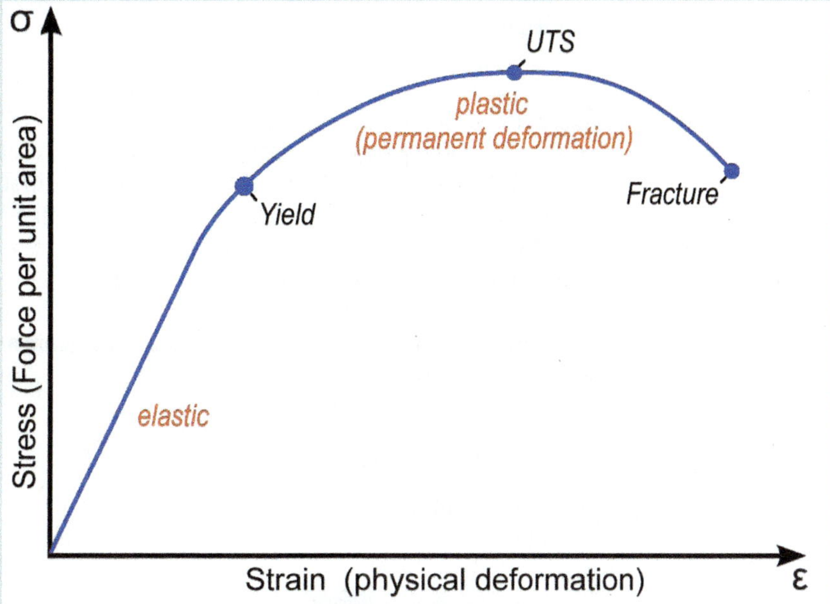

fracture stress

Beyond this, we will see damage until finally we reach the **fracture stress**. Different material has different stress-strain curves.

How do we take these properties into account? Imagine selecting a material for high-speed rail: Emerging high-speed transportation can use magnetics to eliminate friction between the wheels and track. The material we select for the track needs to be ferromagnetic (support magnetism). If we allow for contact between wheels and track, we would look for a material that would have a large elastic region – one that could handle significant stress without deforming. For very small-scale design, we may need to measure the stress and strain to high precision.

When testing a material, the scientist must use the machine most suitable for the job to ensure accuracy and precision of the results. If a material is brittle and fractures easily, then a large machine for highly ductile (stretchy) samples may not be appropriate.

Discuss/Research: When designing a new, thinner, larger cell phone, our design team intends to make the back cover out of a very thin layer of metal. Do we want high or low values of the following: Young's Modulus, ferromagnetism, heat conductivity, and mass?

Research: Say you discovered a new material, test its material properties, and it exhibits qualities never seen before! Look online for venues where you could publish your research.

DESIGN 10: A SIMPLE CHEMICAL TECHNOLOGY: SLIME

Problem Situation

Slime is a gooey substance sold at many toy stores. In its pure state, it is a clear colorless gel that resembles hair gel. Slime can be shaped into a ball and bounced, or flattened to make imprints of comic strips. It can also be stretched if pulled slowly, or it can "fracture" if pulled apart quickly. It has properties such as tensile strength, flexibility, and elasticity. It can be stored in the refrigerator in a tightly sealed plastic bag for about 2 weeks before becoming moldy. Slime is made from a polymer called PVA, polyvinyl alcohol. A polymer is a very large molecule that is made up of many smaller parts.

Your Challenge

You and your team members are to design your own slime that will bounce and stretch.

Materials you will need:

Use digital scales and graduated cylinders to precisely measure each quantity.

- 10 ml white glue
- 12.5 ml of borax solution (4 gm borax dissolved in 100 ml of water)
- 50 ml of polymer solution (4% polyvinyl alcohol, PVA)
- disposable plastic cup
- food coloring
- maximum of 3 ml of borax per formulation
- mixing rod or tongue depressor

1. Clarify the Design Specifications and Constraints

To solve the problem, your design must meet the following specifications and constraints:

- When formed into a ball, the slime should bounce at least 60 cm high.
- When pulled apart, the same ball of slime should stretch at least 20 cm.
- All slime formulations must total 10 ml.

In your engineering notebook, answer the following question in your own words:

1. What are the specifications and constraints the design must meet?

2. Research and Investigate

In order to better complete the design challenge, you need to first gather information to help you build a knowledge base.

Knowledge and Skill Builder 1: Cross-Linking Polymers

Complete this activity with your classmates. Four students stand in line holding hands. Four more students stand in another line behind the first line. These students represent the PVA chains of molecules. Can they move freely?

Two more students go between the two lines of students and hold onto a student's hand in each line. These students represent a borate molecule. Can the lines move freely now? Is there a limit to the number of borate molecules that can be added?

If there were water molecules between the chains, they would be trapped in a three-dimensional network, which is called a gel.

Investigate cross-linked polymers online or using an encyclopedia.

Knowledge and Skill Builder 2: **Preliminary Slime and Gluep Formulations**

The following formulation for slime works, but it may not satisfy the design specifications. You can use this as a basis for a factor analysis.

1. Put 12.5 ml of PVA solution in a disposable cup. You may add one drop of food coloring to better see the liquid.
2. Stir the PVA solution, noting its viscosity, or how easily it drips from the mixing rod.
3. Measure 3.0 m of borax solution and pour it into the PVA solution, stirring vigorously until the gelling is complete.
4. Remove the gel from the cup and knead the mixture—this ensures that the components are well blended.
5. Roll the gel into a ball and place it on the table. Does it maintain its shape?
6. Drop the ball. What happens?
7. Roll the gel into a 6 cm-long cylinder and place it on the table. Slowly pull it from both ends. What is the maximum length reached before breaking?
8. Again, roll the gel into a 6 cm-long cylinder. Quickly pull it apart. What is the maximum length reached? Describe what happened.

The following formulation is for gluep. This is a cross-linked polymer as well.

1. Mix 10 ml of white glue with 10 ml of water (add a drop of food coloring if you wish).
2. Add 7 ml of borax solution and stir vigorously. The gluep will cling around the stick and can be pulled out, usually in one blob.
3. Knead the gluep well to mix the components.
4. Roll it into a ball and place it on the table. Does it maintain its shape?
5. Drop the ball. What happens?
6. Roll the gluep into a 6 cm-long cylinder and place it on the table. Slowly pull it from both ends. What is the maximum length reached before breaking?
7. Again, roll the gluep into a 6 cm-long cylinder. Quickly pull it apart. What is the maximum length reached? Describe what happened.

Knowledge and Skill Builder 3: **Factor Analysis**

Perform a factor analysis on the slime to gather data that allows you make an optimum formulation. You can vary the ratio of PVA and borax solutions. If your teacher allows, you can vary the percent of borax in solution. For example, change it to 5% or 6%.

Test	PVA (ml)	Borax (ml)	Bounce (cm)	Stretch (cm)
1				
2				
3				

Plot the Bounce and Stretch as dependent variables (this will require two line graphs) versus the different tests as independent variables. What conclusions have you reached about the ratio of borax and PVA? Does increasing the borax improve the gel? (Think about Knowledge and Skill Builder 1.)

3. Generate Alternative Designs

Describe two possible solutions to the problem. Remember to consider the specifications and constraints. In your description, indicate what you consider each solution's strengths and weaknesses. Document both of your alternate designs in your engineering notebook or digital portfolio.

4. Choose and Justify the Optimal Solution

From your various alternatives, pick your preferred solution. In your engineering notebook, respond to the following questions:

1. Explain why your solution meets the specifications and constraints.
2. Why is your choice the better alternative?
3. What trade-offs, if any, did you make in selecting this alternative?

5. Develop a Prototype

Make your final gel using the alternative design you chose. Then, in your engineering notebook, perform the following steps and answer the questions posed:

1. Include a copy of the formulation you used as a page in your notebook.
2. In any technological activity, you will use seven resources: people, capital, time, information, energy, material, and tools and machines. Indicate which resources were most important in this activity, and how you made trade-offs among them.

6. Test and Evaluate

Now that you have a prototype, how will you evaluate your design? In your engineering notebook, respond to the following prompts:

1. Develop a test plan. Describe any testing procedures you will use.
2. Indicate how the results will show that the design solves the problem and meets the specifications and constraints.

7. Redesign the Solution

Once the testing period is over, it is time to take a step back and look for areas in which a redesign will improve the initial design. Respond to the following questions in your engineering notebook:

1. If you had to redesign your solution, what problems did you encounter?
2. Did you modify your original design concept? Why?
3. If you were to redesign your model, what changes would you recommend in your new design? Explain your reasoning.
4. What additional trade-offs, if any, would you have to make?

8. Communicate Your Achievements

In your engineering notebook, describe the plan you will use to present your solution to your class, and show what handouts you will use (including PowerPoint slides). Be sure your presentation contains the following elements.

Introduction: What is this design challenge about?
Purpose: Why did you do this investigation?
Procedure: How did you go about designing your slime?
Results: What did you discover? Include relevant tables, charts, and equations – if any. Also, list the volumes and quantities used to the highest degree of precision available (according to your equipment).
Discussion: Comment on your results. What were the limitations you faced while designing?
Conclusion: Summarize the presentation and provide the audience with a sense of finality.

Chapter 10 ASSESSMENT QUESTIONS

1. What is a chemical?
2. Using the periodic table, give the symbol and atomic weight of the following elements: copper, tin, lead, and uranium.
3. In your own words, define the following terms: atom, element, molecule, and compound.
4. How are elements and compounds formed?
5. Why is the chemical structure of molecules important to chemists?
6. Explain the ways in which natural and synthetic chemicals are both different and similar.
7. Explain why different chemical reactions take varying amounts of time to occur.
8. List the kinds of information that are important in chemical technology.
9. What are catalysts? Why are they so important to chemists?
10. Why does the chemical industry spend more on research and development than any other industry?
11. What is the difference between endothermic and exothermic reactions?
12. Describe the chemical process of cracking and how it is used to produce gasoline.
13. Name three synthetic products that were developed during and shortly after World War II.
14. Why were there so many synthetic products developed during World War II? Use specific examples from the previous question in your answer.
15. Explain what a surfactant does and where it might be used.
16. If you were to dilute 10 milliliters of a chemical solution to measure one part per million (ppm), how many liters of water would you need to do this?
17. How is a new chemical product designed? Explain the steps.
18. List the methods of purifying a product.
19. Describe what happens during the formulation process.
20. Explain the difference between batch processing and continuous processing.
21. Research: Find an innovative chemical technology used in either food science, personal protection, or energy efficiency.
    a. Write a 2-page report. Be sure to describe why this is an innovative technology.
    b. Prepare a 2-minute presentation. Be sure to discuss why this is particularly innovative.
22. Research: Compare chemical, mechanical, and physical properties of five materials. What is the difference between these types of properties?
23. Find the stress-strain curve for two of these metals: iron, hardened steel, titanium, or aluminum. Compare the curves: What can you say about the material properties of your two metals based on these curves?
24. Research and present: Two emerging and innovative technologies are custom medicine delivery via nanotechnology and organic electronics. Prepare a 3-page report and a 2-minute presentation as follows:
    - Why is each considered innovative and emerging technology?
    - Describe the state-of-the-art: Are these developed? In use? Where?
    - Describe the unique material, physical, and mechanical properties involved that make these innovative. In other words, what properties are new and unique?

CHAPTER 11
STATISTICS

probability

Probability is associated with assessing the likelihood that an event will or will not occur. For example:

Airplane crash	River breaching a levee	Nuclear reactor accident
Tornado	Failure of equipment	Terminal cancer
Earthquake	Microprocessor failure	Space probe data reception

statistics

Statistics are used for design-concept evaluation because they provide quantitative measures to "things" that behave in a random manner. This evaluation helps us make rational decisions about everything from natural events to manufactured products. Statistics, as well as probability, use numerical evidence to aid decision making in the face of uncertainty. Roles of statistics in engineering include the following:

- Evaluation of new or alternative designs, concepts, and procedures
- Estimation of amount to bid on projects
- Management (human uncertainty, economic uncertainty, and others)
- Determination of degree of acceptable item-to-item variation (quality control)

Often, the best way to analyze an engineering problem is to conduct an experiment. When we take this approach, we face several questions:

- How many tests do we need?
- How confident are we in the results?
- Can we extrapolate the results to other conditions?
- Can we estimate how often the result will lie within a specified range?

There are many other related issues, but addressing all of them requires a separate book. Some of you will take or have taken an entire course in probability or statistics, so for now we just touch on some of the important fundamentals.

- **Repeated tests:** When a test is conducted multiple times, we will not get the same (exact) result each time. For example, use a ruler to measure the length of a particular brand of shoe manufactured by Company X. You can produce a table of values, all of which will be nearly, but not exactly, the same. This is because you make slight errors in measurement, so even if every shoe is identical, there will always be some errors in your measurement. Moreover, every shoe is not exactly identical.
- **Differences in a population:** What is the heart rate of all students in a class? Obviously, not everyone will have the same heart rate. We do expect, however, that everyone's rate will lie between, say, 40 and 140 beats per minute. Through measurement, we can determine this variation. In fact, we may find that the average rate for females and males differs. Statistical procedures help us analyze situations such as this.

- **Manufacturing errors:** Suppose you are manufacturing a run of widgets and a buyer wants all of them to be exactly alike. Obviously, this is impossible, but you can make them almost alike and then tell the buyer how much variation to expect. If you measure each widget as it comes off the assembly line, there will certainly be some variation.
- **Design criteria for products:** When you build a house, you would like it to stand safely for some period of time. For example, you might specify that the house be designed to withstand a windstorm that would occur, on average, every 50 years. For that case, you must be able to calculate the wind speeds associated with such a storm. Statistical methods allow you to do this.

11.1 HISTOGRAMS

To illustrate several common statistical concepts, we use data representing the height of several tenth-grade students. Table 11-1 shows the height, to the nearest inch, of each student in a typical class. Table 11-2 shows the same data, summarized by number of students at each height.

Table 11-1 Student height

Student ID	Height (*H*) [in]
A	67
B	73
C	71
D	69
E	68
F	64
G	70
H	72
I	67
J	71
K	70
L	68
M	66
N	71
O	74
P	71
Q	68
R	72
S	67
T	64
U	75
V	74
W	72

Table 11-2 Summary of height data

Height (*H*) [in]	Number of Students
62	0
63	0
64	2
65	0
66	1
67	3
68	3
69	1
70	2
71	4
72	3
73	1
74	2
75	1
76	0
77	0
Total	23

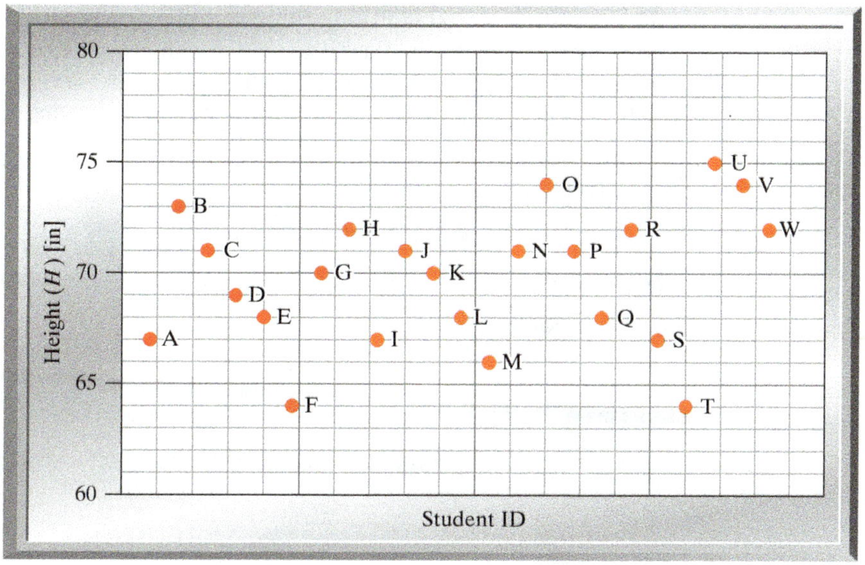

Figure 11-1 Example of student height, shown on scatter plot.

When we graph the values shown in Table 11-1, we end up with a scatter plot with data that is exactly the same: scattered, as shown in Figure 11-1.

Instead of using a scatter plot, we can group the data and plot the group values in a chart similar to a column chart, shown in Figure 11-2. Using the summarized data shown in Table 11-2, we will place two height ranges into a single column or **bin**. The first bin will contain all student-height values less than 62 inches. The next bin will contain student-height values of 62 and 63 inches. The next bin will contain student-height values of 64 and 65 inches, and so on. The x-axis of the graph is the height values; the y-axis is the number of students measured at each height range. Graphs of this nature are called **histograms**. By counting the number of blocks, we find the area under the curve represents the total number of samples taken, in this case, the total number of students (23) observed.

bin

histogram

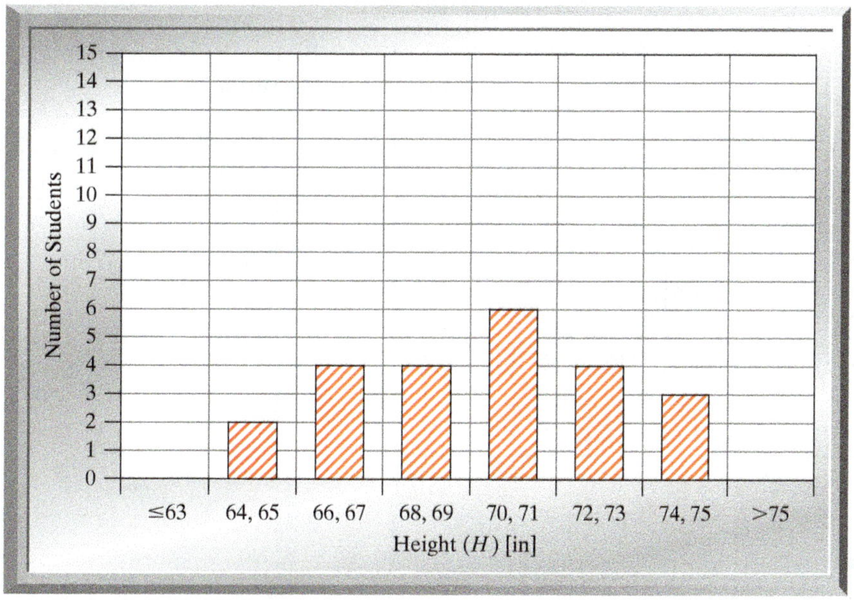

Figure 11-2 Example of student height, shown on a histogram.

Technically, before beginning this example we should have mathematically determined a bin size, rather than arbitrarily grouping the measurements in pairs (62 and 63 in one bin, 64 and 65 in the next bin, etc.). There are several ways to calculate the bin size that will best display the information; below is one method.

Determination of Bin Size

Step One: Determine the number of bins needed.

Number of Bins = Square root of number of data points, rounded to whole number

Step Two: Determine the range of the data.

Range = $X_{max} - X_{min}$

Step Three: Determine the number of items in each bin.

Bin Size = Range divided by Number of Bins, rounded to whole number

Let us apply this to our example.

Step 1: As shown, we have a class of 23 students, so we would need five bins, since the square root of 23 is about 4.8, which rounds to 5. Four would probably also work fine, as would 6. Remember this is just a rule of thumb.
Step 2: The shortest person is 64 inches tall and the tallest is 75, so the range is 11 inches.
Step 3: Dividing the range determined in Step 2 by the number of bins determined in Step 1, we get 2.2, or about 2 inches per bin. On the other hand, we might instead decide to have four bins. If we divide the range by 4, we have 2.75 or 3 inches per bin.

distribution

Depending on the number of bins, we sometimes get two different, but acceptable, bin sizes. By changing the bin size, we can change the appearance of the data spread, or the data **distribution**.

What happens to the student height data if we alter the bin size? The plot on the preceding page shows a 2-inch bin interval, and Figure 11-3(a) shows a 3-inch interval.

In Figure 11-3(b), we have used a 4-inch bin interval, and while it is not what we obtained from the "rule of thumb" (2- or 3-inch intervals), it is still mathematically correct but not as informative as the other two.

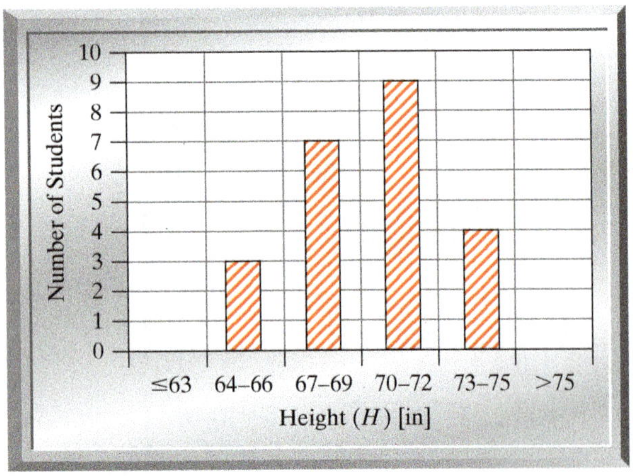

Figure 11-3(a) Bin size of 3.

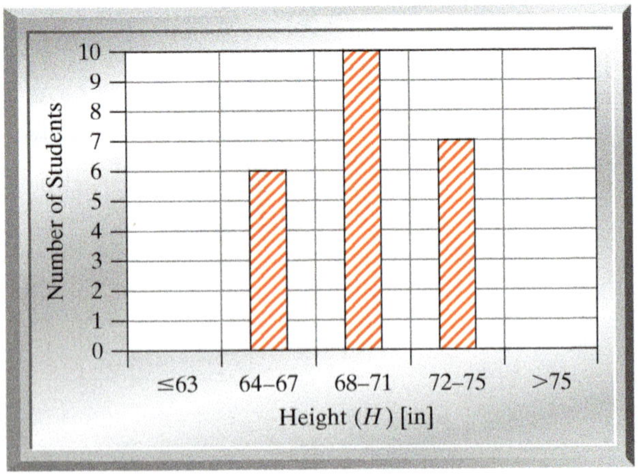

Figure 11-3(b) Bin size of 4.

11.2 STATISTICAL BEHAVIOR

When we have gathered the data and plotted a distribution, the next step is to explain the outcome to others. For convenience, we identify a set of parameters to describe distributions.

average
mean
central tendency

One parameter of a distribution is the average value. The **average**, or **mean**, is an estimate of the value most representative of the population. This is often called the **central tendency** of the data. The computation of the mean ($\overline{X}$) of a data set containing N values is given in the equation below.

$$\text{Mean} = \overline{X} = \frac{1}{N}(X_1 + X_2 + \cdots + X_N) = \frac{1}{N}\sum_{i=1}^{N} X_i$$

In other words, the mean is the sum of all of the values divided by the total number of values.

median

The **median**, another measure of central tendency, is the value between the lower half and the upper half of the population. In other words, if all data points are listed in numerical order, the median is the value exactly in the middle of the list. If the number of data points is odd, the median will be the middle value of the population. If the number of data points is even, however, the median will be the average of the two values at the center. A few examples should clarify this.

Set	Data	Mean	Median
1	1, 2, 3, 4, 5, 6, 7	4	4
2	1, 50, 70, 100	55	60
3	5, 10, 20, 40, 80	31	20
4	50, 50, 50, 50, 50, 1,000	208	50

Review the data shown in set 4. It would seem logical if every data point has a value of 50 except one, the average of the data should be about 50; instead, it is 208! This illustrates the sensitivity of the mean to **extreme values**, or **outliers**. Note that the median is unaffected or only slightly affected. It is for this reason that the mean is insufficient to describe the central tendency of all distributions.

extreme value
outlier

variance
standard deviation

Two other terms are useful in describing a distribution: **variance** and **standard deviation**. Both of these terms quantify how widely a set of values is scattered about the mean. To determine the variance (σ^2), the difference between each point and the mean is determined, and each difference is squared to keep all terms positive. This sum is then divided by one less than the number of data points.

$$\text{Variance} = \sigma^2 = \frac{1}{N-1}((\overline{X} - X_1)^2 + (\overline{X} - X_2)^2 + \cdots + (\overline{X} - X_N)^2)$$

$$= \frac{1}{N-1}\sum_{i}^{N}(\overline{X} - X_i)^2$$

The standard deviation (σ) is found by taking the square root of the variance:

$$\text{Standard deviation} = \sigma = \sqrt{\text{Variance}}$$

Table 11-3 Height data

Height (*H*) [in]	Number of Students
62	0
63	0
64	2
65	0
66	1
67	3
68	3
69	1
70	2
71	4
72	3
73	1
74	2
75	1
76	0
77	0
Total	23

If we again examine the data found in Table 11-3, we can calculate the mean, median, variance, and standard deviation for our height data.

Calculation of the Mean:

Total number of points (N)
= 23 students

The sum of all heights
= (2 students * 64 inches/student) + (1 * 66) + (3 * 67) + (3 * 68) + (1 * 69) + (2 * 70) + (4 * 71) + (3 * 72) + (1 * 73) + (2 * 74) + (1 * 75)
= 1,604 inches

Mean
= 1,604 inches/23 students
= 69.7 inches/student

Calculation of the Median:

Put data in order of value, listing each entry once
64, 64, 66, 67, 67, 67, 68, 68, 68, 69, 70, 70, 71, 71, 71, 71, 72, 72, 72, 73, 74, 74, 75

Find the center value since the total number of students is odd
64, 64, 66, 67, 67, 67, 68, 68, 68, 69, **70**, 70, 71, 71, 71, 71, 72, 72, 72, 73, 74, 74, 75

Median = 70 inches

Calculation of Variance: Note that the variance will have the same units as the variable in question squared, in this case, "inches squared."

$$\textit{Variance} = \frac{1}{23-1}((69.7 - 64)^2 + (69.7 - 64)^2 + (69.7 - 66)^2 + \cdots + (69.7 - 75)^2) = 9.5 \text{ in}^2$$

Calculation of Standard Deviation: The standard deviation has the same units as the variable in question, in this case, "inches."

$$\textit{Standard deviation} = \sqrt{9.5} = 3.08 \text{ in}$$

EXAMPLE 11-1

Consider the following velocity data, listed in units of feet per second. Determine the mean, median, variance, and standard deviation of the data.

1	28	14	32	35	25	14	28	5
16	42	35	26	5	33	35	16	14

Calculation of the Mean:

Total number of points (N) = 18

Sum of all data $(\Sigma X_i) = (1) + (2 * 5) + (3 * 14) + \cdots + (42) = 404$

Mean $= 404/18 = 22.4$ *feet per second*

Calculation of the Median:

Put data in order, listing each entry once.

1, 5, 5, 14, 14, 14, 16, 16, 25, 26, 28, 28, 32, 33, 35, 35, 35, 42

Find the center two values and average them, since total number of entries is even (18).

1, 5, 5, 14, 14, 14, 16, 16, ***25***, ***26***, *28, 28, 32, 33, 35, 35, 35, 42*

Median $= (25 + 26)/2 = 25.5$ *feet per second*

Calculation of Variance:

$$\text{Variance} = \frac{1}{18 - 1}((22.4 - 1)^2 + \cdots + (22.4 - 42)^2) = 147\ (\text{ft/s})^2$$

Calculation of Standard Deviation:

$$\text{Standard deviation} = \sqrt{147} = 12.1 \text{ feet per second}$$

EXAMPLE 11-2

Consider the following energy data, given in units of joules. Determine the mean, median, variance, and standard deviation of the data.

159	837	618	208	971	571	379	220	31

Calculation of the Mean:

Total number of points (N) = 9

Sum of all data $(\Sigma X_i) = 159 + 837 + \cdots + 31 = 3{,}994$

Mean $= 3{,}994/9 = 443.7 = 444$ *joules*

Calculation of the Median:

Put data in order, listing each entry once.

31, 159, 208, 220, 379, 571, 618, 837, 971

Find the center value since the total number of students is odd (9; center value at entry 5).

31, 159, 208, 220, ***379***, *571, 618, 837, 971*

Median = 379 joules

Calculation of Variance:

$$\text{Variance} = \frac{1}{9 - 1}[(444 - 31)^2 + \cdots + (444 - 971)^2] = 105{,}059 \text{ joules}^2$$

Calculation of Standard Deviation:

$$\text{Standard deviation} = \sqrt{105{,}059} = 324 \text{ joules}$$

11.3 DISTRIBUTIONS

From Figures 11-2 and 11-3 on the preceding pages, you can see a similarity in the histogram shape of all three plots. The values start small, increase in size, and then decrease again. In the case of student height, this means that a few people are short, most people have some "average" height, and a few people are tall. This same conclusion is true in many things we can measure. For example:

- If we weigh many standard-size watermelons (neither miniature nor giant), we will find that most weigh between 20 and 30 pounds. A few weigh less than 20 pounds, and a few weigh more than 30 pounds.
- As we look through a dictionary, we find that there are many words with between four and six letters. There are a few with one, two, or three letters and a few with more than six, but clearly most have between four and six letters.
- To improve efficiency in the office, we had an expert monitor the length of phone calls made by the staff. The expert found that most of the time, phone calls lasted between 3 and 5 minutes, but a few were longer than 5 and some others lasted only a minute.

11.3.1 NORMAL DISTRIBUTIONS

We wanted to know how many "flexes" it takes to cause a paper clip to fail, so we asked volunteers to test the bending performance of paper clips by doing the following:

- Unfold the paper clip at the center point so that the resulting wire forms an "S" shape.
- Bend the clip back and forth at the center point until it breaks.
- Record the number of "flexes" required to break the clip.

Using these data, we created Figure 11-4. This is the same as the earlier histogram, but the "boxes" are replaced by a smooth curve through the values.

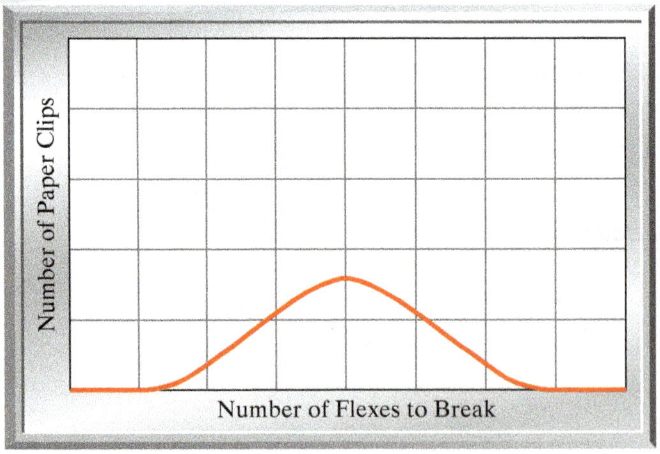

Figure 11-4 Distribution of paper clip failure.

distribution
68-95-99.7 rule

When you are interested in the *shape* of the curve rather than the exact data values, you can replace the bars of the histogram with a smooth curve and rename the graph a **distribution**. A distribution is considered *normal* if the following rules hold true. This is known as the **68-95-99.7 rule**, shown in Figure 11-5.

- 68% of values are within one standard deviation (1σ) of the mean (μ).
- 95% of values are within two standard deviations (2σ).
- 99.7% of values are within three standard deviations (3σ).

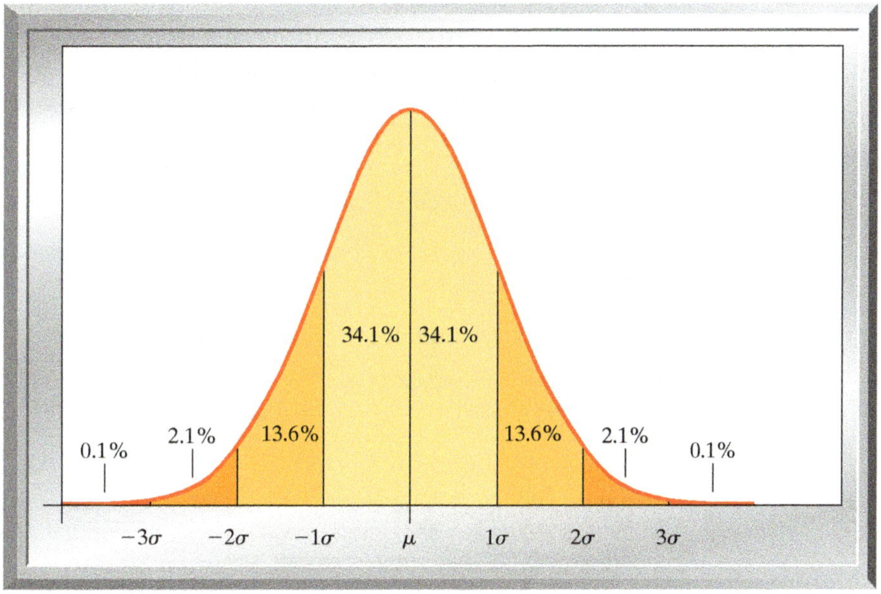

Figure 11-5 "Normal" distribution, showing the 68–95–99.7 rule.

● EXAMPLE 11-3

Suppose we ask a group of students how many states they have visited. The results might appear as shown below.

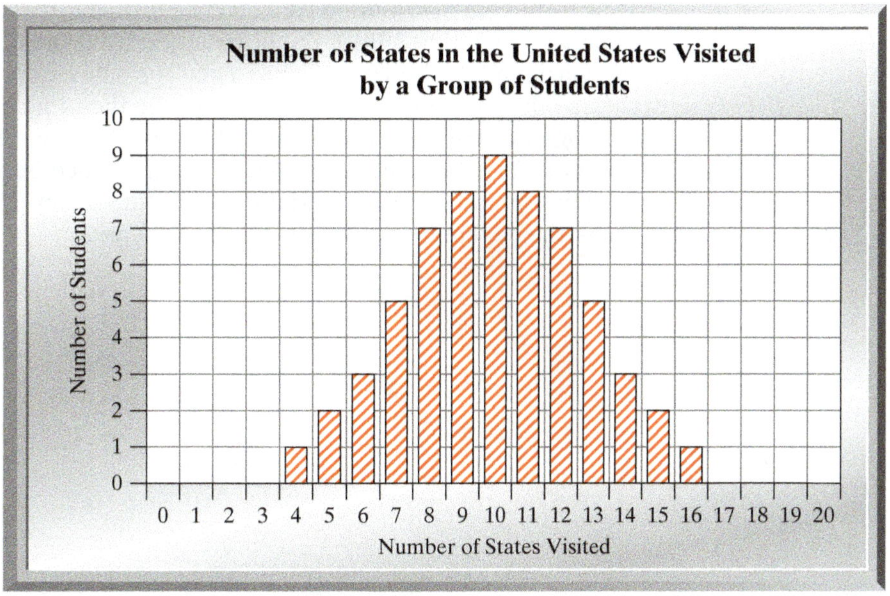

It seems that most have visited between 8 and 12, and that as many have visited more than 10 as have visited fewer than 10. A few have visited as many as 16 states, and all the students have visited at least 4. Let us calculate some values pertinent to this situation.

How many students are there in this group?

To do this we simply add the number of students represented by each bar, or

$$1 + 2 + 3 + 5 + 7 + 8 + 9 + 8 + 7 + 5 + 3 + 2 + 1 = 61 \text{ students}$$

What is the cumulative number of state visits?

*We answer this by totaling the product of the bar height with the number of states represented by the bar. For example, 5 students have visited 7 states, so those 5 students have visited a total of 5 * 7 = 35 states. Or, 8 students have visited 11 states, so those students have visited a total of 88 states. We calculate*

$$1*4 + 2*5 + 3*6 + \cdots + 2*15 + 1*16 = 610 \text{ states}$$

What is the average number of states visited by a student?

Once we have the values from our first two answers, this is straightforward division: the total number of visits divided by the total number of students.

$$610/61 = 10 \text{ states per student}$$

Notice that the value 10 is in the center of the distribution. For distributions that are symmetrical (such as this one), the average value is the one in the center, the one represented by the largest number of occurrences.

Decrease in Variance

Let us examine the shape of the distribution of paper clip failures discussed earlier. How would the distribution change if we brought in a machine that did it "exactly" the same way each time? Both the distribution from the data class and the distribution of the machine are shown in Figure 11-6. The same number of clips was tested in each case, so the areas under each curve must be the same.

This exercise illustrates that distributions that have the same mean (and median) can look very different. In this case, the difference between these two distributions is in their "spread," or their variation about the mean. The effect of using a machine to break the paperclips was a decrease in the variance.

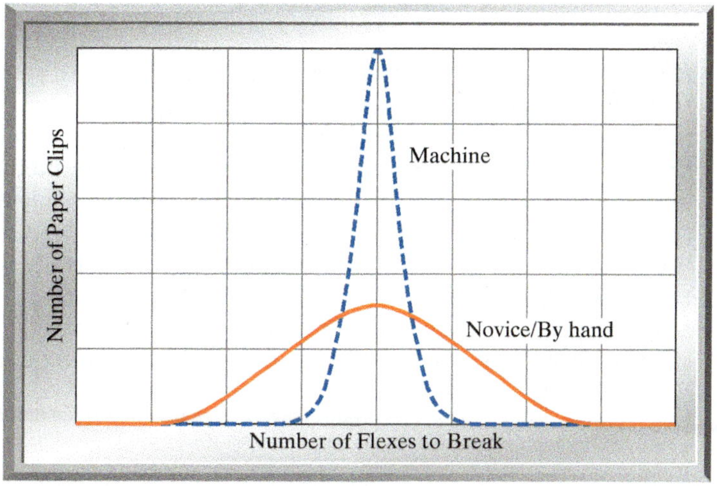

Figure 11-6 Distribution of paper clip failure after a decrease in variance.

Shift in Mean

Redraw the paper clip distribution; then on the same plot, sketch the distribution if each volunteer tested the same number of clips that were manufactured by the same manufacturer as before with the same variance, but were stronger and typically required 10 more flexes to fail. The result is shown in Figure 11-7. The stronger material caused the distribution to shift to the right. Since the variance and number of clips remained the same, the shape and size of the curve remains the same.

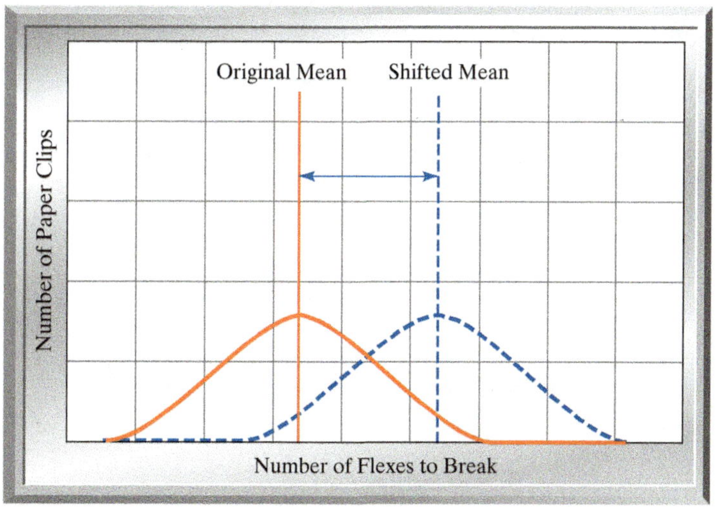

Figure 11-7 Distribution of paper clip failure with a shift.

11.3.2 SKEWED DATA

skewed
positively skewed
negatively skewed
range restriction

It is often easy to place an upper or lower limit on the value of the possible outcome. In these cases, the distribution is no longer symmetric—it is **skewed.** A population is **positively skewed** if the mean has been pulled higher than the median, and **negatively skewed** if the mean has been pulled lower than the median (see Figure 11-8 for an example of a positively skewed graph). You have probably heard news reports that use the median to describe a distribution of income in the United States. The median is used in this case because the distribution is positively skewed. This skew is caused by two factors, the presence of extreme values (millionaires) and the **range restriction**, the latter because income cannot be lower than $0. The extreme values causing the positive skew are not shown on the graph. Most of these would be far off the page to the right.

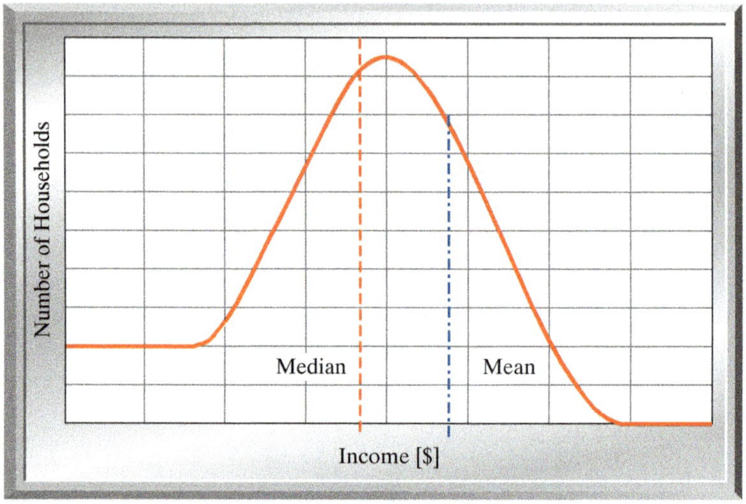

Figure 11-8 Distribution of positively skewed data.

● EXAMPLE 11-4

For each scenario, identify one graph from the following set that best illustrates how the baseline curve would change under the conditions of that scenario. Each graph shows the usual distribution (labeled baseline) and the way the distribution would be modified from the baseline shape (labeled modified) under certain conditions.

The graphs show SAT composite (verbal + quantitative) scores, for which 400 is generally considered to be the minimum possible score and 1,600 is considered to be the maximum possible score.

(a) The designers of the SAT inadvertently made the test more difficult, while the variance of the scores remains the same.

This is shown by Curve (F): Area and variance the same; mean shifted to left.

(b) The variability of scores is reduced by switching to true/false questions, while the average remains the same.

This is shown by Curve (D): Area and mean are the same; distribution is narrower.

(c) A population boom increases the number of students seeking college admission.

This is shown by Curve (B): Area increases, distribution stays the same.

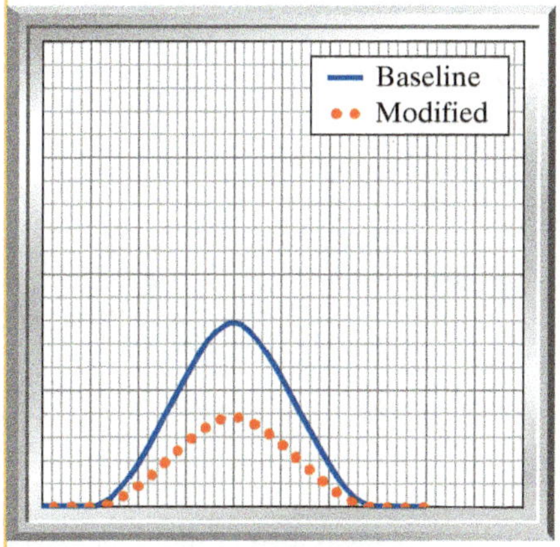

(A)

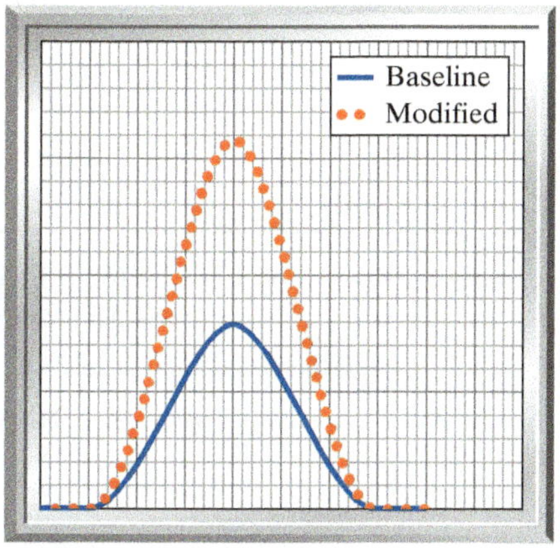

(B)

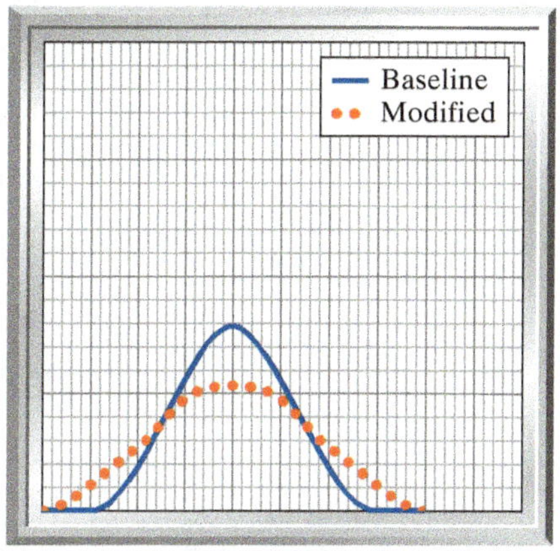

(C)

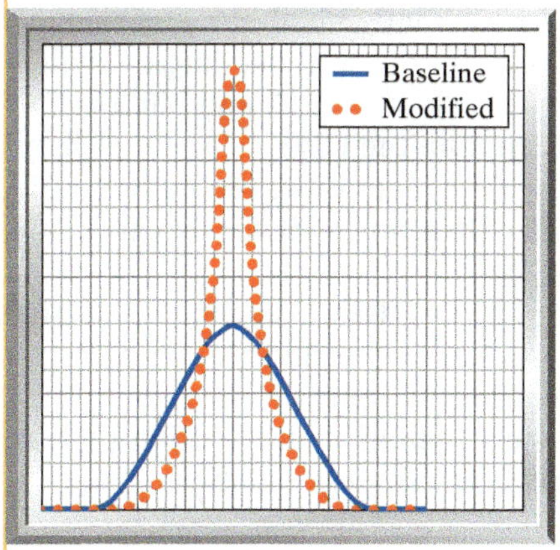

(D)

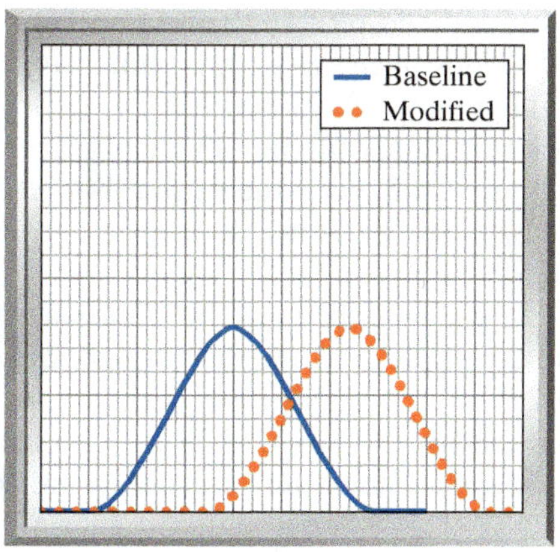

(E)

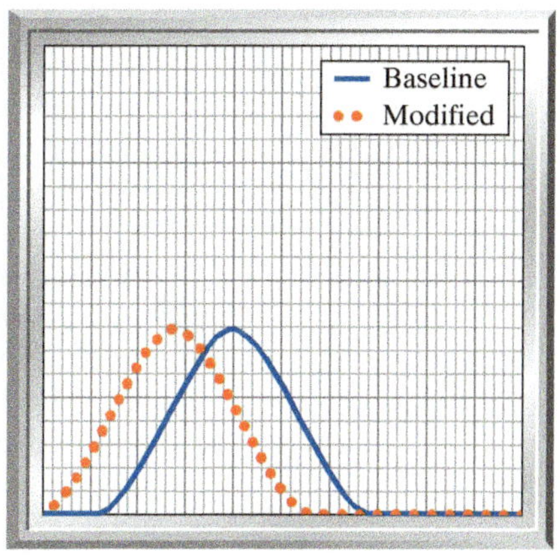

(F)

11.4 CUMULATIVE DISTRIBUTION FUNCTIONS

normalized plot

For the earlier student height plot using two heights per bin, we will graph the bin data but now show the values on the ordinate as a fraction rather than a whole number. To do this, we divide the number of students in each bin of the histogram by the total number of students. If we now add the heights of all the bars in the new plot, they should equal 1. This is called a **normalized plot**, shown in Figure 11-9. We "normalized" the values by dividing by the total number of data points. This graph holds no new information; it is simply a rescaling of the histogram we drew earlier.

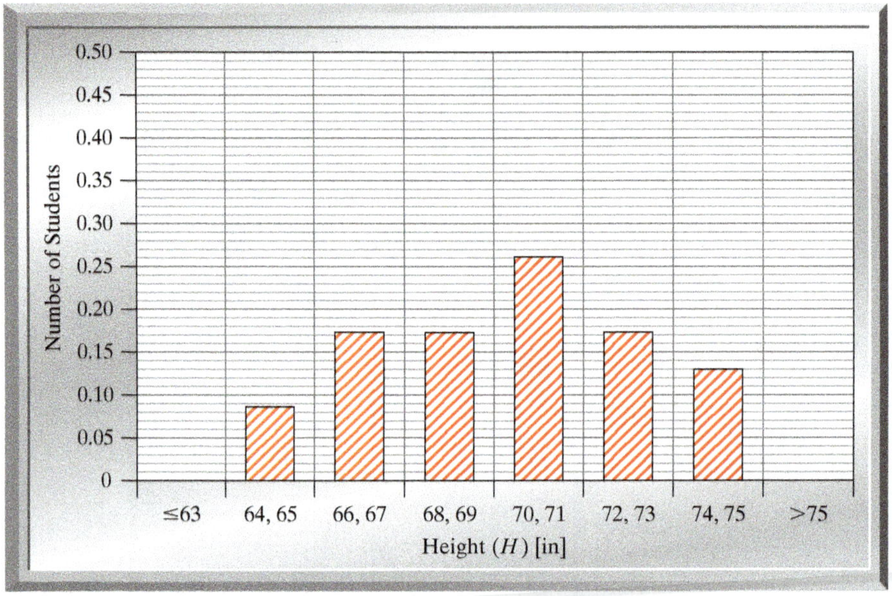

Figure 11-9 Normalized plot of student height, originally shown in Figure 11-2.

cumulative distribution function (CDF)

This plot can be used as an intermediate step to obtain a final plot called a **cumulative distribution function (CDF)**. We derive this plot by summing the values for each bin on the normalized plot from the first bin up to each individual bin. For example, suppose the values in the first three bins were 0, 0.08, and 0.18. By adding the values, we get new "cumulative" values: bin 1 = 0; bin 2 = 0 + 0.08 = 0.08; and bin 3 = 0 + 0.08 + 0.18 = 0.26. It should be obvious that we can obtain the CDF value for each bin by adding the normalized value of that bin to the CDF value of the bin before it. The CDF values are usually shown as percentages rather than fractions, for example, 50% instead of 0.5.

As we move across the plot, the values should go from 0 to 1. Using the height data in the normalized plot below, we have produced a cumulative distribution shown in Figure 11-10. Sometimes, the CDF is shown as a continuous, curved line rather than a column chart. Both the original histogram and the cumulative distribution plot are useful tools in answering questions about the composition of a population.

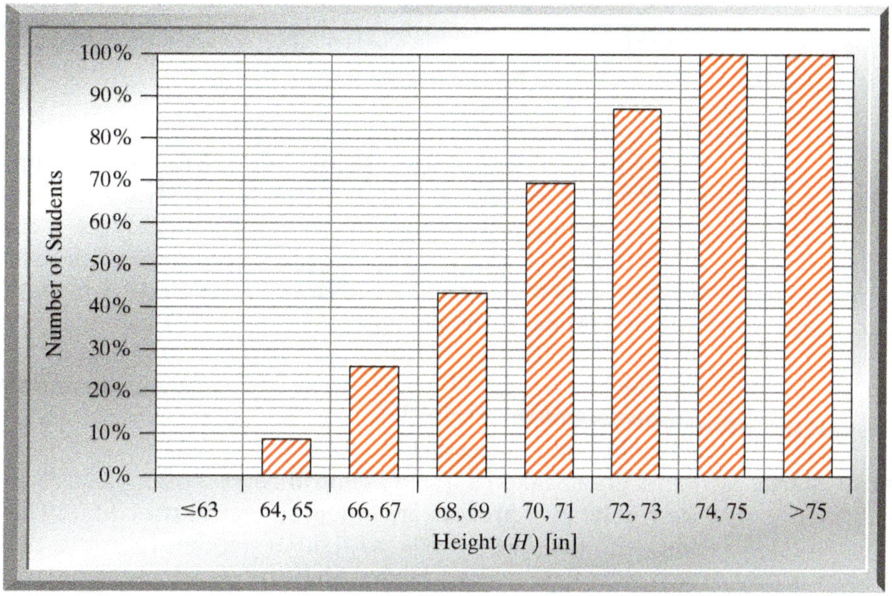

Figure 11-10 CDF of student height, originally shown in Figure 11-2.

EXAMPLE 11-5

Consider the following pressure data, given in units of pascals. Draw the histogram and CDF of the data.

36	9	33	11	23	3
34	39	56	51	39	1
27	25	2	1	53	32
14	41	55	28	29	19
51	15	25	10	35	38

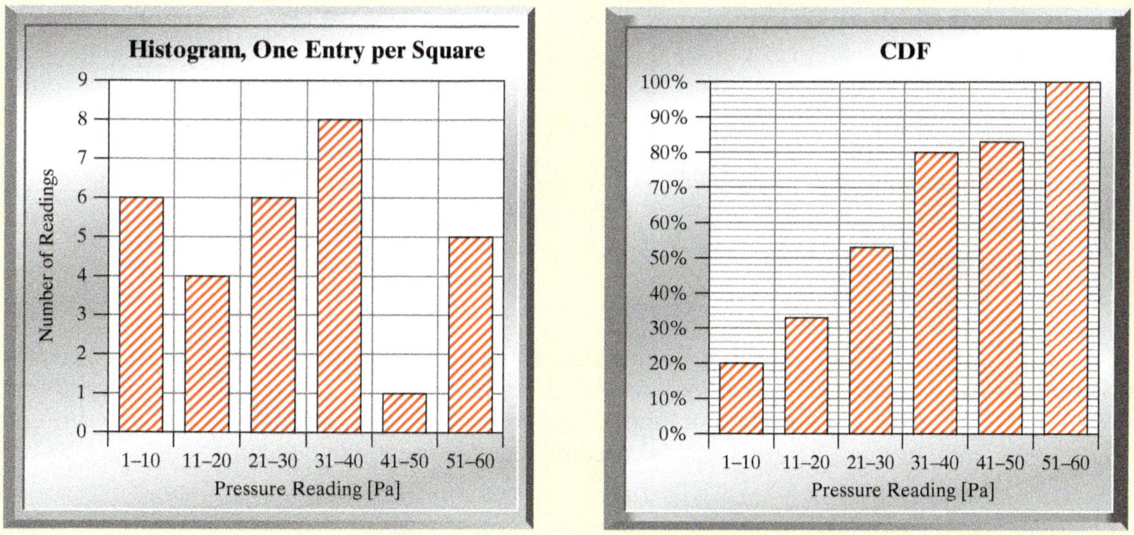

11.5 STATISTICAL PROCESS CONTROL (SPC)

We showed that a histogram such as the one shown in Figure 11-11 visually summarizes how a set of values is distributed. Sometimes, however, we are not only interested in the values themselves, but also in how the distribution changes over time.

For example, as a machine in a factory operates, it may slowly (or occasionally quickly) lose proper alignment or calibration due to wear, vibrations, and so on. If a machine was making bolts with a mean length of 1 inch and a standard deviation of 0.01 inch when it first began operating, after it had made 100,000 bolts, the alignment may have drifted so that the mean was only 0.95 inches with a standard deviation of 0.02 inch. This may be unacceptable to the customer purchasing the bolts, so the parameters of the process need to be monitored over time to make sure the machine is readjusted as necessary.

quality control chart

A graph called a **quality control chart** is often used to show how close to the mean the results of a process are when measured over time. The graph is usually a scatter plot, with the x-axis shown as time or another indicator that would change with time, such as batch number. Figure 11-12 shows a sample control chart, the mean, and the standard deviation.

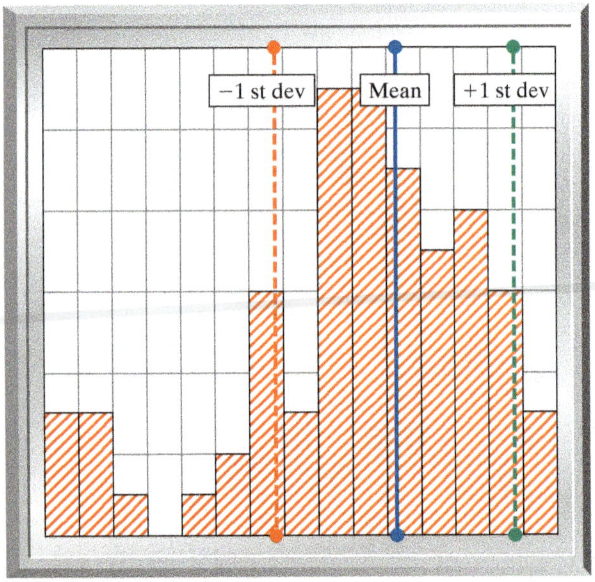

Figure 11-11 Sample histogram.

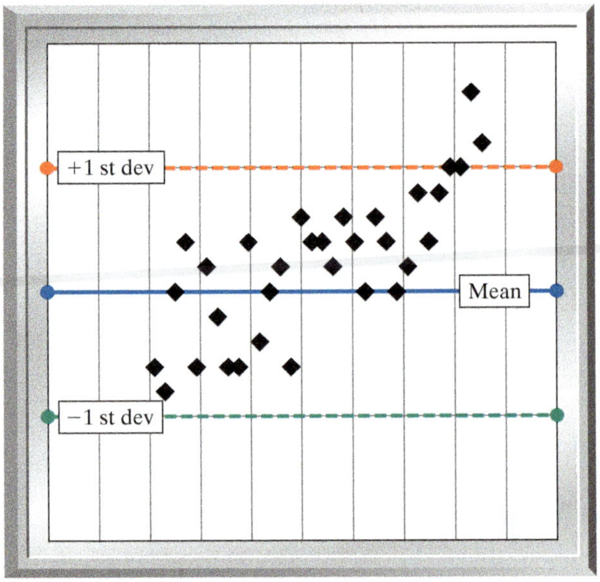

Figure 11-12 Sample quality chart.

Statistical Process Control (SPC)

When tracking a manufacturing process, engineers are often concerned with whether the process is "in control," or behaving as expected. **Statistical Process Control (SPC)** is a method of monitoring, controlling, and improving a manufacturing process. In some situations, the desired mean and acceptable deviation limits may be preset for a variety of reasons (chemistry, safety, etc.). Often, the upper or lower limits of control are determined by the desired end result. For example:

- The reactor temperature must not rise above 85 degrees Celsius or the reactant will vaporize.
- The injection pressure should be between 50 and 75 kilopascals to ensure that the part is molded properly.
- A bolt must be machined to ±0.02 inches to fit properly in a chair leg.

An engineer will study how the process relates to the control limits and will make adjustments to the process accordingly. To discuss whether a process is in control, we can divide a chart into zones, shown in Figure 11-13, to create a control chart. The mean is determined either by the desired end result (the iron content of the product must average 84%) or by the process itself (the reactor temperature should average 70 degrees Fahrenheit for an optimum reaction to occur). The standard deviation is most often determined by experimentation. For example:

- The purity range for this product is ±0.0005%.
- The standard deviation for the reactor temperature must not exceed 5 degrees Fahrenheit or the reaction will create unwanted by-products.
- The standard deviation of the current gain of the transistors being produced must be less than 15.

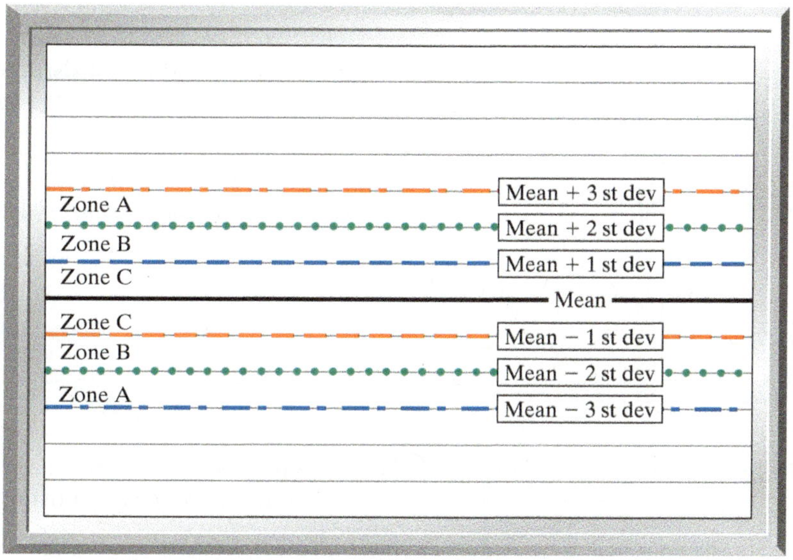

Figure 11-13 Standard deviation ranges.

Eight Ways to Be Out of Control

Nelson Rules

A variety of conditions can indicate that a process is out of control. First published by Lloyd S. Nelson in the October 1984 issue of the *Journal of Quality Technology*, the **Nelson Rules** are listed on the following pages, with examples and graphs. In the graphs, solid points indicate the rule violations. The actual conditions may vary slightly from company to company, but most take the same standard form. For example, a company may operate with Rule 3 stated as seven or eight points in a row instead of six.

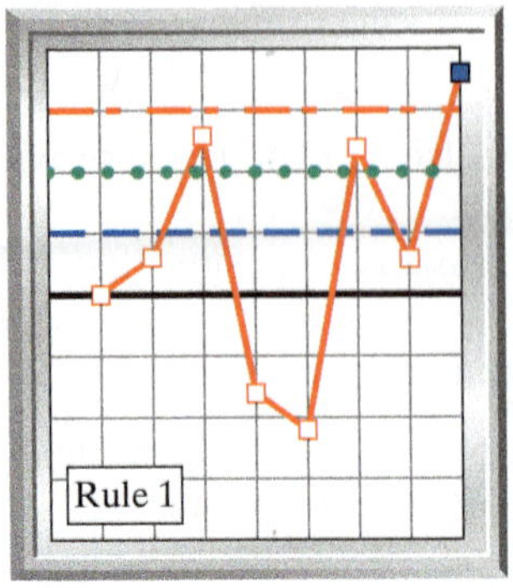

1. **A point falls anywhere beyond Zone A.** The value is more than three standard deviations away from the mean. May occur on either side of the mean.

   Example: The mean temperature of a reactor is 85 degrees Celsius with a standard deviation of 5 degrees Celsius. If the temperature exceeds 100 degrees Celsius, the reactor vessel may explode. If the temperature falls below 70 degrees Celsius, the reaction cannot proceed properly.

2. **Nine points in a row occur on the same side of the mean.** The actual value seems to be drifting away from the mean.

   Example: The percentage of boron in a semiconductor should be 250 parts per billion. Nine consecutive samples have boron contents less than this value. The machine incorporating the boron into the semiconductor material may need to be cleaned or recalibrated.

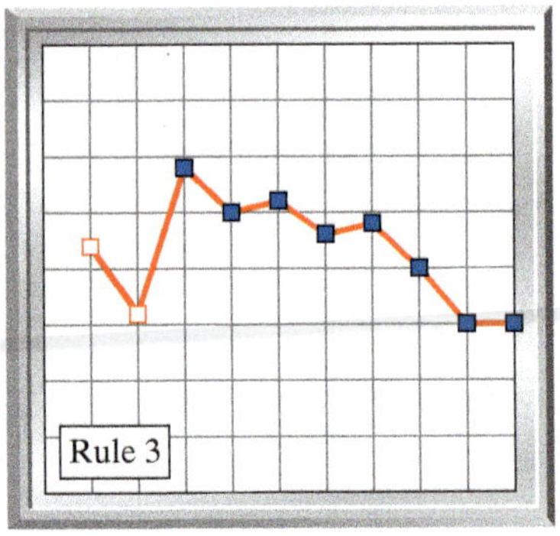

3. **Six points occur with a consistently increasing or decreasing trend.** If this pattern continues, the values will eventually become unacceptable.

   Example: The shaft length of a part is increasing with each successive sample; perhaps the grinding wheel needs to be changed.

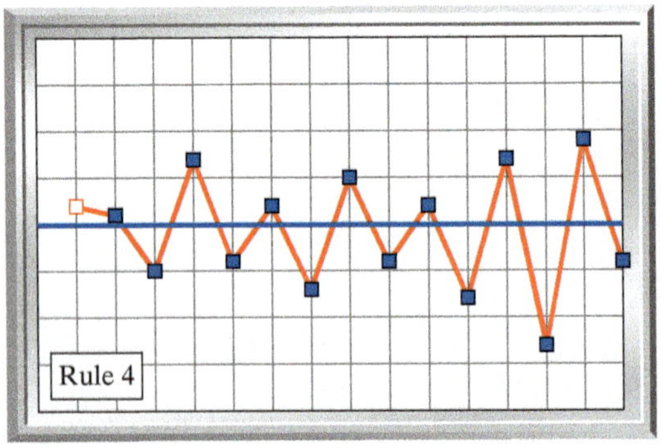

4. **Fourteen points in a row alternate from one side of the mean to the other.** The process is unstable.

   Example: The control system for a crane errs from one side to the other. This may indicate a sensor failure or the need to reprogram the controller.

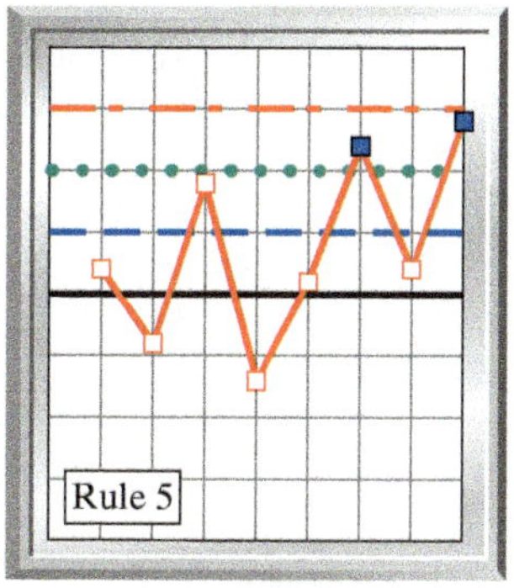

5. **Two out of three points in a row occur in Zone A.** The process is close to the upper limit; take preventive measures now.

Example: A robot that is spot-welding parts in an automobile is coming close to the edge of the material being welded. It probably needs attention.

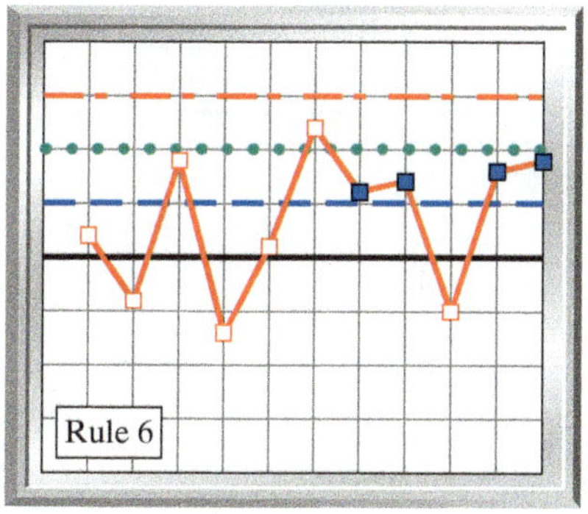

6. **Four out of five points in a row occur in Zone B.** The process is very close to the upper limit; take preventive measures now.

Example: Four out of five customers in the bank teller queue have waited more than one standard deviation to be helped. Perhaps another teller is needed.

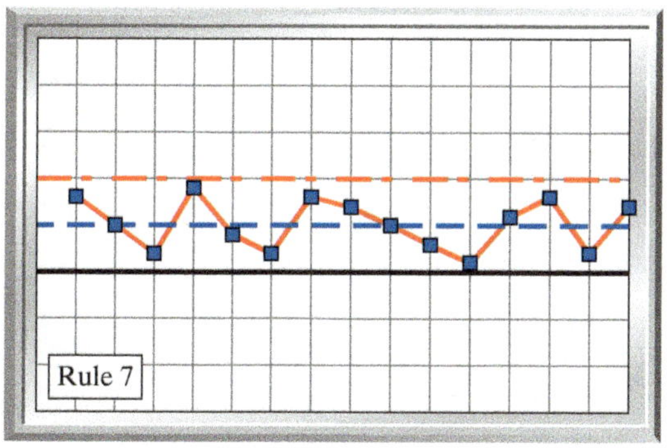

7. **Fifteen points in a row occur in Zone C; points can occur on either side of the mean.** The process is running too perfectly; in many applications the restrictions can be loosened to save time and money.

Example: The thickness of all washers being manufactured for quarter-inch bolts is within 0.0005 inch of the desired mean. Very few applications require washers with such close tolerances. Perhaps the process could be set to process the washers faster.

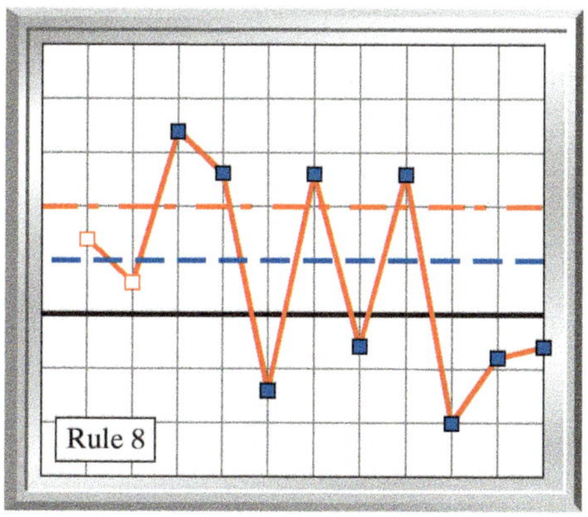

8. **Eight points in a row occur beyond Zone C; points can occur on either side of the mean.** The process does not run close enough to the mean; the parts are never quite on target; may indicate a need for a process adjustment.

Example: The postmark machine in a regional postal distribution center is stamping the envelopes too high or too low; it probably needs attention.

EXAMPLE 11-6

The data shown in the table were collected from a manufacturing process that makes bolts.

Assume the process specifies an average bolt length of 10 inches, with a standard deviation of 0.25 inches. Is this process under statistical control?

Part	Length (*L*) [in]	Part	Length (*L*) [in]
1	10.00	11	10.25
2	10.25	12	10.65
3	10.65	13	9.50
4	9.50	14	9.36
5	9.36	15	9.25
6	9.00	16	10.50
7	10.50	17	10.20
8	10.20	18	9.80
9	9.80	19	10.45
10	10.00	20	10.10

The control chart for these data is shown below. Rule 1 is violated since a point falls outside of Zone A (part 6) and Rule 5 is violated as parts 14 and 15 fall inside Zone A. The process is not in statistical control.

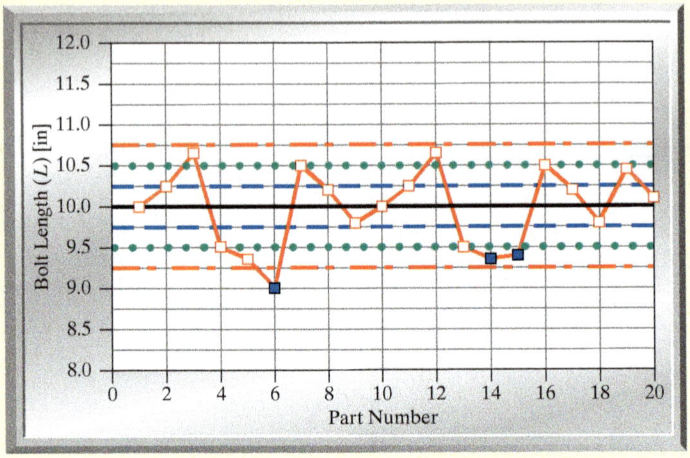

Chapter 11 ASSESSMENT QUESTIONS

1. The table below lists the number of computer chips rejected for defects during random testing over the course of a week on a manufacturing line. Four samples of 20 parts are pulled each day. Use the following data to generate a histogram and CDF in Excel.

1	1	8	0	2	0
0	2	10	1	3	2
0	1	12	0	2	1
1	6	15	0	0	
3	8	1	2	5	

2. For the following mass data given in units of kilograms, determine the mean, median, variance, and standard deviation.

8	7	9	11	16
12	2	9	10	9

3. This exercise includes the measurement of a distributed quantity and the graphical presentation of the results. You are to determine how many flexes it takes to cause a paper clip to fail.

Test the bending performance of 20 paper clips by doing the following:

- Unfold the paper clip at the center point so that the resulting wire forms an "S" shape.
- Bend the clip back and forth at the center point until it breaks.
- Record the number of flexes required to break the clip.

On a copy of the table on the following page, record the raw data for the paper clips you break. Then, summarize the data for the team by adding up how many clips broke at each number of flexes. Each team member should contribute 20 data points unless otherwise instructed by your teacher. Create a histogram with an appropriate bin size and a CDF, using the data collected and creating the graphs:

(a) by hand
(b) Research: using Excel
(c) Research: using MATLAB

Paper clip flexing data

Paper Clip	Flexes to Break
1	
2	
3	
4	
5	
6	
7	
8	
9	
10	
11	
12	
13	
14	
15	
16	
17	
18	
19	
20	

Summary of data

No. of Flexes	No. of Clips

4. For the following pressure data, recorded in units of pound-force per square inch, answer the following questions.

1	14	2	15	10
6	3	1	18	

(a) What is the mean of the data?
(b) What is the median of the data?
(c) What is the variance of the data?
(d) What is the standard deviation of the data?

5. A technician tested two temperature probes by inserting their probes in boiling water, recording the readings, removing and drying the probes, and repeating the process. The results are shown below, giving temperature reading in degrees Celsius.

(a) What is the mean of each probe?
(b) What is the median of each probe?
(c) What is the variance of each probe?
(d) What is the standard deviation of each probe?

Probe 1	87.5	86.5	88	89.5	87	88.5	89
Probe 2	95.5	100	101.5	97.5	90.5	91.5	103.5

6. For each graph shown below, decide if the mean, variance, or population size has changed.

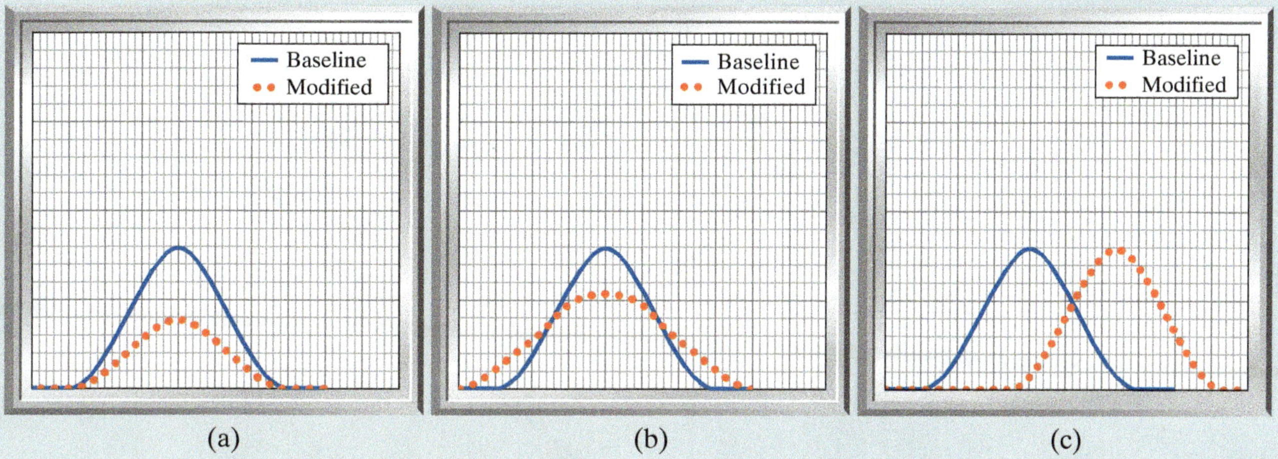

7. One of the NAE Grand Challenges for Engineeering is **Develop Carbon Sequestration Methods**. According to the NAE website: "In pre-industrial times, every million molecules of air contained about 280 molecules of carbon dioxide. Today that proportion exceeds 380 molecules per million, and it continues to climb. Evidence is mounting that carbon dioxide's heat-trapping power has already started to boost average global temperatures. If carbon dioxide levels continue upward, further warming could have dire consequences, resulting from rising sea levels, agriculture disruptions, and stronger storms (e.g., hurricanes) striking more often."

The Mauna Loa Carbon Dioxide Record is the longest continuous record of atmospheric concentrations of carbon dioxide (CO_2), the chief greenhouse gas responsible for global climate warming. These data are modeled as the Keeling Curve, a graph showing the variation in concentration of atmospheric CO_2 based on measurements taken at the Mauna Loa Observatory in Hawaii under the supervision of Charles David Keeling. It is often called the most important geophysical record on Earth and has been instrumental in showing that mankind is changing the composition of the atmosphere through the combustion of fossil fuels.

The Keeling Curve also shows a cyclic variation in each year corresponding to the seasonal change in the uptake of CO_2 by the world's land vegetation. Most of this vegetation is in the northern hemisphere, where most of the land is located. The level decreases from northern spring onward as new plant growth takes CO_2 out of the atmosphere through photosynthesis and rises again in the northern fall as plants and leaves die off and decay to release the gas back into the atmosphere.

Data and wording for this problem set were obtained from: www.esrl.noaa.gov/gmd/ccgg/trends/. Additional information on the Mauna Loa Observatory can be found at: http://scrippsco2.ucsd.edu/.

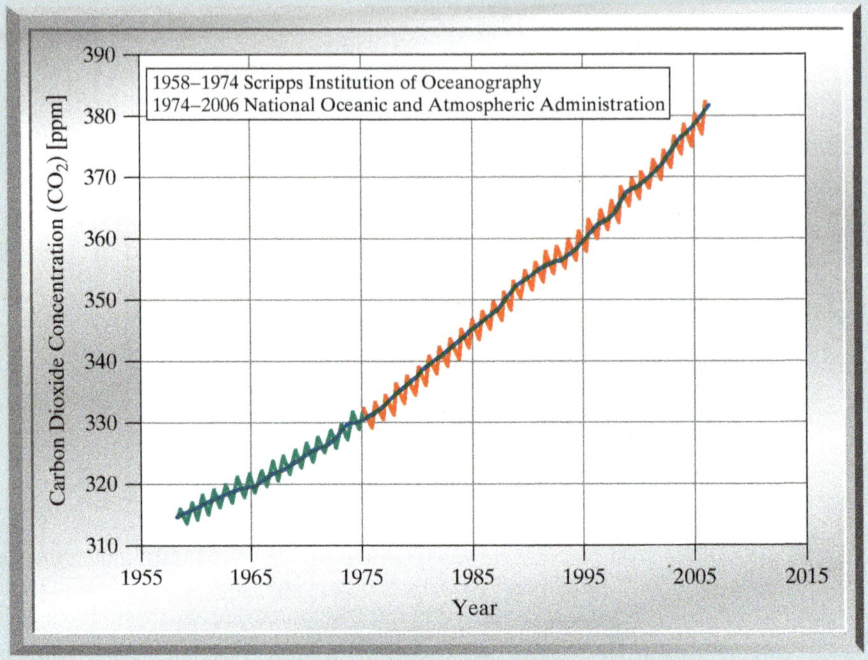

Examine the increase in monthly CO_2 emissions for 2009, taken from the Mauna Loa data set. All values given are in parts per million [ppm] CO_2 as the difference between the December 2008 and the monthly 2009 reading.

1.38	4.64	−0.77	3.23	2.20	0.45
1.87	3.89	−1.16	3.92	0.37	1.73

(a) What is the mean of these data?
(b) What is the median of these data?
(c) The variance of the data set shown here is 3.5 parts per million squared [ppm^2]. What is the standard deviation of these data?
(d) The estimated annual growth rates for Mauna Loa are close, but not identical, to the global growth rates. The standard deviation of the differences is 0.26 parts per million per year [ppm/year]. What is the variance?

8. You use the data from the Mauna Loa observatory in the previous question to create the following histogram and CDF. These data reflect the observed yearly increase in CO_2 emissions for a period of 51 years. The annual mean rate of growth of CO_2 in a given year is the difference in concentration between the end of December and the start of January of that year. If used as an average for the globe, it would represent the sum of all CO_2 added to, and removed from, the atmosphere during the year by human activities and by natural processes.

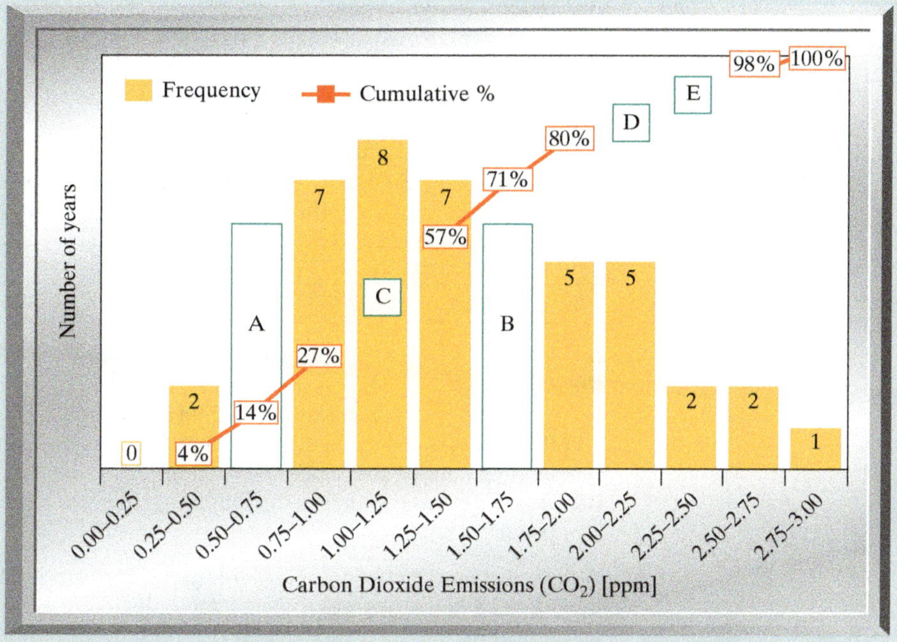

(a) What is the value of point A?
(b) What is the value of point B?
(c) What is the value of point C?
(d) What is the value of point D?
(e) What is the value of point E?

NOTE

In the graph, A and B are *not* drawn to scale, and the locations of C, D, and E are approximate. In other words, you cannot *guess* the value based upon the graph—you must *calculate* the value.

9. Polyetheretherketone (PEEK)™ are polymers that are resistant to both organic and aqueous environments; they are used in bearings, piston parts, and pumps. Several tests were conducted to determine the ultimate tensile strengths in units of megapascals [MPa]. The following CDF shows results from 320 points.

NOTE

In the graph, the locations of A–E are approximate. In other words, you cannot *guess* the value based upon the graph—you must *calculate* the value.

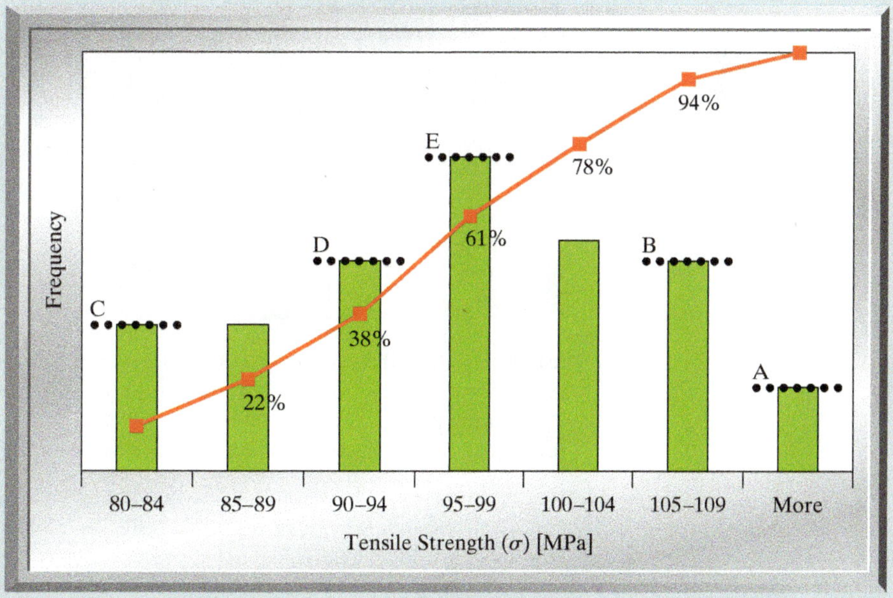

(a) What is the frequency value of A on the chart?
(b) What is the frequency value of B on the chart?
(c) What is the frequency value of C on the chart?
(d) What is the frequency value of D on the chart?
(e) What is the frequency value of E on the chart?

10. A technician tested a temperature probe by inserting it in boiling acetic acid (theoretical boiling point is 118 degrees Fahrenheit), recording the readings, removing and drying the probe, and repeating the process. The data are shown in the following table.

Temperature, Probe 1 (T) [°F]	120	118	105	115	105	120	125
Temperature, Probe 2 (T) [°F]	100	90	95	105	90	Missing Point	

(a) Determine the mean of Probe 1.
(b) Determine the median of Probe 1.
(c) A second probe was tested, yielding a mean of 95 degrees Fahrenheit and a median of 92.5 degrees Fahrenheit. If the data from Probe 2 are as shown above, determine the missing data point.
(d) If a probe has a standard deviation of 8 degrees Fahrenheit, what is the variance of the probe?

11. A technician tested two temperature probes by inserting them in boiling water (theoretical boiling point is 100 degrees Celsius), recording the readings, removing and drying the probe, and repeating the process. The CDF for both probes is shown below.

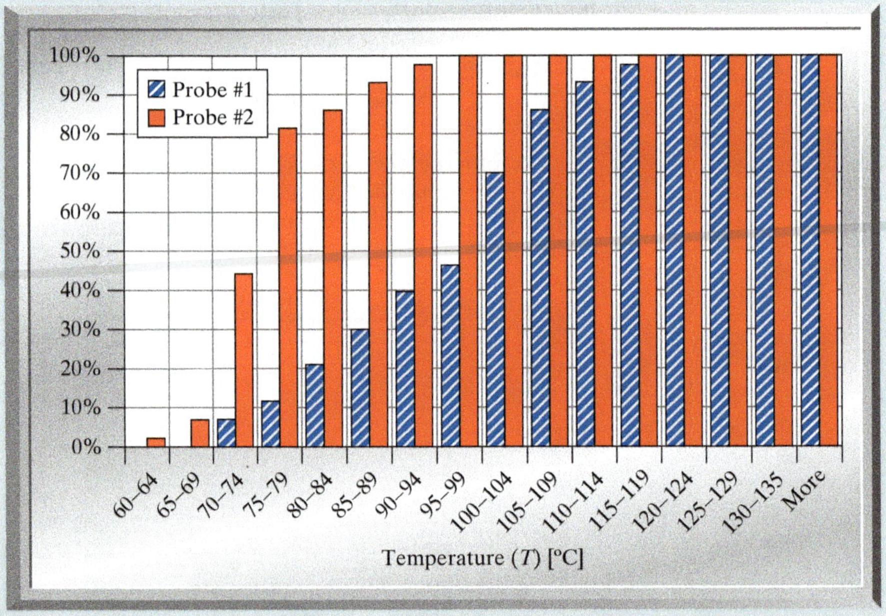

(a) Using this information, which probe would you choose to use? Explain why.
(b) How would you use the probe you chose above to ensure that you found the correct boiling point?
(c) Which probe has the higher standard deviation?

12. During the month of November, the heating system in your apartment appeared to be broken. To prove this, you record the following daily high temperatures in degrees Fahrenheit, taken every other day:

58	54	60	58	55	55	60	60
86	80	85	85	82	85	91	93

(a) Calculate the mean and median of the data.

(b) Draw the associated histogram and CDF for the data by hand; plot temperature on the x-axis and days on the y-axis.

(c) When you take your complaint to the apartment manager, he fails to see the problem; according to his heating bill, your apartment had an average temperature of 72 degrees Fahrenheit. Show both the mean and the median on the graph drawn for part (b). Which is a better presentation of the data, part (a) or part (b)? Justify.

13. The pressure in a water filter is monitored in a chemical plant. The filter should operate at 18 pounds-force per square inch [psi], with a standard deviation of ±2 psi. Analyze the data shown to determine if the filter is behaving as expected (the process is in control; the filter does not require any attention) or if the filter required attention (the process is out of control; the filter should be cleaned). Refer to the Nelson rules to explain your conclusion, and include the time the violations occur.

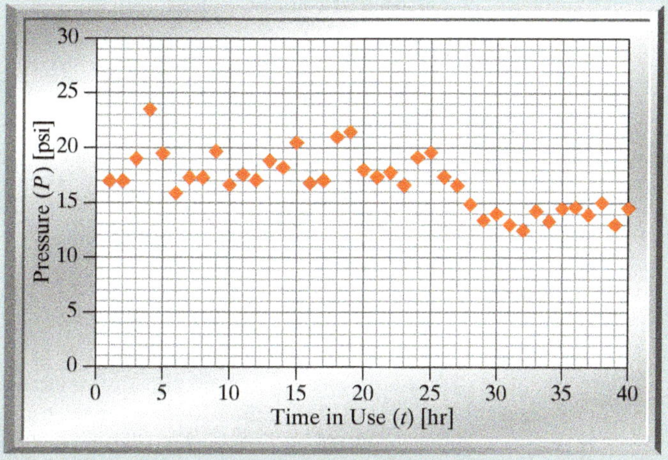

14. The data below were collected from a manufacturing process for making plastic cylinders. According to the specifications, the cylinder diameter should be 100 inches (the average diameter is 100 inches) and the standard deviation is ±5 inches.

Graph the data on a control chart. Be sure to clearly indicate the "Zones" of control.

Using the eight SPC rules, determine whether the process is in statistical control. If it is not in statistical control, indicate which rule or rules are violated and list the part numbers that violate that rule.

Part	Diameter (*D*) [in]	Part	Diameter (*D*) [in]
1	100	10	96
2	106	11	89
3	103	12	89
4	99	13	87
5	90	14	92
6	95	15	94
7	105	16	87
8	107	17	96
9	97	18	98

15. The data below were collected from a manufacturing process involving reactor temperature measured in degrees Celsius. The following values are desired: average = 100 degrees Celsius; standard deviation = ±10 degrees Celsius.

Graph the data on a control chart. Be sure to clearly indicate the "Zones" of control.

Using the eight SPC rules, determine whether the process is in statistical control. If it is not in statistical control, indicate which "rule" or "rules" are violated and list the part numbers that violate that rule.

Reading No.	Temperature (T) [°C]	Reading No.	Temperature (T) [°C]
1	100	11	103
2	105	12	101
3	106	13	100
4	97	14	98
5	98	15	97
6	95	16	96
7	101	17	104
8	100	18	102
9	96	19	95
10	105	20	101

16. The data below were collected from a manufacturing process involving reactor temperature measured in degrees Celsius. The following values are desired: average = 100 degrees Celsius; standard deviation = ±5 degrees Celsius.

Graph the data on a control chart. Be sure to clearly indicate the "Zones" of control.

Using the eight SPC rules, determine whether the process is in statistical control. If it is not in statistical control, indicate which "rule" or "rules" are violated and list the part numbers that violate that rule.

Reading No.	Temperature (T) [°C]	Reading No.	Temperature (T) [°C]
1	101.0	11	97.5
2	103.5	12	100.0
3	98.5	13	92.0
4	100.5	14	97.0
5	96.5	15	103.0
6	102.5	16	103.0
7	108.0	17	88.0
8	100.0	18	100.5
9	102.0	19	102.5
10	104.0	20	98.5

CHAPTER 12

A SCENARIO IN THE FUTURE - REVISITED

The year is 2027. You still love your job as a mechanical engineer at a multinational engineering firm, although you continue to be surprised that you don't work exclusively with other mechanical engineers. You remember, during school, that you thought you would work with mechanical engineers in a cubicle, rarely making human contact. You are surprised by how important it is to work with a team – and that almost every project involves teamwork. You also write quite a bit and make the occasional presentation. When you were in speech class, you thought that speech you made on "How to Grill the Perfect Grilled Cheese Sandwich" was probably the last time you would have to speak to a group.

You continue to enjoy the creativity and innovation you experience regularly. Even though every design hasn't been perfect, the failures allowed you and your team to learn and improve.

You continue to hang out with some people on the team, including the people who went on that trip a few years ago. (You might recall your adventure from Chapter 0.)

In fact, you remember the radioactive water lifter you built on that trip a few years ago, and you wonder – had we used some design process, could we have built a better lifter?

12.1 HAPPY NATURE NATIONAL PARK

You may fondly recall the time your team visited Happy Nature National Park and discovered that the pristine water wasn't quite suitable for drinking. In fact, one lake was found to have radioactive fish and contaminated water. You successfully informed the government, and the lake has since been cleaned.

Figure 12-1 Happy Nature National Park.

Since you and your team members all have vacation time coming up once again, you decide to revisit Happy Nature. "Remember the time we rented the cabin? Let's do it again!" Your team members agree and the plans are set.

Your team of three engineers – once again – packs up and hits the road. You make the drive to Happy Nature. The plan: Enjoy this vacation!

It's déjà vu: When you arrive at the cabin, a note on the door reassures you that this park is now SO CLEAN that there is no need for separate drinking water! The fine print on the note mentions that a few brave and inspiring engineers did discover a little contamination, but it has since been cleaned up.

A few empty bottles and a map to the local stream where you can fill them were left on the counter. Your cabin is still equipped with a beautiful, flat, study table. While it brings back memories of the trip a few years ago, you are pretty confident. They MUST have cleaned the park this time, right?

Figure 12-2 Fresh, clean water.

The water is clear and seems perfectly safe! One member of your team likes to fish, one likes to hike, one prefers to swim – and there is time for all of it. Thirsty? Enjoy a drink of the perfectly clear and clean water.

12.2 TROUBLE IN PARADISE?

Not this time.

Things are going well. However, one member of the team really misses the design aspects of engineering. She has a suggestion: Let's redesign the radioactive water-lifter, but this time, let's just call it a water lifter!

12.3 ENGINEERING DESIGN

Your task: The team must design and build a device to lift the fresh and clean water bottle into a radiation-proof container.

The team decides to build this device once again. The team members scour the cabin in search of supplies. They find a toolbox and assorted other material, including paper clips, a spool of wire, tape, rubber bands, and more.

They also find a toolbox with a note saying:

> You can use these tools, but you may NOT include any of these tools in a design you build.

12.4 YOUR TASK

In teams of 3 or 4, gather the supplies listed in our story.

Material list:

(Note: Your teacher may modify the first part of this list if necessary; as long as each team has the same set of supplies, the design goes on.)

5 cardboard strips: 12" long, 2.5" wide
5 rubber bands
8 clothespins
25 paper clips
20 craft sticks
6 pencils
8 binder clips
3 feet masking tape
3 feet wire
3 feet string or twine

Also (not to be incorporated into the water-lifter):

Scissors
Wire clippers
Ruler
Water bottle (at least ¾ full)
Bucket (preferably smaller, but large enough to hold a water bottle)

This time, spend **20 minutes** designing a water-lifter able to lift a full water bottle into the air and place it in a bucket. After designing, you have **10 minutes** to build your device. The operator may not get any part of his or her body within 18" of the water bottle because we want to have the same experience as we did a few years ago. You may test while you build.

After the activity, discuss as a class:

1. a) Was this activity "engineering"? Why or why not?

   b) Was it "engineering" last time?

2. Did the teams exhibit effective teamwork this time? Did you improve over last time?

3. Did the design step lead to an improved device?

APPENDIX A
FORMULAS, VALUES, SYMBOLS, AND MATLAB FUNCTIONS

CONVERSIONS

Angle

1 rad	= 57.3 deg
π rad	**= 180 deg**

Area

1 acre	= 4047 m^2
	= 0.00156 mi^2

Energy

1 J	= 0.239 cal
	= 9.48×10^{-4} BTU
	= 0.7376 ft lb_f
1 kW h	**= 3,600,000 J**

Force

1 N	= 0.225 lb_f
	= 1 E 5 dyne
1 kip	**= 1,000 lb_f**

Length

1 m	= 3.28 ft
1 km	= 0.621 mi
1 in	**= 2.54 cm**
1 mi	**= 5280 ft**
1 yd	**= 3 ft**

Mass

1 kg	= 2.205 lb_m
1 slug	= 32.2 lb_m
1 ton	**= 2,000 lb_m**

Power

1 W	= 3.412 BTU/h
	= 0.00134 hp
	= 14.34 cal/min
	= 0.7376 ft lb_f/s

Pressure

1 atm	= 1.01325 bar
	= 33.9 ft H_2O
	= 29.92 in Hg
	= 760 mm Hg
	= 101,325 Pa
	= 14.7 psi

Time

1 d	**= 24 h**
1 h	**= 60 min**
1 min	**= 60 s**
1 yr	= 365 d

Temperature

1 K	**= 1 °C**
	= 1.8 °F
	= 1.8 °R

Volume

1 L	= 0.264 gal
	= 0.0353 ft^3
	= 33.8 fl oz
1 mL	**= 1 cm^3 = 1 cc**

Named Units

1 F	= 1 A s/V
1 H	= 1 V s/A
1 Hz	= 1 s^{-1}
1 J	= 1 N m

1 N	= 1 kg m/s^2
1 P	= 1 g/(cm s)
1 Pa	= 1 N/m^2
1 St	= 1 cm^2/s

1 V	= 1 W/A
1 W	= 1 J/s
1 Ω	= 1 V/A

Conversions shown in bold text above indicate exact conversions

SI PREFIXES

Numbers Less Than One		
Power of 10	**Prefix**	**Prefix Abbreviation**
10^{-1}	deci-	d
10^{-2}	centi-	c
10^{-3}	milli-	m
10^{-6}	micro-	μ
10^{-9}	nano-	n
10^{-12}	pico-	p
10^{-15}	femto-	f
10^{-18}	atto-	a
10^{-21}	zepto-	z
10^{-24}	yocto-	y

Example: 1 millimeter [mm] = 1×10^{-3} meters [m]

Numbers Greater Than One		
Power of 10	**Prefix**	**Prefix Abbreviation**
10^{1}	deca-	da
10^{2}	hecto-	h
10^{3}	kilo-	k
10^{6}	Mega-	M
10^{9}	Giga-	G
10^{12}	Tera-	T
10^{15}	Peta-	P
10^{18}	Exa-	E
10^{21}	Zetta-	Z
10^{24}	Yotta-	Y

Example: 1 Megajoule [MJ] = 1×10^{6} joules [J]

FUNDAMENTAL DIMENSIONS [BASE SI UNITS]

I	electric current [A]	M	mass [kg]	T	time [s]
J	light intensity [cd]	N	amount of substance [mol]	Θ	temperature [K]
L	length [m]				

COMMON DERIVED UNITS IN THE SI SYSTEM

Dimension	SI Unit	Base SI Units = Dimensions	Derived from
Force (F)	newton [N]	$1\ \mathrm{N} = 1\dfrac{\mathrm{kg\ m}}{\mathrm{s}^2} \{=\} \dfrac{\mathrm{M\ L}}{\mathrm{T}^2}$	F = m a Force = mass times acceleration
Energy (E)	joule [J]	$1\ \mathrm{J} = 1\ \mathrm{N\ m} = 1\dfrac{\mathrm{kg\ m}^2}{\mathrm{s}^2} \{=\} \dfrac{\mathrm{M\ L}^2}{\mathrm{T}^2}$	E = F d Energy = force times distance
Power (P)	watt [W]	$1\ \mathrm{W} = 1\dfrac{\mathrm{J}}{\mathrm{s}} = 1\dfrac{\mathrm{kg\ m}^2}{\mathrm{s}^3} \{=\} \dfrac{\mathrm{M\ L}^2}{\mathrm{T}^3}$	P = E/t Power = energy per time
Pressure (P)	pascal [Pa]	$1\ \mathrm{Pa} = 1\dfrac{\mathrm{N}}{\mathrm{m}^2} = 1\dfrac{\mathrm{kg}}{\mathrm{m\ s}^2} \{=\} \dfrac{\mathrm{M}}{(\mathrm{L\ T}^2)}$	P = F/A Pressure = force per area
Voltage (V)	volt [V]	$1\ \mathrm{V} = 1\dfrac{\mathrm{W}}{\mathrm{A}} = 1\dfrac{\mathrm{kg\ m}^2}{\mathrm{s}^3\ \mathrm{A}} \{=\} \dfrac{\mathrm{M\ L}^2}{(\mathrm{T}^3\ \mathrm{I})}$	V = P/I Voltage = power per current

PHYSICAL CONSTANTS [VALUE AND UNITS]

c	speed of light in a vacuum	$= 3 \times 10^8$ m/s
e	Euler's number (base of natural logarithm)	= 2.71828 . . .
	elementary charge of an electron	$= 1.602 \times 10^{-19}$ C
φ	Golden ratio	= 1.61803 . . .
F	Faraday's constant	$= 9.65 \times 10^4$ C/mol
g	acceleration due to gravity	$= 9.8$ m/s^2 on Earth
G	gravitational constant	$= 6.67 \times 10^{-11}$ (N m^2)/kg^2
h	Planck constant	$= 6.62 \times 10^{-34}$ Js
k	Boltzmann constant	$= 1.38 \times 10^{-23}$ J/K
k_e	Coulomb's constant	$= 8.98755 \times 10^9$ N m^2/C^2
N_A	Avogadro's number	$= 6.022 \times 10^{23}$ mol^{-1}
π	ratio of circle circumference to diameter	= 3.14159 . . .
R	gas constant	= 8314 (Pa L)/(mol K) = 0.08206 (atm L)/(mol K)
σ	Stefan-Boltzmann constant	$= 5.67 \times 10^{-8}$ W/(m^2 K^4)

Greek Letters [Name, Greek Form, English Form, Common Uses]			
alpha	A, α	A, a	α angle, angular acceleration, Neper frequency, thermal diffusivity
beta	B, β	B, b	β bulk expansion coefficient, coefficient of thermal expansion, transistor current, gain
gamma	Γ, γ	G, g	γ specific weight; Euler's constant
delta	Δ, δ	D, d	Δ change; tip deflection of a beam δ small change, impulse function
epsilon	E, ε	E, e	ε dielectric constant, pipe roughness, porosity, strain
zeta	Z, ζ	Z, z	ζ dampening ratio of spring system
eta	H, η	H, h	η efficiency
theta	Θ, θ	Q, q	Θ dimensionless time θ thermal resistance, Euler angle
iota	I, ι	I, i	ι imaginary unit, consecutive integers
kappa	K, κ	K, k	κ permeability, thermal conductivity
lambda	Λ, λ	L, l	λ wavelength, arrival rate, eigenvalue
mu	M, μ	M, m	μ SI prefix micro (10^{-6}), statistical mean, dynamic viscosity, coefficient of friction
nu	N, ν	N, n	ν frequency, kinematic viscosity, Poisson's ratio, specific volume
xi	Ξ, ξ	X, x	ξ permeability, damping coefficient in circuit
omicron	O, o	O, o	
pi	Π, π	P, p	Π product π 3.141592... (circumference/diameter of circle), osmotic pressure
rho	P, ρ	R, r	ρ density, resistivity
sigma	Σ, σ	S, s	Σ summation σ conductivity, standard deviation, surface tension, stress, Stefan-Boltzmann constant
tau	T, τ	T, t	τ time constant, shear stress, torque
upsilon	Y, υ	U, u	
phi	Φ, ϕ or φ	F, f	Φ potential ϕ, φ Euler angle, Golden ratio
chi	X, χ	C, c	χ electronegativity
psi	Ψ, ψ	Y, y	ψ Euler angle
omega	Ω, ω	W, w	Ω unit of ohm ω angular frequency, angular velocity

Nomenclature [common units]

a	acceleration [m/s^2]
A	altitude [m]
	area [m^2]
C	capacitance [F]
	circumference [m]
	concentration [g/L]
C_d	coefficient of drag [-]
C_p	specific heat [J/(g °C)]
d	distance [m]
D	diameter [m]
E	energy [J]
	Young's modulus [Pa]
F	force [N]
f	frequency [Hz]
	friction factor [-]
G	conductance [S]
h	heat transfer coefficient [W/(m^2 °C)]
H	height [m]
I	current [A]
	moment of inertia [kg m^2]
k	spring constant [N/m]
	thermal conductivity [W/(m °C)]
K	beam stiffness [N/m]
KE	kinetic energy [J]
L	inductance [H]
	length [m]
m	mass [kg]
$\dot{m}$	mass flowrate [kg/s]
Ma	Mach Number [-]
n	number of moles [mol]
	shaft speed [Hz]
P	perimeter [m]
	power [W]
	pressure [Pa]
	vertical load of beam [N]
PE	potential energy [J]
Q	electric charge [C]
	thermal energy [cal]
	volumetric flowrate [m^3/s]
r	radius [m]
R	resistance [Ω]
Re	Reynolds Number [-]
S	arc length [m]
SG	specific gravity [-]
T	period [s]
	temperature [K]
t	time [s]
v	velocity [m/s]
V	volume [m^3]
	voltage [V]
w	weight [N]
W	width [m]
	work [J]
x	displacement [m]
α	neper frequency [Hz]
	thermal diffusivity [m^2/s]
β	coefficient of thermal expansion [$°C^{-1}$]
γ	specific weight [N/m^3]
Δ	delta; change in quantity
	tip deflection of beam [m]
ε	pipe roughness [mm]
	strain [-]
η	efficiency [-]
θ	angle [rad]
μ	coefficient of friction [-]
	dynamic viscosity [cP]
ν	kinematic viscosity [St]
ρ	mass density [kg/m^3]
	resistivity [Ω m]
σ	stress [Pa]
	surface tension [J/m^2]
τ	shear stress [Pa]
ω	angular frequency [rad/s]

Unit Abbreviations

°C	degree Celsius
°F	degree Fahrenheit
°R	degree Rankine
A	ampere
amu	atomic mass unit
atm	atmosphere
AU	astronomical unit
BTU	British thermal unit
C	coulomb
cal	calorie
cc	cubic centimeter
cd	candela
cP	centipoise
d	day
Da	dalton
deg, °	degree
eV	electron volt
F	farad
fl oz	fluid ounce
ft H_2O	feet of water
ft	foot
g	gram
gal	gallon
gpm	gallons per minute
H	henry
h	hour
hp	horsepower
Hz	hertz
in Hg	inches of mercury
in	inch
J	joule
K	kelvin
kg	kilogram
kWh	kilowatt-hour
L	liter
lb_f	pounds-force
lb_m	pounds-mass
m	meter
mi	mile
min	minute
mm Hg	millimeters mercury
mol	mole
mph	miles per hour
N	newton
Pa	pascal
psi	pound-force per square inch
rad	radian
s	second
S	seimens
St	stokes
t	ton
tbsp	tablespoon
tsp	teaspoon
V	volt
W	watt
yd	yard
yr	year
Ω	ohm

Miscellaneous Equations

Antoine equation

$$P = 10^{\left(A - \frac{B}{T+C}\right)}$$

Cantilever Beam Deflection

$$P = K\,\Delta$$

Discharging Capacitor

$$V = V_0\, e^{-t/RC}$$

Moore's Law

$$T = T_0\, 2^{\frac{t}{2}}$$

Oscillating spring

$$f = \frac{1}{2\pi}\sqrt{\frac{k}{m}}$$

Poiseuille's Equation

$$R = \frac{8\mu L}{\pi r^4}$$

Pythagoras' Theorem

$$d = \sqrt{(x_2 - x_1)^2 + (y_2 - y_1)^2}$$

Resonant frequency for one inductor and one capacitor

$$f_R = \frac{1}{2\pi\sqrt{L\,C}}$$

Shockley Equation

$$I = I_0\left(e^{\frac{V_D}{nV_T}} - 1\right)$$

Torricelli's Law

$$v = \sqrt{2\,g\,H}$$

van der Waals Equation

$$\left(P + \frac{an^2}{V^2}\right)(V - b\,n) = n\,R\,T$$

Bingham plastic

$$\tau = \mu\frac{\Delta v}{\Delta y} + \tau_0$$

Dimensionless Number Equations

Coefficient of Drag

$$C_d = \frac{F}{\frac{1}{2}\rho v^2 A_p}$$

Friction factor: Laminar Flow

$$f = \frac{64}{Re}$$

Friction factor: Blasius Approximation

$$f = 0.316\,(Re)^{-1/4}$$

Friction factor: Colebrook

$$\frac{1}{\sqrt{f}} = -2\,\log\left(\frac{\varepsilon/D}{3.7} + \frac{2.51}{Re\sqrt{f}}\right)$$

Mach Number

$$Ma = \frac{speed_{object}}{speed_{sound\ in\ air}}$$

Power Number

$$N_P = \frac{P}{\rho n^3 D^5}$$

Reynolds Number: Pipe Flow

$$Re = \frac{\rho D v}{\mu}$$

Reynolds Number: Use with Coefficient of Drag

$$Re = \frac{D_p \rho v}{\mu}$$

Reynolds Number: Use with Power Number

$$Re = \frac{D^2 n \rho}{\mu}$$

Statistics Equations

Mean

$$\bar{X} = \frac{1}{N}(X_1 + X_2 + \cdots + X_N) = \frac{1}{N}\sum_{j=1}^{N} X_j$$

Median

Place data in ascending order.
If N = odd, median = central value.
If N = even, median = mean of two central values.

Variance

$$V_X^2 = \frac{1}{N-1}((\bar{X} - X_1)^2 + \cdots + (\bar{X} - X_N)^2) = \frac{1}{N-1}\sum_{j=1}^{N}(\bar{X} - X_j)^2$$

Standard Deviation

$$SD_X = \sqrt{V_X^2}$$

Equations

Current
$Q = I t$

Conductance
$G = 1/R$

Power: Electric
$P = V I = I^2 R = V^2/R$

Electric charge
$Q = CV$

Inductance
$V = L\, dI/dt$

Capacitors: Parallel
$C_{eff} = C_1 + C_2 + \cdots + C_{N-1} + C_N$

Capacitors: Series
$$C_{eff} = \left(\frac{1}{C_1} + \frac{1}{C_2} + \cdots + \frac{1}{C_{N-1}} + \frac{1}{C_N}\right)^{-1}$$

Capacitance: Energy Stored
$E_C = 1/2\, C\, V^2$

Inductance: Energy Stored
$E_L = 1/2\, L\, I^2$

Inductors: Series
$L_{eff} = L_1 + L_2 + \cdots + L_{N-1} + L_N$

Inductors: Parallel
$$L_{eff} = \left(\frac{1}{L_1} + \frac{1}{L_2} + \cdots + \frac{1}{L_{N-1}} + \frac{1}{L_N}\right)^{-1}$$

Circuits: Ohm's Law
$V = I R$

Circuits: Parallel Resistors
$$R_{eff} = \left(\frac{1}{R_1} + \frac{1}{R_2} + \cdots + \frac{1}{R_{N-1}} + \frac{1}{R_N}\right)^{-1}$$

Circuits: Series Resistors
$R_{eff} = R_1 + R_2 + \cdots + R_{N-1} + R_N$

Density
$\rho = m/V$

Efficiency
$\eta = \text{output/input}$

Work
$W = F\, \Delta x$

Potential Energy
$PE = m\, g\, \Delta H$

Kinetic Energy, translational
$KE_T = \frac{1}{2} m (v_f^2 - v_i^2)$

Kinetic Energy, rotational
$KE_R = \frac{1}{2} I (\omega_f^2 - \omega_i^2)$

Kinetic Energy, total
$KE = KE_T + KE_R$

Thermal Energy
$Q = m\, C_p\, \Delta T$

Force: Newton's 2nd Law
$F = m\, a$

Interpolation
$$\frac{(A - A_1)}{(A_2 - A_1)} = \frac{(B - B_1)}{(B_2 - B_1)}$$

Power
$P = E/t$

Pressure
$P = F/A$

Pressure, Hydrostatic: Pascal's Law
$P_{hydro} = \rho\, g\, H$

Pressure, Total
$P_{total} = P_{surface} + \rho\, g\, H$

Pressure: Ideal Gas Law
$P V = n R T$

Specific Gravity
$SG = \rho_{object}/\rho_{water}$

Specific Weight
$\gamma = w/V$

Springs: Hooke's Law
$F = k\, x$

Springs: Parallel
$k_{eff} = k_1 + k_2 + \cdots + k_{N-1} + k_N$

Springs: Series
$$k_{eff} = \left(\frac{1}{k_1} + \frac{1}{k_2} + \cdots + \frac{1}{k_{N-1}} + \frac{1}{k_N}\right)^{-1}$$

Temperature
$$\frac{T\,[°F] - 32}{180} = \frac{T\,[°C] - 0}{100}$$
$T\,[K] = T\,[°C] + 273$
$T\,[°R] = T\,[°F] + 460$

Temperature: Newton's Law of Cooling
$T = T_0\, e^{-kt}$

Viscosity, Dynamic: Newton's Law of Viscosity
$\tau = \mu\, (\Delta v/\Delta y)$

Viscosity, Kinematic
$\nu = \mu/\rho$

Weight
$w = m\, g$

Young's Modulus
$\sigma = E\, \varepsilon$

Geometric Formulas

Rectangle

Area = a b
Perimeter = 2 a + 2 b

Rectangular Parallelepiped

Volume = a b c
Surface area = 2 (a b + a c + b c)

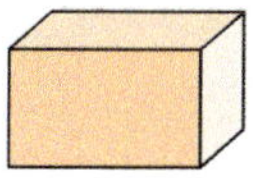

Circle

Area = πr^2
Perimeter = $2\pi r$

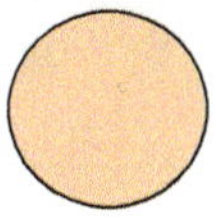

Sphere

Volume = $4/3\ \pi r^3$
Surface area = $4\ \pi r^2$

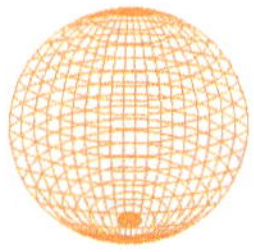

Triangle

Area = $\frac{1}{2}$ b H

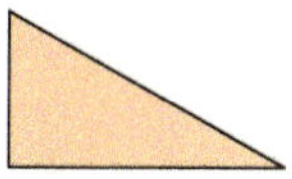

Right Circular Cone

Volume = $1/3\ \pi r^2 H$

Right Circular Cylinder

Volume = $\pi r^2 H$
Lateral Surface Area = $2\pi r H$

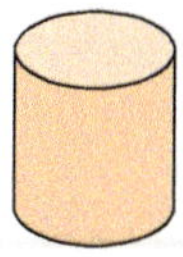

MATLAB SPECIAL CHARACTER REFERENCE

These special characters are available for use plots (title, legend, xlabel, ylabel, etc.).

Character Sequence	Symbol	Character Sequence	Symbol	Character Sequence	Symbol
\alpha	α	\upsilon	υ	\infty	∞
\angle	∠	\phi	ϕ	\clubsuit	♣
\ast	*	\chi	χ	\diamondsuit	♦
\beta	β	\psi	ψ	\heartsuit	♥
\gamma	γ	\omega	ω	\spadesuit	♠
\delta	δ	\Gamma	Γ	\leftrightarrow	↔
\epsilon	ε	\Delta	Δ	\leftarrow	←
\zeta	ζ	\Theta	Θ	\Leftarrow	⇐
\eta	η	\Lambda	Λ	\uparrow	↑
\theta	Θ	\Xi	Ξ	\rightarrow	→
\vartheta	ϑ	\Pi	Π	\Rightarrow	⇒
\iota	ι	\Sigma	Σ	\downarrow	↓
\kappa	κ	\Upsilon	γ	\circ	∘
\lambda	λ	\Phi	Φ	\pm	±
\mu	μ	\Psi	Ψ	\geq	≥
\nu	ν	\Omega	Ω	\propto	∝
\xi	ξ	\forall	∀	\partial	∂
\pi	π	\exists	∃	\bullet	•
\rho	ρ	\ni	∋	\div	÷
\sigma	σ	\cong	≅	\neq	≠
\varsigma	ς	\approx	≈	\aleph	ℵ
\tau	τ	\Re	ℜ	\wp	℘
\equiv	≡	\oplus	⊕	\oslash	∅
\Im	ℑ	\cup	∪	\supseteq	⊇
\otimes	⊗	\subseteq	⊆	\subset	⊂
\cap	∩	\in	∈	\o	o
\supset	⊃	\lceil		\nabla	∇
\int	∫	\cdot	·	\ldots	...
\rfloor	⌊	\neg	¬	\prime	′
\lfloor	⌋	\times	×	\0	∅
\perp	⊥	\surd	√	\mid	∣
\wedge	∧	\varpi	ϖ	\copyright	©
\rceil	⌉	\rangle	〉		
\vee	∨	\sim	∼		
\langle	〈	\leq	≤		

Original version: http://www.mathworks.com/help/techdoc/ref/text_props.html#String
Note: Greek letters appear in italics throughout this text but will appear as shown on this table in Matlab.

MATLAB LINE STYLES, MARKERS, AND COLORS

Line style specifiers

Specifier	Line Style
'–'	Solid line (default)
'--'	Dashed line
':'	Dotted line
'–.'	Dash-dot line
'none'	No line

Marker specifiers

Specifier	Marker Type
'+'	Plus sign
'o'	Circle
'*'	Asterisk
'.'	Point
'x'	Cross
'square' or 's'	Square
'diamond' or 'd'	Diamond
'^'	Upward-pointing triangle
'V'	Downward-pointing triangle
'>'	Right-pointing triangle
'<'	Left-pointing triangle
'pentagram' or 'p'	Five-pointed star (pentagram)
'hexagram' or 'h'''	Six-pointed star (hexagram)
'none'	No marker (default)

Color specifiers

Specifier	Color
r	Red
g	Green
b	Blue
c	Cyan
m	Magenta
y	Yellow
k	Black
w	White

Original source: http://www.mathworks.com/help/techdoc/ref/linespec.html

MATLAB GRAPHING PROPERTIES

Most graphing functions have the ability to apply special properties for additional customization of the appearance. This section contains a list of some of the different property names and property values available in MATLAB. Note the property name and property values are placed inside single quotes, with exceptions listed.

Plotting Properties:

Functions: **plot**

Example

```
plot(X,Y,'s','MarkerSize',12)
```

Property Name	Purpose	Property Values	Default
LineWidth	To change the width of lines on a graph	Numerical values, without single quotes	0.5
MarkerEdgeColor	To change the border color of all markers	Same as BackgroundColor	Default marker color
MarkerFaceColor	To change the fill color of all markers	Same as BackgroundColor	Default marker color
MarkerSize	To change the size of all markers	Numerical values, without single quotes	2.0

Text Properties:

Functions: **text, xlabel, ylabel, title**

Example

```
text(100,200,'Example Text','EdgeColor','red')
title('Example Title','FontWeight','bold','FontSize',16)
```

Property Name	Purpose	Property Values	Default
BackgroundColor	To change the background color of the textbox created by the function	yellow green magenta blue cyan white red black	transparent
Color	To change the color of the text created by the function	Same as BackgroundColor	black
EdgeColor	To change the color of the border around the textbox created by the function	Same as BackgroundColor	none
FontAngle	To change the font angle	normal italic oblique	normal
FontSize	To change the font size (in points)	Numerical values, without single quotes	10.0

Note: By default, all font size are given in the unit of points (e.g., 12 pt font)

Text Properties (continued)

Property Name	Purpose	Property Values	Default
`FontWeight`	To change the font weight	`light` `demi` `normal` `bold`	`normal`

`HorizontalAlignment`	To change the horizontal alignment of the text within the textbox	`left`
		`center`
		`right`
			`center`

`LineStyle`	To change the style of the line bordering the textbox. Enabled when `EdgeColor` changed to a visible color	`-`
		`--`
		`:`
		`-.`
		`none`
			By default, the style is `-`, but not visible until `EdgeColor` changed

`LineWidth`	To change the width (in unit of points) of the line bordering the textbox. Enabled when `EdgeColor` changed to a visible color	Numerical values, without single quotes	0.5
`Margin`	To change the amount of space (in unit of pixels) between the text and the line bordering the textbox. Enabled when `EdgeColor` changed to a visible color	Numerical values, without single quotes	2.0

`VerticalAlignment`	To change the vertical alignment of the text within the textbox	`top`
		`cap`
		`middle`
		`baseline`
		`bottom`
			`bottom`

ASCII code

Character	A	B	C	D	E	F	G	H	I	J	K	L	M
Value	65	66	67	68	69	70	71	72	73	74	75	76	77

Character	N	O	P	Q	R	S	T	U	V	W	X	Y	Z
Value	78	79	80	81	82	83	84	85	86	87	88	89	90

Character	a	b	c	d	e	f	g	h	i	j	k	l	m
Value	97	98	99	100	101	102	103	104	105	106	107	108	109

Character	n	o	p	q	r	s	t	u	v	w	x	y	z
Value	110	111	112	113	114	115	116	117	118	119	120	121	122

Character	0	1	2	3	4	5	6	7	8	9
Value	48	49	50	51	52	53	54	55	56	57

APPENDIX B TOLERANCES

OVERVIEW

Tolerances on dimensions are necessary to specify the acceptable variability in the dimension of a part. Tolerances can be indicated for all dimensions using a general note or they can be specified for each individual dimension. Tolerances are based on the function of the part or are used to ensure that mating parts fit together. However, in many cases the manufacturing process determines the tolerance. The surface finish may also be specified on a drawing to indicate the roughness of the surface. To further minimize ambiguity in engineering drawings, geometric dimensioning and tolerancing is used to specify the form of a feature, such as flatness or roundness, or to indicate the ideal position of a feature.

B.1 WHY TOLERANCES?

In 1815, it was hoped that muskets could be produced in a number of different U. S. Government and private armories so that the parts of the musket would be interchangeable. Before that time, muskets were individually handcrafted, so parts from one musket rarely fit another musket. Because engineering drawings were not commonly used, a number of "perfect" muskets were made according to master gauges and jigs. **Gauges** are devices used to check the individual dimensions of a part. For instance, the diameter of a shaft can be checked using a gauge consisting of two holes: one of the maximum allowable diameter of the shaft and one of the minimum allowable diameter. If the shaft fits through the larger hole, but not the smaller hole, then it is within the allowable dimensions for the shaft. **Jigs** are devices used to hold a workpiece and guide the tool to ensure repeatable machining. The master gauges and jigs were sent to the armories with the instruction that "no deviations from the pattern were to be allowed." In 1824, this concept of interchangeable parts was tested by disassembling a hundred rifles from several different armories, mixing the parts, and then reassembling the parts at random. Most reassembled rifles worked as they should, proving the concept and value of interchangeable parts.

gauges

jigs

The interchangeability of parts is so common now that it is hard to imagine anything different. The concept allows parts made in various locations by different manufacturers to be successfully assembled and to function properly as an assembly. Although engineering drawings rather than "master parts" came into use in the mid-19th century, it was the two World Wars in the 20th century that brought about the development of methods of showing the acceptable variation in a dimension on a drawing, or **tolerancing**.

tolerancing

The need to control precisely the geometry of a part arises from the part's function. For instance, the cross-sectional geometry of an airfoil, such as the wing of a jet aircraft or a turbine blade, must be accurately controlled to ensure aerodynamic efficiency. The more commonplace need for controlling geometry results from the requirement for parts to fit together. But it is quite difficult to make every part the exact size that is specified in a dimension because of slight differences in tool size, machine tool wear, human operator error, and other factors. As a consequence, tolerances are used in engineering drawings to specify the limit in the variation between mating parts and to provide guidelines to control the manufacturing process. The **tolerance** is the amount that a specific dimension is permitted to vary, or the difference between the maximum and minimum limits of a dimension.

tolerance

Consider a dimension on a drawing specified as 3.750 ± .003. This dimension indicates that the part has a **nominal size**, or general size, of $3\frac{3}{4}$ inches (usually expressed as a common fraction). The **basic size**, or theoretical size for the application of the tolerance, is 3.750 inches (expressed as a decimal). The tolerance of the dimension is .006 inch. Thus, an acceptable machined part can have an **actual size** ranging from 3.747 to 3.753 inches, which are the minimum and maximum limits of the dimension. If the actual size is smaller than 3.747 or larger than 3.753, then the part is not acceptable.

nominal size
basic size
actual size

Of course, it is desirable to make the part as close as possible to the basic size, sometimes called the **target size**. But the more accuracy needed in a machined part, the higher is the manufacturing cost. Furthermore, the method of machining a part limits the tolerance that can be specified. For instance, the tolerance in drilling a .500-inch hole is between .002 and .005 depending on the quality of the drill press and drill bit. To require a tolerance smaller than .002 would require an additional machining process, such as reaming. Consequently, tolerances can play a large role in the cost of manufacturing a part. Therefore, the tolerances should not be specified tighter than necessary for the product to function properly.

target size

An increased awareness of the importance of tolerances in manufacturing has led to an approach in which the deviation of the actual size of a part from the target size has a cost in terms of reworking of the part, scrap, customer dissatisfaction, and poor reliability. An approach to handle this is to design a product so that it is less sensitive to manufacturing tolerances. The idea is to design the part intelligently so that easy-to-manufacture tolerances, rather than exceptionally tight tolerances, are implemented. Consequently, proper fit between mating parts comes about because of good design, instead of depending on tight tolerances. This approach is known as **robust design**

robust design

B.2 DISPLAYING TOLERANCES ON DRAWINGS

Every dimension on a drawing should have a tolerance. Tolerances can be displayed on engineering drawings in several ways depending on the situation.

General tolerances can be given in a note on the drawing or in the title block. An example would be the note ALL DIMENSIONS TO BE HELD TO ± .003". Thus, a dimension of .375 would have a minimum limit of .372 and a maximum limit of .378. Often general tolerances are specified in terms of the number of digits following the decimal point in a dimension. In this case, a note would appear on the drawing, such as

general tolerances

UNLESS OTHERWISE SPECIFIED

TOL: .XX = ±.010
.XXX = ±.005
HOLES = ±.002
ANGLES = ±.5 DEG

Thus, a dimension of .50 on a drawing would indicate an actual size between .490 and .510. A dimension of .500 would indicate an actual size between .495 and .505. For metric dimension, a similar form for general tolerances would be X.X METRIC = ± .08.

limit dimensions

Two other methods of displaying tolerances on a drawing are shown in Figure B-1. The maximum and minimum sizes are specified directly when using **limit dimensions** to specify the tolerance. Usually the upper limit is placed above the lower limit. While this method clearly shows the limits of the dimension, many engineers, designers, and machinists think in terms of the nominal or basic size plus a tolerance, neither of which is immediately evident using limit dimensions. Thus, a more common approach is to use **plus and minus dimensions** where the basic size is shown followed by the tolerance values. If the plus and minus tolerance values are identical, then the basic dimension is followed by a plus-or-minus sign and half of the numerical value of the tolerance. Otherwise, the plus and minus tolerances are indicated separately with the plus above the minus. A **bilateral tolerance**, such as the tolerance for the 1.500-inch dimension in Figure B-1, varies equally in both directions from the basic size. A **unilateral tolerance**, such as the tolerance for the 1.250 diameter of the shaft in Figure B-1, varies in only one direction from the basic size. In both cases, the number of digits to the right of the decimal point for the dimension should be the same as the number of digits to the right of the decimal point for the tolerance.

plus and minus dimensions

bilateral tolerance
unilateral tolerance

Sometimes it is not necessary to specify both limits of a tolerance, and only one limit dimension is needed. MIN or MAX is placed after a numerical dimension to indicate that it is a **single-limit dimension**. The depth of holes, length of threads, and radii of corner rounds are often specified using single-limit dimensions.

single-limit dimension

Many times several methods of representing tolerances are used on a single drawing. For instance, general tolerances may be listed in the title block, and plus and minus dimensions or limit dimensions may be used for a few dimensions to which the general tolerances do not apply.

An important consideration in applying tolerances is the effect of one tolerance on another, especially because tolerances are cumulative. This is known as **tolerance accumulation** or **tolerance stackup** for a chain of dimensions. In the example in Figure B-2, two dimensioning schemes are shown. Above the part, **chain dimensioning** shows the distance from one feature to the next. This would be an appropriate dimensioning scheme if the distance from one hole to the next was critical, because the distance between holes must be between .995 and 1.005. But if the distances of the holes from the left edge of the part were critical, a large tolerance would accumulate. The third hole would be 2.985 from the left edge if all of the dimensions happened to be at the minimum limit of the dimension, or 3.015 from the left edge if all dimensions were at the maximum limit. Thus, the actual dimension of the third hole from the left edge using chain dimensioning is 3.000 ± .015. This tolerance accumulation can be avoided using **datum dimensioning**, where the dimensions are given with respect to a datum, in this case the left edge of the part as shown below the part in Figure B-2. Using this scheme, the third hole is specified to be between 2.995 and 3.005 from the left edge. Of course, the distance between the second and third holes could be as small as .990 if the

tolerance accumulation
tolerance stackup
chain dimensioning

datum dimensioning

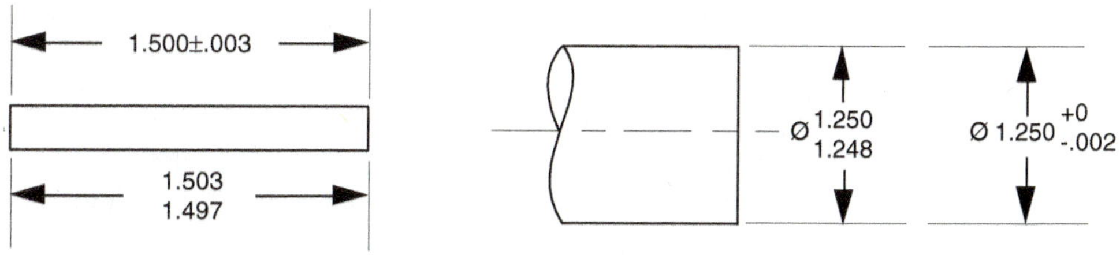

Figure B-1 Plus and minus tolerances and limit tolerances.

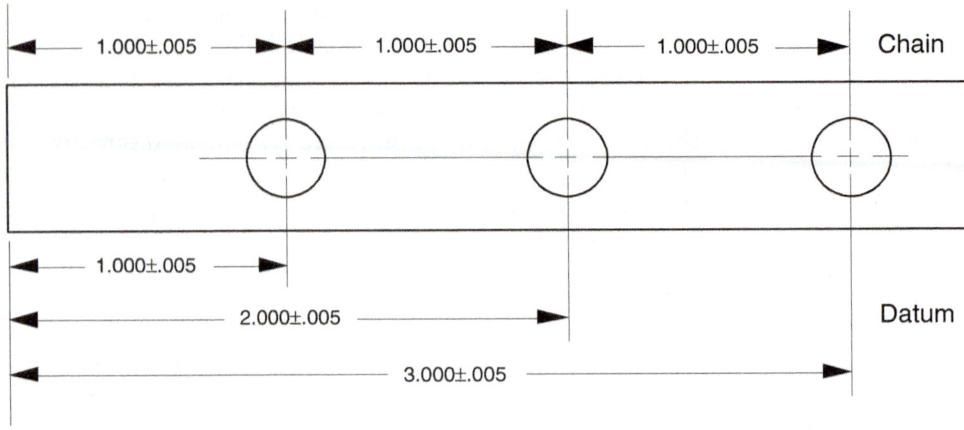

Figure B-2 Chain dimensioning and datum dimensioning.

second hole is at 2.005 and the third hole is at 2.995. Alternatively, the distance between the second and third holes could be as large as 1.010 if the holes are at 1.995 and 3.005. Consequently, the decision about which way to dimension the holes depends on whether the distance between holes is critical or whether the distance from each hole to the left edge is critical.

B.3 HOW TO DETERMINE TOLERANCES

One of the most difficult aspects of dimensioning and tolerancing for inexperienced engineers and designers is to determine what tolerance to specify on a drawing. In some cases a specific tolerance is quite clear based on the function of the design. For instance, if a $\frac{3}{4}$ inch shaft must rotate at moderate speed in a hole, the tolerances on the shaft and the hole can be determined based on standard classes of fits that are available in many engineering and machinist handbooks. In this particular case the hole could be between .7500 and .7512, while the shaft could have a diameter of .7484 to .7492. A shaft any bigger or a hole any smaller would result in too tight of a clearance for the shaft to rotate freely, or could result in the shaft being too big for the hole. A smaller shaft or larger hole than the specified limits would result in a sloppy fit between the shaft and the hole. The point here is not the details of how the tolerances are determined. That is too specific to individual cases for this discussion and is presented in many handbooks and texts. Instead, the point is that the function of the system that is being designed drives the tolerances. In this case, *the tolerances drive the manufacturing process* used to machine the hole and the shaft.

But a more common situation is the opposite: *The manufacturing process drives the tolerance.* If the only machine tool that is available to form a hole is a drill press, the $\frac{3}{4}$ inch hole that is cut will be between .748 and .754 in diameter based on expected tolerances for the drilling operation. But this means that the hole in the example just presented would have too much variability to ensure that the shaft rotates in it properly. Of course, if only one shaft-hole assembly was being made, the hole could be drilled first and measured. Then the shaft could be turned on a lathe to a diameter just slightly smaller to achieve the fit described. However, this approach will clearly not work for mass-produced, interchangeable parts. Either a more accurate machining method is needed to create the hole or the design must be altered to avoid drilling a hole altogether. (In this example, a standard-size, off-the shelf bearing could be inserted into a larger drilled hole to mate with the shaft.)

When the manufacturing processes that are available drive the tolerances, the tolerances on the drawing must reflect the manufacturing process that is used to make the part, and the design must be altered to ensure that the part can be made with the available machining processes. Guidelines are available for the tolerance range of various common machining processes, including the following:

lapping — ***Lapping:*** A process that produces a very smooth and accurate surface by rubbing the surface of the workpiece against a mating form, often using a fine abrasive between the surfaces.

honing — ***Honing:*** A process in which a honing stone made of fine abrasive is used to form a surface on a workpiece.

grinding — ***Grinding:*** A process that removes a small amount of material from a workpiece with a rapidly rotating grinding wheel.

broaching — ***Broaching:*** A process similar to filing in which a cutting tool is reciprocated along its axis to remove a small amount of material from a workpiece.

reaming — ***Reaming:*** Removal of a small amount of material from a drilled hole using a rotating cutting tool (reamer).

turning — ***Turning:*** Axisymmetric removal of material from a rotating workpiece by feeding a cutting tool into the workpiece using a lathe.

boring — ***Boring:*** Axisymmetric removal of material from a workpiece using a tool with a single cutting surface. Either the workpiece or the tool may be rotating.

milling — ***Milling:*** Removal of material by feeding a workpiece into a rotating cutting tool.

drilling — ***Drilling:*** Creating a hole by feeding the end of a sharpened cylindrical cutting tool (drill bit) into the material.

stamping **punching** — ***Stamping and Punching:*** Cutting material using a punch and a die to shear the material, much like a paper hole-punch.

The tolerance typically depends on the size of the feature, as well as the quality of the machine tool, sharpness of the cutting tool, and skill of the operator. Figure B-3 indicates the range of tolerances to be expected for various machining operations. To use the table shown in the figure, first consider the size of the feature in the left column. Then go below the sizes in the left column to the type of machining operation used. The bar associated with that process indicates the range of tolerances in the row associated with the feature size. For instance, consider a feature that has a dimension of 2 inches. If broaching were used to create that feature, the chart indicates that the total tolerance would be .0004 to .0015, depending on the machining conditions. Thus, this dimension could be specified as 2.0000 ± .0002 up to 2.0000 ± .0008 (rounding up). If the same piece were created on a milling machine, the tolerances would be .0025 to .010, indicating a dimension of 2.0000 ± .0013 up to 2.000 ± .005. If broaching were not available, we would have to be satisfied with the tolerances of a milling machine and adjust the design accordingly, if the milling tolerances are too large. If both broaching and milling were available, the criticality of the tolerance and the cost of manufacturing would drive the tolerance that would be specified on the drawing. (Broaching, a mass-production process, is usually more expensive than milling, so milling would be preferred if the milling tolerances are acceptable.) Tolerances for metric dimensioning can be determined by converting the values in Figure B-3 to millimeters and rounding off the result to one less place to the right of the decimal point.

Size (in.)	Total Tolerance (in.)								
0.000-0.599	0.00015	0.0002	0.0003	0.0005	0.0008	0.0012	0.002	0.003	0.005
0.600-0.999	0.00015	0.00025	0.0004	0.0006	0.001	0.0015	0.0025	0.004	0.006
1.000-1.499	0.0002	0.0003	0.0005	0.0008	0.0012	0.002	0.003	0.005	0.008
1.500-2.799	0.00025	0.0004	0.0006	0.001	0.0015	0.0025	0.004	0.006	0.010
2.800-4.499	0.0003	0.0005	0.0008	0.0012	0.002	0.003	0.005	0.008	0.012
4.500-7.799	0.0004	0.0006	0.001	0.0015	0.0025	0.004	0.006	0.010	0.015
7.800-13.599	0.0005	0.0008	0.0012	0.002	0.003	0.005	0.008	0.012	0.025

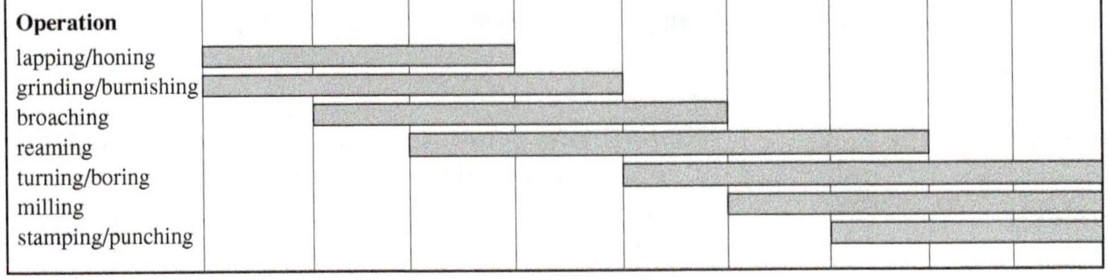

Figure B-3 Range of tolerances for various machining operations.

Typical tolerances on drills depend on the drill size. Standard drill sizes are fractional in increments of $\frac{1}{64}$ inch and range from $\frac{1}{16}$ inch to 4 inches. In addition, drill sizes are specified according to an alphanumeric code. The coded drill sizes range from 80 (.0135 inch) to 1 (.2280 inch) and from A (.234 inch) to Z (.413 inch). The dimensions for the coded drill sizes are widely available in engineering and machinist handbooks. In addition, the proper drill sizes for drilling holes that will be tapped are listed in these handbooks. The standard tolerances for drilled holes are given in Table B-1.

Table B-1 Standard Tolerances for Drilled Holes

Drill Size (In.)	Tolerance (In.)	
	Plus	Minus
0.0135 (80)–0.185 (13)	0.003	0.002
0.1875–0.246 (D)	0.004	0.002
0.250–0.750	0.005	0.002
0.7656–1.000	0.007	0.003
1.0156–2.000	0.010	0.004
2.0312–3.500	0.015	0.005

B.4 FITS

The degree of tightness or looseness between mating parts is known as the fit. This concept is most easily explained in terms of a shaft and a hole, although the concept can be applied to other mating parts in which an internal member is designed to fit within an external member. For example, a shaft that is .750 inch in diameter will fit into a hole that is .751 inch in diameter with .001 inch of clearance, but it will be too large for a hole that is .749 inch in diameter. But the situation is more complicated than this because the tolerances on the diameters of both the shaft and the hole play roles in the fit. Consider the example shown in Figure B-4 of a hole and a shaft, each with a basic diameter of .750 inch.

clearance fit

In a **clearance fit**, the internal component fits in the mating part so that there is always a space or gap between the parts. In Figure B-4, the largest shaft size for a clearance fit is .748 inch and the smallest hole size is .750 inch, so there is always a space or **clearance** of at least +.002 inch. This space is called the **allowance**. It is always positive for a clearance fit. The gap may be even larger. The smallest shaft in the largest hole results in an allowance of +.006 inch. A clearance fit is used when a shaft is meant to rotate or slide within the hole.

clearance
allowance

interference fit

An **interference fit** occurs when the internal component is larger than the mating external member, so that the two parts will interfere with one another if they are assembled. For the interference fit in Figure B-4, the smallest shaft, which is .753 inch, is too big to fit into the largest hole, which is only .752 inch. In this case, the allowance is −.001 inch, a negative number to indicate interference. This is known as the **least material condition**, or LMC, for which the least material for the shaft and the block with the hole are present. In other words, both the shaft and the portion of the block left over after the hole is cut are as small as possible. For the block, this corresponds to the largest hole. At the opposite extreme is the **maximum material condition**, or MMC, for which the maximum amount of material is present for the shaft and the block with the hole. For this condition, which corresponds to the largest shaft (.755 inch) and the smallest hole (.750 inch), the allowance is −.005 inch. To assemble these parts requires either using force to press the shaft into the hole, which is called a **press fit**, or using heating or cooling to stretch the hole or shrink the shaft. The latter are called **shrink fits**. They are often accomplished by heating the hole to expand it or cooling the shaft to shrink it, assembling the shaft in the hole, and then allowing the assembly to return to room temperature. Interference fits can be quite useful in design applications to mechanically fasten parts together without the use of fasteners or adhesive.

least material condition

maximum material condition

press fit
shrink fits

transition fit

A **transition fit** results when the tolerances are such that the fit can be either a clearance fit or an interference fit when the parts are assembled. The transition fit in Figure B-4 is a clearance fit if the shaft has a dimension of .749 and the hole is .752 (the least material condition), but it is an interference fit if the shaft is .751 and the hole is .750 (the maximum material condition). Transition fits are sometimes used to position or locate one part with respect to another.

ANSI, ISO, and other organizations have issued engineering standards for a variety of types of shaft-in-hole fits to aid engineers in designing for particular applications. For instance, "running and sliding fits" prescribe the clearances for situations in which a shaft should rotate (running) or slide in a hole. There are a variety of running and sliding fits ranging from "close sliding fits" in which the clearance is quite small so that there is no perceptible play between the parts to "loose running fits" in which the clearance is quite large when accuracy is not essential. The engineering

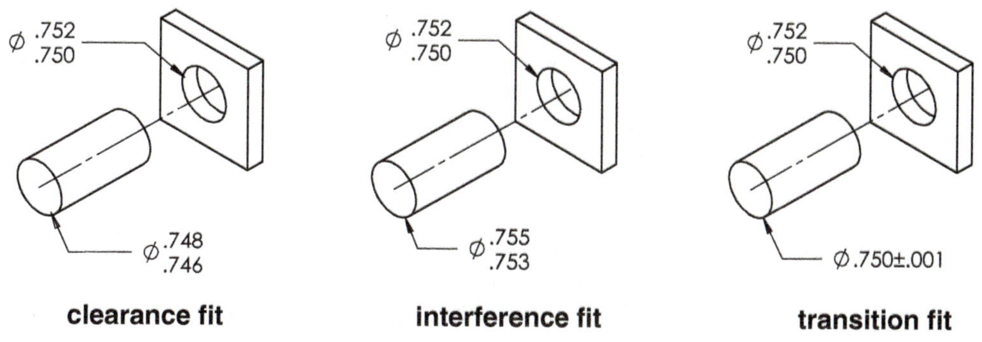

Figure B-4 Clearance, interference, and transition fits.

standard indicates the optimal clearance to be specified for the particular type of running or sliding condition that is required. There are also engineering standards for "locational fits," which are used to position one part with respect to another, and for "force and shrink fits" to fasten parts together.

B.5 SURFACE FINISH

roughness

waviness

lay

All surfaces are rough and irregular on a microscopic scale. The roughness can be described in terms of asperities (peaks or ridges) and valleys. The irregularities of a surface are classified into **roughness** and **waviness**. Roughness describes the smallscale, somewhat random irregularity of a surface, while waviness describes a more regular variation in the surface that typically has a wavelength of 0.8 mm or more. The surface can be thought of as a roughness superimposed on waviness. The roughness and waviness can be measured in several ways, but the most common is with a profilometer, an instrument that drags a fine-pointed stylus over the surface and records the height of the stylus. The roughness is measured in terms of the roughness average (R_a). The roughness average is the average height of peaks above or valleys below the local surface height. More precisely, it is defined as the average of the absolute values of the deviation of the local surface height from the average height of the surface. This is equivalent to half the average of the peak-to-valley height. In addition, the **lay**, or predominant direction of the marks left by the machine tools, may be important in describing the surface.

The surface finish resulting from several machining and forming processes are shown in Figure B-5. This figure shows the range of possible surface roughness averages (R_a) in both micrometers ($1\,\mu\text{m} = 10^{-6}\,\text{m}$) and microinches ($1\,\mu\text{in} = 10^{-6}\,\text{in}$). For instance, surfaces formed by drilling are likely to have a roughness from 12.5 μm down to 0.8 μm (500 μin down to 32 μin). Typically, the roughness is in the middle of the specified range, about 6.3 μm down to 1.6 μm for the case of drilling. In addition to the machining processes described in Section B.3, the following processes are also included in Figure B-5

sand casting

forging

permanent mold casting

Sawing: Cutting material by moving a toothed blade through the material.

Polishing: Removal of scratches and tool marks on a surface using a belt or rotating wheel of soft material such as cloth or leather, often with a fine abrasive.

Sand Casting: Forming a part by pouring molten metal into a mold made of sand.

Forging: Shaping of metal to a desired form by pressing or hammering, usually with the metal hot.

Permanent Mold Casting: Forming a part by pouring molten metal into a permanent mold.

To put the roughness averages in perspective, consider that a 0.10-μm roughness is a mirrorlike surface, free from visible marks of any kind. A 1.6-μm roughness is a high-quality, smooth machined finish. A 6.3-μm roughness is the ordinary finish from typical machining operations.

The surface finish may be specified on a drawing using the special symbols touching the surface or with a leader line to the surface as shown in Figure B-6. For the top surface, the roughness is specified as 1.2-μm corresponding to a high-quality, smooth machine finish. The surface finish symbol on the right side of the object includes a horizontal bar that indicates that material removal by machining is required for this surface. The surface

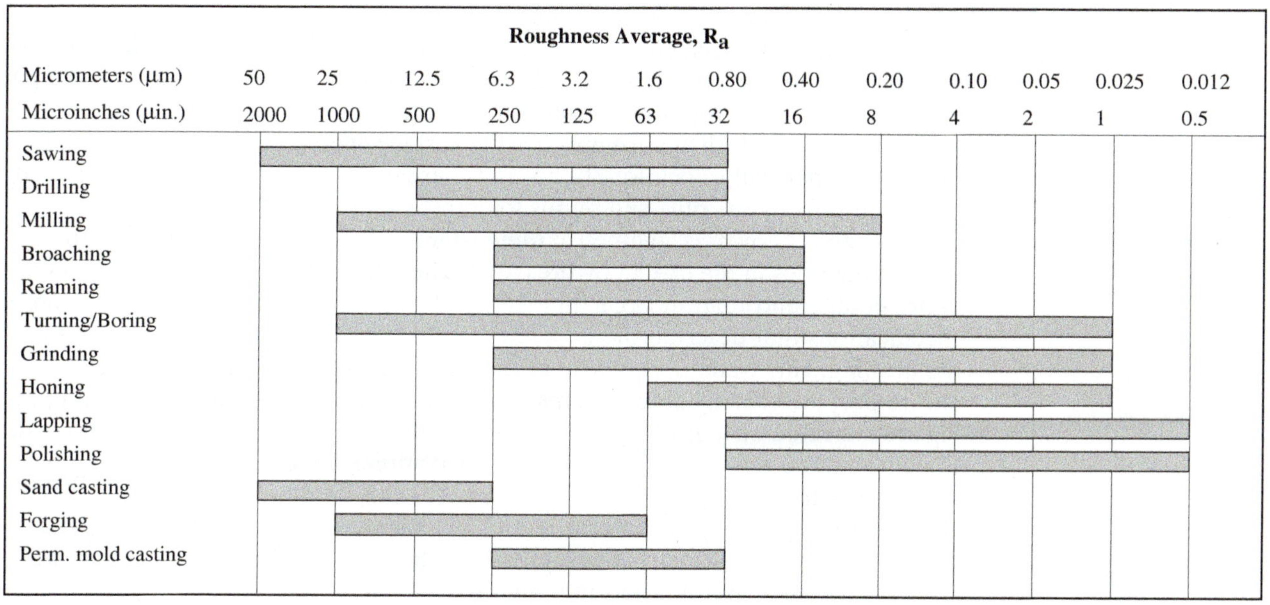

Figure B-5 Surface roughness for various machining operations.

finish symbol on the left side of the object indicates that material removal from this side is prohibited. This surface must be the one resulting from the original process used to produce the surface, such as casting, forging, or injection molding.

Other notations can be added to the surface roughness symbol. A number to the left of the surface finish mark describes the material removal allowance indicating the amount of stock material to be removed by machining. A symbol to the right of the surface roughness mark indicates the lay, or predominant direction of machining marks: perpendicular, parallel, circular, or radial. Other notations can be used to specify the waviness. Details of surface finish notation are available in engineering and machinist handbooks.

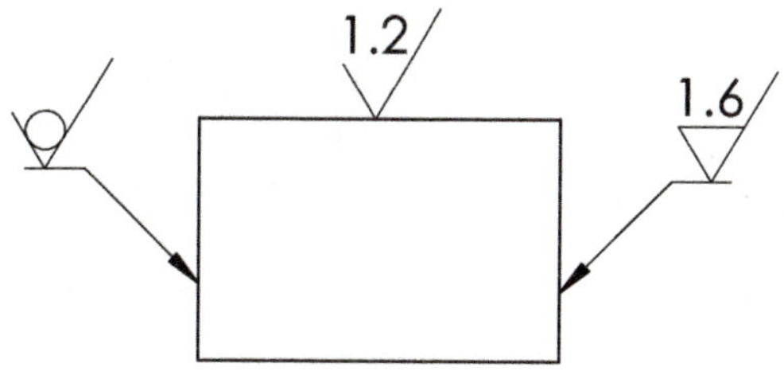

Figure B-6 Surface roughness symbols.

B.6 GEOMETRIC DIMENSIONING AND TOLERANCING

Ideally, engineering drawings should be unambiguous in their interpretation. In many cases, stating dimensions and tolerances provides adequate information so that the part can be manufactured to ensure interchangeability of parts and optimal performance. But sometimes the traditional dimensioning using the plus and minus scheme does not adequately describe the geometry of the part. For instance, consider the part shown in Figure B-7. The dimension shown in the drawing could have several possible interpretations. For instance, the dimension at the top edge could be 14.9 mm while the dimension at the bottom edge could be 15.1 mm, resulting in a trapezoidal part instead of a rectangular part. Or the right edge of the part could be bowed so that the part is 15.1 mm wide at the top and bottom, but 14.9 mm wide at the middle. As a result, the part would not fit flush against another flat surface even though the part is 15 ±.1 mm wide at any particular location. In both of these cases, the geometry of the part is not described fully by the plus and minus tolerances on the dimension.

geometric dimensioning and tolerancing

A system known as **geometric dimensioning and tolerancing** is used to prevent such ambiguities in the design by presenting the design intent more clearly. Geometric dimensioning and tolerancing supplements the traditional dimensioning system with specifications that describe the geometry of the part to enhance the interpretation of the drawing. This system is used to indicate dimensions critical to a part's function and ensure part interchangeability. Items related to geometric dimensioning and tolerancing are indicated on a drawing by a feature symbol consisting of a frame or box around the information describing the geometry. A complete discussion of geometric dimensioning and tolerancing is beyond the scope of this book—there are entire books on the subject. Nevertheless, several of the key aspects and applications of geometric dimensioning and tolerancing are presented here to familiarize the reader with the basic terminology and symbols that are used.

form tolerances
straightness
flatness
circularity
roundness
cylindricity
profile tolerances

The various types of tolerances and the characteristics that they describe are indicated in Figure B-8. Each has its own symbol that is used on a drawing. Perhaps the easiest type of geometric tolerances to understand are the **form tolerances**, which describe the shape of a single feature. For instance, straightness indicates the limits on how much a surface or axis can bow with respect to a straight line. Figure B-9 shows the **straightness** specification for a surface. A feature control box with the straightness symbol (a horizontal bar) and a numerical dimension are connected by a leader line to the surface. The interpretation is that the upper surface must be straight enough so that all points on the surface are within the tolerance zone of 0.010 mm. An example of a **flatness** specification is shown in Figure B-10. The parallelogram symbol with a numerical dimension in a feature control box indicates that the planar surface must lie between two parallel planes 0.20 mm apart. Other single feature specifications include **circularity** or **roundness** (for circular features to be within a tolerance zone defined by two concentric circles), and **cylindricity** (for cylindrical features to be within a tolerance zone defined by two concentric cylinders). Closely related to form tolerances are **profile tolerances**, which define the acceptable deviation of the outline of an object (profile) from that specified in the drawing. These tolerances define a zone on either side of the true profile shown on the drawing within which the line or surface should remain. Profile tolerances can be applied to either a line or a surface.

orientation tolerances
parallelism

Orientation tolerances prescribe the relation between features. **Parallelism** controls the degree to which a surface is parallel to a reference plane, or datum plane.

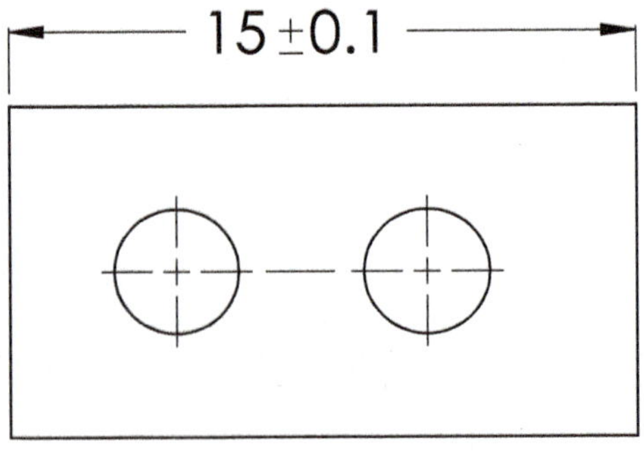

Figure B-7 Ambiguity can result from dimensioning as shown.

	TYPE OF TOLERANCE	CHARACTERISTIC	SYMBOL
FOR INDIVIDUAL FEATURES	FORM	STRAIGHTNESS	—
		FLATNESS	⏥
		CIRCULARITY (ROUNDNESS)	○
		CYLINDRICITY	⌭
FOR INDIVIDUAL OR RELATED FEATURES	PROFILE	PROFILE OF A LINE	⌒
		PROFILE OF A SURFACE	⌓
FOR RELATED FEATURES	ORIENTATION	ANGULARITY	∠
		PERPENDICULARITY	⊥
		PARALLELISM	//
	LOCATION	POSITION	⌖
		CONCENTRICITY	◎
		SYMMETRY	⌯
	RUNOUT	CIRCULAR RUNOUT	↗
		TOTAL RUNOUT	⌰

Figure B-8 Types of tolerances and their symbols.

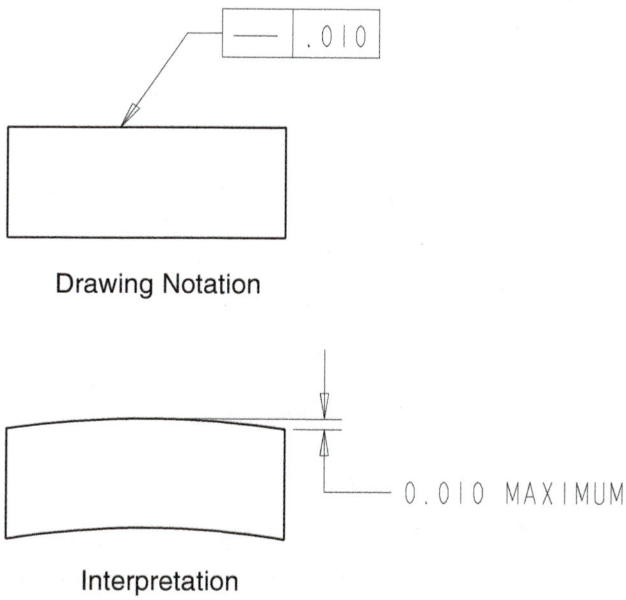

Figure B-9 Straightness specification.

In Figure B-11, the horizontal datum plane that coincides with the bottom edge of the part is identified with a box around B attached to the bottom edge. The parallelism feature symbol (parallel lines), the numerical tolerance, and the reference datum plane are included in a feature control box with a leader line to the surface to which the tolerance applies. In this case, the upper surface must be within 0.10 mm of parallel to datum plane B. **Perpendicularity** with respect to a datum plane is indicated in a similar

perpendicularity

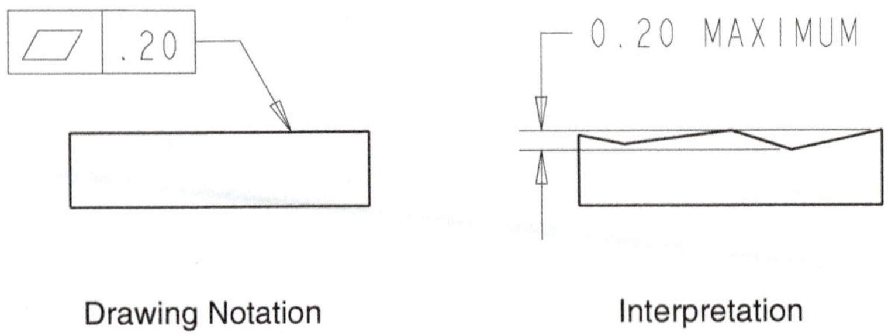

Figure B-10 Flatness specification.

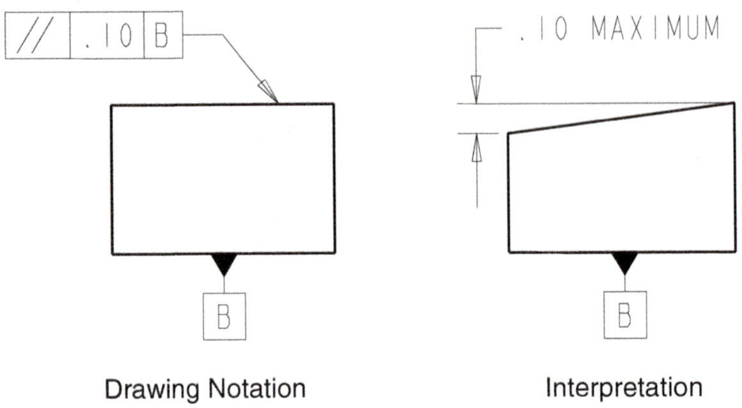

Figure B-11 Parallelism specification.

way, as shown in Figure B-12. Here the vertical plane must be perpendicular to datum plane A to within 0.20 mm. In this example, an alternative notation for specifying datum plane A is used. The datum plane symbol used in Figure B-12 is common, but the datum plane symbol used in Figure B-11 and Figure B-13 is preferred. The datum plane symbol can be applied in several different ways as shown in Figure B-11 and Figure B-13. Several other relations between features can be specified, including the tolerance zone for **angularity** from a datum plane or axis, the tolerance zone for **concentricity** of a surface of revolution with respect to an axis, the tolerance zone for **symmetry** with respect to a datum plane, and the tolerance zone for **runout**, which describes the circularity with respect to an axis of revolution.

angularity
concentricity
symmetry
runout

The most difficult concept in geometric dimensioning and tolerancing is the concept of **positional tolerances**. It is demonstrated most easily in indicating the position of a hole, as shown in Figure B-13a. In this case, the ultimate goal is to have a circular open area (hole) that is at least 25.0 mm in diameter with its center exactly at the position indicated in the drawing. Note that even though we want a 25.0-mm hole, the hole is dimensioned as 25.3 mm in the drawing. The reason for this will shortly become evident.

positional tolerances

Rather than describing the position of the hole with a plus-or-minus tolerance, the ideal position of the hole is prescribed. The boxes around the 40 mm and 60 mm dimensions indicate that they are basic **true-position dimensions**, which describe the perfect position of the center of the hole. Then the feature control box below the dimension of the hole shows how close the center of the hole should be to the true position. In this case, the hole should be within a circular tolerance zone with a diameter of 0.1 mm of the true position (indicated by the plus in the circle within the feature control box followed by ∅.1). The reference datum planes are also included in the feature control box. Datum plane A is the surface of the part in the plane of the

true-position dimensions

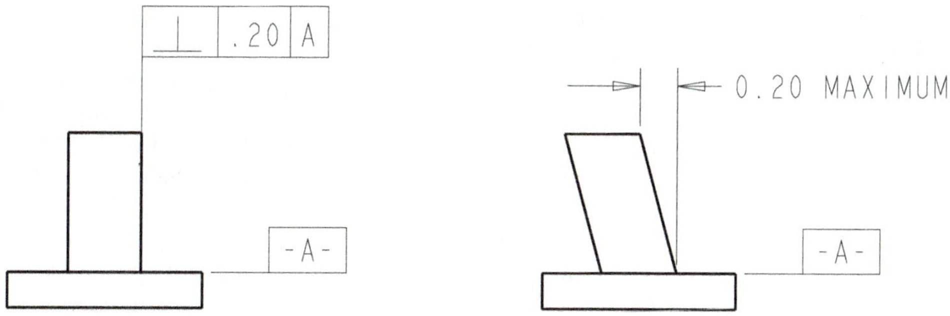

Figure B-12
Perpendicularity specification.

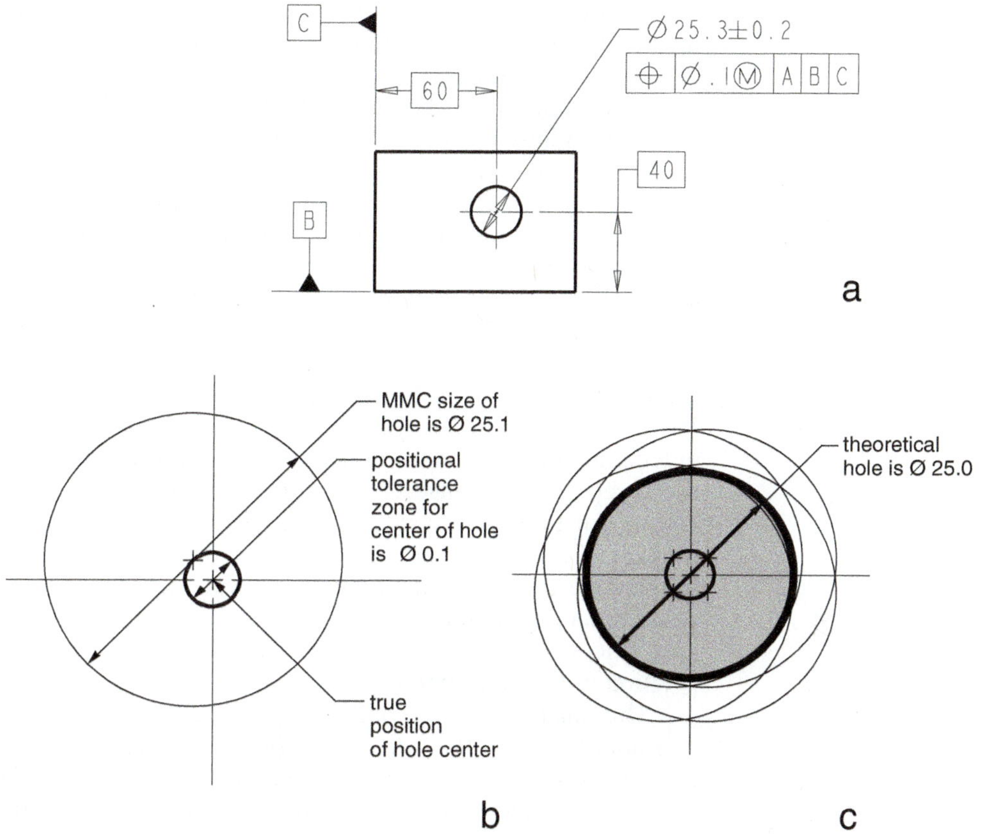

Figure B-13 Positional tolerances.

page. Datum planes B and C are references for the true-position dimensions of 40 mm and 60 mm. These three mutually perpendicular datum planes provide a reference point for the positional tolerance of the hole.

The positional tolerance zone for the center of the hole is shown in Figure B-13b. The "true position of the hole center" is based on the 40 and 60 mm dimensions. Around this center is a circle with a diameter of 0.1 mm (not drawn to scale in Figure B-13b) that defines the "positional tolerance zone." The center of the actual hole must be within this circle. For instance, an acceptable hole would be a 25.1-mmdiameter hole with its center on the edge of this circular tolerance zone, as shown in Figure B-13b.

maximum material condition

The circled M in the feature control box in Figure B-13a indicates that the 0.1-mm circular positional tolerance zone is applied at the **maximum material condition** (MMC). For a hole, the maximum material condition is the smallest acceptable dimension of the hole, because this is the situation in which the most material would be present. For the part in Figure B-13a, the diameter of the hole for the maximum material condition is 25.1 mm, given a dimension of 25.3 ± 0.2 for the diameter of the hole. Thus, the circle representing a hole with its center on the edge of the positional tolerance zone has a diameter of 25.1 mm, as shown in Figure B-13b. Of course, the 25.1-mm-diameter hole could be anywhere within the positional tolerance zone or on its edge. Three more 25.1-mm-diameter circles are drawn in Figure B-13c in addition to the original 25.1-mm circle. The centers of all four circles are on the edge of the positional tolerance zone. Note that there is a circle that lies inside all of the 25.1-mm-diameter circles, which is called the **theoretical hole**. It is relatively straightforward to find the diameter of the theoretical hole, based on the positional tolerance zone for the center of the hole and the MMC size of the hole. The radius of the theoretical hole is the distance from the true position of the hole center to the nearest portion of any one of the MMC circles. Here the theoretical hole has a diameter of 25.0 mm. In fact, the theoretical hole always has a diameter equal to the diameter for the maximum material condition minus the diameter of the positional tolerance (in this case, 25.1 mm − 0.1 mm = 25.0 mm).

theoretical hole

The key result is that if the hole diameter is within the specified tolerance (25.3 ± .2 mm) and the center of the hole is within the positional tolerance zone (.1 mm), there will always be a circular opening in the part with a diameter of at least 25.0 mm. Now assume that the hole must mate with a circular pin that is positioned at the true center of the hole and is exactly 25.0 mm in diameter. Given the dimensioning scheme in Figure B-13a, the pin will always align with the theoretical hole and fit through the theoretical hole, even if the actual hole is not positioned so that its center is exactly at the true position of the hole center. Of course, the theoretical hole is based on the maximum material condition (smallest hole allowed). If the hole is at the large end of the acceptable tolerance, 25.5 mm in diameter, circles drawn in Figure B-13c will be larger. But the theoretical hole will still be open for the pin. Likewise, if the circles drawn in Figure B-13c are not on the edge of the positional tolerance zone, but inside of the zone, the entire theoretical hole will still be open for the pin. Thus, defining the position and diameter of the hole as in Figure B-13a forces the existence of an open area (theoretical hole) that is at least 25.0 mm in diameter and centered at the true position of the hole center.

The maximum material condition reflects the practical aspect of assembly. Consider two parts that are bolted together with several bolts. If the bolt holes are not accurately positioned, not all of the bolts can be inserted. Either drilling out the holes to make them larger or using smaller diameter bolts fixes the problem. The idea of the maximum material condition is to capture these trade-offs. The maximum material condition indicates the inner boundary (the theoretical hole) for the actual hole. The boundary of the actual hole must always be outside of this inner boundary. In this way, a bolt or pin that is positioned at the true position of the hole will always fit through the actual hole. Likewise, the least material condition, or largest hole size, can be used to define a theoretical circle within which the actual hole will be. The boundary for the actual hole must lie inside of the theoretical circle as defined by the LMC. Further information regarding geometric dimensioning and tolerancing is provided in engineering handbooks and graphics textbooks.

APPENDIX C
TECHNOLOGY TIMELINE

5400 B.C.

3500 B.C.

221 B.C.

Fire and the development of early stone tools	Stone Age 1,000,000 B.C.
Early oil lamps that burned fish and animal oil	12,000 B.C.
Writing (cuneiform and hieroglyphics)	Agricultural Age 3500 B.C.
Tools made from metal; development of the wheel	Bronze Age 3000 B.C.
Smelting of iron; development of the Greek alphabet	Iron Age 1200 B.C.
Irrigation and sewers in Rome	700 B.C.
Construction of the Great Wall of China	221 B.C.
First high-level aqueduct in Rome; use of cement	144 B.C.
Water wheels	100 B.C.
Sailing ships that could travel with and against the wind	A.D. 500
Chinese use first paper currency	A.D. 1024
Windmills	A.D. 1200
Chinese use gunpowder in war against Mongols	A.D. 1281
Age of Exploration	A.D. 1400–1600
Gutenberg invents movable type	A.D. 1450

1600 Magnetism discovered by William Gilbert

1610 Galileo's telescope

1628 William Harvey discovers circulation of the blood

1700 Newton's Laws

A.D. 1700

1712 Newcomen's steam engine

1724 Fahrenheit thermometer invented

1752 Benjamin Franklin proves lightning is electricity

A.D. 1752

Industrial Age 1760–1840 Industrial Revolution in Britain

1769 James Watt's steam engine

1770 Hargreaves's spinning jenny patented

1784 Wrought-iron process by Henry Cort

1785 Cartwright's loom

1789 LaVoisier's theory of chemical combustion

1793 Eli Whitney's cotton

1804 Jacquard's loom

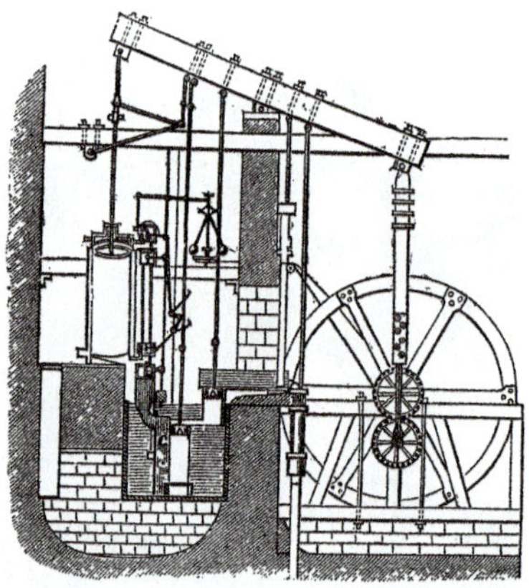

A.D. 1769

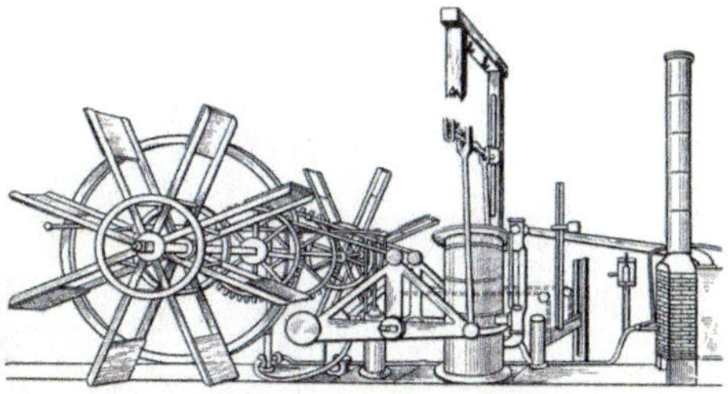

1807

1843

1903

Kodak camera (1888: Model shown is from 1934)

Robert Fulton's steamboat	1807
First textile mill in United States, in Rhode Island	1822
The steam locomotive	1829
Bessemer process for making steel	1835
Photography invented	1839
Morse begins telegraph line between Baltimore, MD, and Washington, DC	1843
Alexander Graham Bell invents the telephone	1876
First successful gasoline-driven car, by Carl Benz in Germany	1885
Daimler's internal combustion engine automobile	1887
First Kodak handheld camera	1888
Diesel engine invented by Dr. Rudolf Diesel	1893
Marconi sends first transatlantic radio signals	1901
Wright brothers' airplane	1903
Albert Einstein proposes his theory of relativity	1905
Model T Ford built using mass production	1908

1909 Beginning of age of plastics

1915 SONAR invented

1921 Hybrid corn greatly improves crop yields

1923 Automatic traffic light patented by Garrett A. Morgan

1924 First round-the-world flight; Kleenex tissues

1925 First successful experiments with hydroponics

1926 First television demonstration; pop-up toaster

1928 Penicillin invented by Alexander Fleming

1929 Electroencephalograph (EEG); foam rubber

1931 Empire State Building in New York City

1931

1934 The electron microscope

1935 Radar developed; nylon patented; B-17 bomber produced

1938 The ballpoint pen is patented

1939 First jet aircraft flies in Germany

1942 First color snapshots

1950

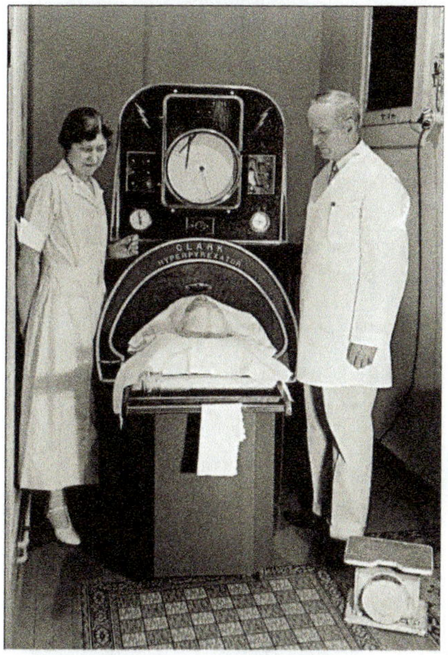

1952

First automatic digital computer, the Mark I — 1944

ENIAC computer built using 18,000 vacuum tubes — 1945

Microwave oven; first Honda motorcycle; development of the transistor at Bell Laboratories — 1947

Power steering; start of regular color TV broadcasts; first credit card (Diner's Club) — 1950

Salk polio vaccine; 3-D movies — 1952

Nuclear-powered submarine "Nautilus"; photocell developed — 1954

Nuclear-powered electricity generation; fiber optics — 1955

First desktop computer (Burroughs E-101) — 1956

Sputnik — 1957

First U.S. satellite — Information Age 1958

Russia sends unmanned spacecraft to the moon (Luna 2); Xerox copier — 1959

Lasers; light-emitting diodes (LEDs); weather satellites — 1960

China tests nuclear bomb; IBM word processors — 1964

First space walk — 1965

Electronic fuel injection; Russians make successful soft landing on the moon (Luna 9) — 1966

1969 American manned moon landing; first Concorde SST flight

1971 Microprocessor developed by Intel; Mariner 9 orbits Mars

1972 CAT scan; photography from satellites (Landsat); videodiscs

1973 Genetic engineering; Skylab orbiting space station launched

1974 Toronto Communications Tower (tallest building in the world)

1976 Birth of the Apple computer; electronic cameras

1977 Trans-Alaska pipeline system completed

1979 Skylab, the orbiting space station, falls back to Earth

1980 Solar-powered aircraft

1981 First launch of a reusable space vehicle (space shuttle *Columbia*)

1982 Artificial heart

1984 Compact disc player

1986 DNA fingerprinting

1990 NASA's Hubble Space Telescope deployed by space shuttle *Discovery*

1997 Ebay online auction and shopping website

1969

1977

1984

2002

First human chromosome completely sequenced — 1999

Xbox video game console — 2001

Worldwide introduction of the Toyota Prius hybrid car — 2001

iPod and iTunes — 2001

First production of Segway electric vehicle — 2002

MySpace website launched — 2003

World of Warcraft massively multiplayer online game released — 2004

Nintendo DS — 2004

2004 Facebook social networking website launched

2005 Sony PlayStation Portable released in North America

2005 YouTube.com launches

2007 iPhone

2007 Google search engine becomes the most visited website

2008 Mapping and sequencing of eight human genomes

2008 Tesla Roadster and Chevrolet Volt electric cars developed

2010 3D Television

2013 PlayStation 4 (PS4)

2014 Driverless cars

2015 Drone use for package delivery proposed

2004

2008

2015

APPENDIX D

TECHNOLOGY STUDENT ASSOCIATION (TSA)

The Technology Student Association (TSA) is a national organization of students engaged in science, technology, engineering, and mathematics (STEM). Open to students who are currently enrolled in or who have completed technology education courses, TSA's membership includes over 235,000 middle and high school students in approximately 2,000 schools spanning 49 states. TSA is supported by educators, parents, and business leaders who believe in the need for a technologically literate society. Members learn through exciting competitive events, leadership opportunities, and much more. The diversity of activities makes TSA a positive experience for every student. From engineers to business managers, our alumni credit TSA with a positive influence on their lives.

Chapters

TSA chapters take the study of STEM (science, technology, engineering, and mathematics) beyond the classroom and give students the chance to pursue academic challenges among friends with similar goals and interests. Together, chapter members work on competitive events, attend conferences at the state and national levels, and have a good time raising funds to get there. Chapter organization develops leadership, as members may become officers within their state and then run nationally. Our chapters are committed to a national service project and are among the most service-oriented groups in the community.

To start a TSA chapter, all that is needed is a motivated teacher and an eager group of students. An expensive array of high-tech equipment is not necessary.

Competitions

Imagine an activity so captivating that your students spend hours working on it after school for weeks at a time. That's what happens when TSA members engage in TSA's competitions. Only TSA members have the opportunity to compete at exciting state conferences and then at the annual National Conference. Expert judging by technology educators and industry representatives inspires the best from participants. Members are rewarded not only with medals or trophies, but also with memories of the camaraderie and the challenge of a state or national TSA conference.

Middle School and High School Level Competitions

TSA offers competitions at both the middle and high school levels. Students can work individually or as a team in such areas as agriculture and biotechnology, Website design, dragster design, flight, geospatial technology, CAD with 3D modeling, digital video production, extemporaneous speaking, structural design and engineering, and scientific visualization.

Each year TSA brings together leading technology educators to revise, review, and develop competitions at the middle or high school level. Every competitive event is correlated to STEM concepts, and the criteria for evaluating each competition is detailed. TSA competitions (approximately 30 on the middle school level and 30 on the high school level) can be implemented at the local, state, and national level.

Once affiliated, your chapter has access to Total TSA, which includes full guidelines for competitions. In addition, Total USA includes correlations to STEM and Common Core State standards, leadership activities, with suggestions for using the activities with specific events, and membership materials. The membership section provides the information you need to know to run and maintain your chapter.

Many state TSA delegations host an annual competitive event conference, as well as a leadership conference. These conferences give student members the opportunity to test their technology skills; many students then choose to compete at the national level conference. Membership in TSA is required in order to participate at any TSA-sanctioned event at the local, state, or national level.

Whether working on an activity at the local level, or competing at the state or national level, all students who use TSA materials benefit from being a TSA member. Opportunities exist for chapter, state, and national leadership roles through elected offices and working committees. Networking among chapters is encouraged.

Curriculum Integration–Meeting STEM Criteria Through TSA Participation

Education directives at the federal and state levels focus on providing avenues for initiatives such as STEM (Science, Technology, Engineering, and Mathematics) education. TSA helps teachers meet the criteria for STEM education goals through exciting programs that include technology activities, competitions, leadership, and teamwork for students. In addition, all TSA activities are correlated to the Standards for Technological Literacy, as set forth by the International Technology and Engineering Educators Association (ITEEA). Teachers can use TSA's activities during the school day in their classroom and extend the learning beyond the classroom through a TSA chapter.

Advantages of TSA Membership

Perhaps the most important benefit of TSA membership is the inspiration and enthusiasm that students gain from receiving recognition for applying their knowledge. Demonstrating skills in a forum beyond school walls motivates students, especially when there is a large, like-minded audience of peers who support them. Other benefits include:

- Enhancing technological literacy and leadership skills
- Working on complex activities from start to finish
- Competing on local, state, and national levels
- Developing leadership skills
- Meeting and working with business and industry leaders
- Attending organized meetings
- Networking with students and teachers with similar interests from all over the United States
- Contributing time and effort to a national service project
- Receiving national membership services, including the online TSA newsletter, *School Scene*
- Receiving recognition through TSA's honor society, scholarships, and achievement programs
- Participating in a national organization that is helping to shape the future of technology education

TSA Chapter Advisors

Serving as a TSA chapter advisor at your school means more than having a "technology club." It means your school is a part of a national organization that promotes and recognizes excellence in technology students through applied science, technology, engineering, and mathematics (STEM) competitions and leadership activities.

For most TSA advisors, TSA is not the only organization with which they are involved. Many also coach sports, serve as class advisors, or work on other school activities. So why do they also advise a TSA chapter? Most would tell you that they do so because they feel TSA offers:

- Leadership opportunities
- Teamwork experiences
- Challenging competitions
- A sense of belonging to a group of students interested in STEM areas

How Much Time Does It Take?

There is no question that being an advisor for a TSA chapter (or any other student activity) takes time and commitment. How much time is a function of several variables:

- **Co-curricular or extra-curricular**. TSA is designed so that it can be a part of a technology classroom, or can be an after-school activity.
- **Program of work**. This is the basis of chapter activities for the year. The program of work is designed to accommodate the amount of time the advisor and members are willing to devote to the endeavor. Considerations for the program of work include: competitions, fundraising if needed, service project, and social events.
- **Organization**. Organizing your technology program around TSA can actually save time. TSA is designed to be student run through elected officers; this is often a great classroom management tool.

More information about the Technology Student Association and how to start a chapter can be found at www.tsaweb.org or by contacting the national TSA office via email at general@tsaweb.org or toll free at 888-860-9010.

TESTS OF ENGINEERING APTITUDE, MATHEMATICS, AND SCIENCE (TEAMS)

TEAMS is a one-day competition for students in middle and high school that allows them to apply their knowledge of skills in science, technology, engineering, and mathematics (STEM) to issues facing our global society. Focused on an annual theme, original academic and innovative concepts are developed for the competition based on the National Academy of Engineering's Grand Challenges. Engaging in the challenges requires critical job-readiness skills such as teamwork, analytical thinking, and multi-dimensional problem solving.

How TEAMS works:

- Teams of 4 to 8 students work together to answer multiple choice and short essay questions based on a National Academy of Engineering Grand Challenge theme.
- The competitions occur at a host university or at your own school during the competition window in February and March each year.
- Recognition includes participation certificates and state and national awards.
- Top ranking teams qualify to compete in the national TEAMS competitions, held each year during the national TSA conference at the end of June.

Why teachers value TEAMS:

- The TEAMS competition is creative, exciting, and theme-based, with problems designed to be solved in teams – the way engineers work.
- TEAMS resources and materials for competitions are designed and tested by experienced, credentialed engineering professors and professionals.
- No need for extra materials, equipment, computers, or supplies.
- Competitions can occur at your own school – no need to travel.

Through TEAMS, your students gain:

- Problem solving abilities
- Collaboration and team-building experience
- An understanding of the role of engineering in solving societal, environmental, community, and global problems
- An opportunity to explore engineering as a profession
- Exposure to STEM disciplines outside the classroom

The TEAMS competition is managed by the Technology Student Association (TSA). More information about TEAMS can be found at teams.tsaweb.org, via email teams@tsaweb.org, or toll free at 888-860-9010.

GLOSSARY

A

ABET Accreditation Board for Engineering and Technology

accrediting evaluating

accuracy a measure of how close a calculation or measurement (or an average of a group of measurements) is to the actual value

actions any executable steps in an algorithm that do not require a decision to be made

aesthetics the way something looks and how that affects people's feelings

agile manufacturing an organizational strategy in which decisions are made quickly and a greater variety of products is produced quickly and inexpensively

Agricultural Era a period in history during which most people lived off the land

algorithm step-by-step description of accomplishing a task

aligned parts or sections of drawings that directly line up with each other

alloy a mixture of two or more metals

ampere the unit of measurement for electrical current

analytics the use of data to determine actions or events

anthropometry the science of measuring people

aqueduct a channel built to carry water

array a list or group of similar variables

artificial intelligence any use of machines to imitate human intelligence

assembly drawings drawings that show how the components of a design fit together

assembly line a system in which each worker performs a specific operation in assembling an item as it is passed along, often on a slowly moving belt, from one workstation to the next

assignment operator tells the computer to put a value into a variable

automation a system in which machines are controlled automatically

auxiliary view a pictorial drawing that shows a straight-on view of one surface of an object

average an estimate of the value most representative of a population

B

base feature a solid model that is roughly the size and shape of the part that is to be modeled

batch process a process in which all materials are included to complete one operation

bill of materials list of material required for a component or assembly, number required in the assembly, and other information

bin a single column in a histogram

biomass vegetation and animal wastes

block comment a block of non-executed text between the special block characters %{ and %} that allow the user to type full length paragraphs as a comment without needing to type the % symbol before each sentence

brainstorming a method of coming up with alternative designs

British thermal unit the amount of heat required to raise the temperature of one pound-mass of water by one degree Fahrenheit

Bronze Age the period during which people began to craft tools and weapons from bronze rather than from stone, from about 3000 B.C. to about 1200 B.C.

C

calorie the amount of heat required to raise the temperature of one gram of water by one degree Celsius

capacitance units of coulombs per volt

catalysts certain chemical materials that speed up a chemical reaction

centerlines thin, long-dash/short-dash lines that mark axes of rotationally symmetric parts or features

central tendency an estimate of the value most representative of a population

charge known as (Q), it is generally quantified using the derived unit coulomb [C]

chemical a substance that contains any of the elements found in the periodic table

closed-loop system a system that can respond and change its inputs or outputs

Code of Ethics description of right vs. wrong
comment item placed in a sketch to explain certain pieces
comparator an instrument that compares the actual output of a system with the desired output or result
compound a substance that is formed by the combination of two or more elements
computer-aided design (CAD) computer-aided design technology that allows the user to create a design on the computer screen and then send it directly to a machine tool
computer-aided manufacturing (CAM) a software application that uses computers to control factory machines
computer-integrated manufacturing (CIM) uses computers not only for design and manufacturing, but also for business needs
concise brief and comprehensive
conduct material property defining how easily heat or electrical current flow
constraint-based solid modeling overcomes the weakness of constructive solid geometry modeling by making the modeling process more intuitive
constraints limitations, boundaries on a solution; the limits imposed on a design solution
construction lines lines used to guide other lines that are not actually part of the component being drawn
constructive solid geometry based on the combination of geometric primitives such as right rectangular prisms (blocks), right triangular prisms (wedges), spheres, cones, and cylinders
continuous process a process in which materials are constantly added to the reactor, where they are mixed and heated according to the formulation, continuously producing the desired product
contour dimensioning placing dimensions of a feature in the view where the feature occurs in its most characteristic shape and where it is visible (as opposed to hidden)
controller a device that turns a process on or off, or that changes it in some way
coulomb a charge of one coulomb represents the total charge on approximately 6.24 * 1018 protons
Coulomb's Law states that the actual force exerted on a charged object varies with both the amount of charge on each object (Q1 and Q2) and the distance (r) between the charges
counter variable keeps track of the number of times we have repeated a calculation or decision
cracking a chemical process in which a complex chemical compound is broken (cracked) into simpler compounds
craft approach the method of making products one at a time, from start to finish
criteria specifications
critical path the minimum amount of time to complete a project
cross section a two-dimensional profile of a part
cumulative distribution function (CDF) a plot in which the values for each bin on the normalized plot are summed from the first bin up to each individual bin

D

debugging a process to identify and correct any errors in the code of a project
decimal places the number of digits to the right of the decimal point
decisions any executable steps in an algorithm that require the answer to a question with "Yes" or "No"
density the amount of mass per unit of volume
dependability a dependable person is one who can be relied upon
dependencies show that a specific task has to complete before a new task begins
descriptive geometry the mathematical principles of projection. These principles form the basis of engineering graphics today.
design a process that takes an idea and implements it physically
design challenge states the problem and lists the specifications for a design
design intent the incorporation of "intelligence" into a design
design process an iterative series of steps engineers use to arrive at the solution to a problem
detail drawings document the detailed design of individual components using orthographic views
diesel engine an engine in which diesel fuel is injected into the cylinder at the end of the compression step
dimension a measurable physical idea; it generally consists solely of a word description with no numbers
dimension lines thin lines with arrowheads at each end that indicate sizes in the drawing
dimensional constraints used to specify distances between items in a solid model
distribution data spread
documentation human readable notes within the code of a program
durable last a long time

E

efficiency a measure of how much of a quantity, typically energy or power, is lost in a process

electric charge positive or negative charges

electric current a movement of electric charge in a wire or other solid material

electrical generator a device that converts mechanical energy into electric energy

elegant elegant code uses the simplest possible algorithm and design needed to solve the problem at hand efficiently

element a pure substance made up of only one type of atom

endothermic reaction a reaction that absorbs energy in the form of heat

energy the ability to do work

energy of translational motion a form of energy possessed by an object in motion

engineering a profession that draws upon knowledge of mathematics and science to create ways to economically use available materials and the forces of nature to benefit the human race through the process of design

engineering graphics a highly stylized way of presenting images of parts or assemblies

entrepreneur a person who comes up with a good idea and uses that idea to make money

ergonomics also called human factors engineering; deals with designing products so they can be used easily and comfortably

error checking allows for the creation of robust solutions to problems that will not lead to incorrect or unstable answers

ethical having to do with right and wrong as defined by an external source, such as within an Engineering Code of Ethics

ethical dilemma a difficult decision about whether something is right or wrong

ethics ethical decisions are those based on a right vs. wrong as defined for a population by an external source

exothermic reaction a reaction that releases energy in the form of heat

exploded view reveals details of individual parts, along with showing the order in which they are assembled

exponential rate of change extraordinarily rapid developments; shows a sharp curve on a graph

extension lines thin lines that extend from the object to the dimension line to indicate which feature is associated with the dimension

external combustion fuel is burned in a chamber outside the engine, providing heat to another liquid or gas

extreme value minimum and maximum values of data

extruded stretched

F

factory system a system in which people made goods using machines rather than by hand

farads measure of capacitance; one farad equals one coulomb per volt

feature-based modeling attempts to make the modeling process more efficient by creating and modifying geometric features of a solid model in a way that represents how geometries are created using common manufacturing processes

features characteristics of a device, drawing, part, or assembly

feedback information about the output of a system that can be used to adjust it

feedback loop a return to an earlier location in an algorithm with updated values of variables

Fermi problems problems in which so much information is missing that a solution seems impossible

flexible manufacturing the efficient production of small numbers of products

flowchart a graphical representation of a written algorithm that describes the sequence of actions, decisions, and path of a process

foreshortening lines of an object are changed to produce the illusion of the part extending into the third dimension

formulation a systematic method of creating a result

fossil fuels energy-rich substances from the remains of plants and animals that lived millions of years ago

fracture stress amount of stress applied when material fractures

functionality the ability of a product, system, or process to fulfill its intended purpose over its desired life span

functions smaller programs referenced within a larger program

G

Gantt chart a project schedule, listing each task and showing how long it takes, and assigning tasks to show who has responsibility for tasks

gas chromatograph tests the purity of a substance by heating a product past its boiling point to create gases

gas turbine engine an engine which heats fuel and air, causing a turbine blade to spin, which can then drive a generator to generate electricity

gasoline engine an engine in which a spark ignites the fuel

geometric constraints used to define positional relationships between entities in the model in terms of the geometry

geothermal energy thermal energy that is stored below Earth's surface

Grand Challenges a list of 14 challenges for the twenty-first century, each of which has the potential to transform the way we live, work, and play

graphical algorithms a pictorial representation of instructions required to solve a problem

H

hazardous materials materials that can be harmful to health or the environment but are still in their original, usable form

hazardous waste waste that is potentially harmful to health or the environment

henrys unit of measure of inductance

hidden lines thin, dashed lines that represent features that are hidden behind surfaces in the current view

histogram plot using bars to represent a distribution of numerical data

history-based modeling states that the order in which parts are created is critical

horsepower a unit of power originally used to quantify how the steam engine could replace the work done by a horse

hybrid electric vehicles combine the engine of a conventional vehicle with the battery and electric motor of an electric vehicle

I

ideation the generation of concepts or ideas to solve a design problem

index variable keeps track of the number of times we have repeated a calculation or decision

inductance a wire in a changing magnetic field such as a collapsing one will have a current induced: this is inductance

inductor component that resists sudden changes to current

Industrial Revolution a period, which began in Britain during the late 1700s and lasted until the mid-1900s, in which human and animal muscle power was replaced by machines

INF a special MATLAB reserved word representing infinity

infinite feedback loop a nonterminating feedback loop

Information Age a period that continues today in which the changes caused by machinery and factory production gradually gave way to a new type of technology: information

informed design process a process providing a way to make decisions without complete knowledge and then reassess those decisions later on

input the command given to a system; the known quantities used in a design or program

interchangeable manufactured to be exactly alike so that any one can be substituted for any other

interference when physical parts interfere with the operation of other physical parts

internal combustion engine an engine in which the combustion process takes place inside the engine; when the fuel is ignited, chemical energy is converted to thermal energy and then to mechanical energy

Iron Age the era when iron came into common use, began around 1200 B.C. in the Middle East and about 450 B.C. in Great Britain

isometric projection simplifies the pictorial perperspective. In an isometric projection, parallel lines remain parallel.

J

jet engine gas turbine attached to the wing of an airplane

joule the amount of energy required to move an object one meter using one Newton of force

just-in-time manufacturing uses careful scheduling to keep materials and products in storage for as little time as possible

K

kinetic energy energy that results from motion

L

leader lines thin, solid lines terminated with arrowheads that are used to indicate a feature with which a dimension or note is associated
line of sight angle at which an object is viewed
linear rate of change a rate that looks like a straight line on a graph

M

manufacturing the process of making goods by hand or machine
mass a fundamental property of materials, so it is imperative to have a standard definition of the scientific unit, kilogram
mass production the production of goods in large quantities by groups of workers in factories
Material Safety Data Sheet (MSDS) paperwork explaining how hazardous materials are to be used and stored
mathematical model a model that uses equations to describe how a product will function or perform
MATLAB a programming language: short for matrix laboratory
mean an estimate of the value most representative of a population
median the value between the lower half and the upper half of a population
metric a measurement system based on powers of 10; a measurable quantity used to assess performance
M-Files MATLAB program files
model a computer representation of an object or part
Modulus of Elasticity the slope of the line, λ
molecule the smallest unit of a chemical compound, made of two or more atoms that act as a unit
Moore's Law states that the number of transistors within a computer processor doubles every two years
morals moral decisions are those that apply to an individual
multimeter device capable of measuring current, voltage, or resistance
multiview projection a collection of flat 2-D drawings of the different sides of an object

N

nanotechnology the creation of materials, devices, and systems using individual atoms and molecules
natural product includes substances that appear in nature (such as edible plants and fruits) as well as substances that can be made from natural ingredients (such as wine and soap)
negatively skewed a distribution in which the mean has been pulled lower than the median
Nelson Rules a variety of conditions that indicate that a process is out of control
nondurable designed to be consumed in a short period of time
nonrenewable energy source a source of energy that cannot be replaced once it has been used up
normal distribution distribution which follows the 68-95-99.7 rule
normalized plot a plot in which the values are divided by the total number of data points
nuclear fission the splitting of an atom's nucleus
nuclear fusion a process in which the nuclei of two atoms are forced together to form a new nucleus, releasing a large amount of energy

O

oblique projection places the principal face of the object parallel to the plane of the paper with the axes in the plane of the paper perpendicular to one another
Ohm's Law an equation that describes the relationship between current, voltage, and resistance: $V = R*I$
open source the licensing to the software and internal workings of the platform are free
open-loop system a system that has no way of monitoring and adjusting itself
order of search the sequence MATLAB goes through when you type a name in either the Command Window or in a program or function
orders of magnitude often used when comparing things of very different scales, such as a small rock and a planet
orthographic projection typically shows three views of an object; each view shows a different side
outlier an extreme value
output what is produced by a system; the actual result
overdimensioned having one redundant dimension

P

parallel projection type of drawing in which lines along edges are parallel and do not converge
parametric parameters of the model may be modified to change its geometry
parent-child relation an aspect of solid modeling that states that the child cannot exist without the parent existing first
parts list includes the part name, identification number, material, number required in the assembly, and other information
Paste Special pasting text (or something else) into a document, spreadsheet, or other file along with other information, such as format
periodic table lists all the known elements in a way that reflects the special properties of each
perspective projection drawn so that parallel lines converge in the distance
pharmaceutical products that include medicines, vaccines, and drugs used for healthcare
photovoltaic cell a solar cell that can turn sunlight into electricity
pictorial perspective this type of drawing presents an object much as it would look to the human eye or in a photograph
positively skewed a distribution in which the mean has been pulled higher than the median
potential energy energy that is stored
power the amount of work done during a given period of time
power grid interconnected network of electrical transmission delivering power from suppliers to consumers
precedence of lines when two lines in a particular view coincide, this dictates which line is shown: visible lines have precedence over hidden lines, which have precedence over centerlines
precision a combination of accuracy and repeatability
primitive modeling based on the combination of geometric primitives such as right rectangular prisms (blocks), right triangular prisms (wedges), spheres, cones, and cylinders
probability associated with assessing the likelihood that an event will or will not occur
process the action part of a system
productivity refers to how quickly and cost-effectively a product is made
programming the process of expressing an algorithm in a language that a computer can interpret; the process in which a problem is formulated in a way that a computer can understand and solve
programs a set of instructions given to a computer to perform a specific task
project management prioritizing tasks and following schedules to produce efficient results
projectors parallel lines along the edges of a cube
prototype a full-scale, fully operational version of a solution
punctuality being on time and ready when expected
purify a process done to a substance so that it is made of only one kind of matter

Q

quality degree of excellence
quality circles groups of workers and managers who meet to discuss problems and ways of improving production
quality control chart used to show how close to the mean the results of a process are when measured over time

R

radian the angle at the center of a circle formed by an arc (S) equal in length to the radius (r) of that circle
radioactive waste waste material containing unstable atoms that give off radiation
range restriction rule which prohibits some values, usually based on characteristics of the data
reference dimension provided only for information to help in interpreting the drawing
reliability a reliable person can be relied upon for accuracy and honesty each and every time
rendering a virtual light source is assumed to be near the object to provide a three-dimensional lighting effect
renewable energy sources any source of energy that can be replaced
repeatability a measure of how close together multiple measurements of the same parameter are, whether or not they are close to the actual value
resistance opposition to the flow of current through a material
responsibility taking ownership of one's role and deliverables within a project
revolving extruding around an axis
risk/benefit trade-off a common trade-off made in solving large problems; to obtain desired benefits, designers must allow for risk while trying to keep the risk as low as possible

robotics the area of artificial intelligence that designs and builds robots

rotational kinetic energy states that a rotating object has energy whether it is translating (moving along a path) or not

S

safety factor a multiple of some criteria used in a design

scale indicates the ratio of the size of the drawing to the size of the actual object

scale model a model that is either larger or smaller than the final product

scope the entire description of a project

sensor a monitor of a system's output

seven technological resources people, capital, time, information, energy, materials, and tools and machines

shading a virtual light source is assumed to be near the object to provide a three-dimensional lighting effect

SI prefix In using one of these when reporting numbers, we avoid scientific notation or long strings of zeros.

significant figures the digits considered reliable as a result of measurement or calculation

silhouette lines define the outer edges, or silhouette, of the part in the image plane as viewed looking toward the inclined surface in the direction of the arrow indicating the line of sight

68-95-99.7 rule a rule that determines if a distribution is normal: 68% of values are within one standard deviation of the mean, 95% are within 2 standard deviations and 99.7% are within 3 standard deviations

sketch Arduino's name for files containing code—with a file extension of .ino

skewed a distribution that is not symmetric

smelting a process in which iron was made from iron ore by melting the ore to take out the impurities

solar energy energy from the sun

solid modeling represents an object in the virtual environment just as it exists in reality, having volume as well as surfaces and edges

SOLVEM stands for Sketch, Observations or Objectives, List, Variables, and Equations Manipulation

specifications the performance requirements that a solution must fulfill

spectrometer measures the light energy absorbed or transmitted through a sample

speech recognition digitizing the sound waves of speech and then converting them to the basic sounds that are used in language

speech synthesis converting text to speech by breaking down the written words into basic sounds of the language and then converting the sounds into digital audio

stakeholders who might be affected by the various alternative courses of action that might be implemented

standard deviation found by taking the square root of the variance

Statistical Process Control (SPC) a method of monitoring, controlling, and improving a manufacturing process

statistics provide quantitative measures to "things" that behave in a random manner

stepwise scope all of the known and unknown information at that point in the procedure

stiffness the slope of the line, λ

Stone Age the period during which people used stone to make tools, from about 1 million B.C. to about 3000 B.C.

strain the ratio of the change in length to the original length

stress the applied load pulling the material apart

stress strain curve plots the strain (the ratio of the change in length to the original length) against the stress (the applied load pulling the material apart)

subsystems smaller systems that often make up other, larger systems

surface model defines the object in terms of surfaces such as plates (flat) and shells (curved) in addition to edges

surfactant surface-active agent, normally made from plant and animal fats

symmetry lines thin, long-dash/short-dash lines that mark axes of rotationally symmetric parts or features

synthesize create molecules for a particular material by combining two or more substances in a chemical reaction

synthetic product made in a laboratory using specific chemical processes

system a group of interrelated components designed collectively to achieve a desired goal

T

tensile test plots the stress strain curve of a material

thermal energy the energy of the atoms that make up a substance

3-D model the most general model used in CAD software; the CAD software electronically stores the entire three-dimensional shape of the object

3-D projection a drawing that looks like a visual representation of an object: drawn "as it looks" with lines that converge in the distance

time management managing the available time for a project or job effectively

title block records important information about a drawing; normally appears in the lower right corner of the drawing

turbine a circular device with blades generally used to generate electricity

2 1/2-D model has a third dimension that is recognized by the CAD software, but the third dimension is simply an extrusion of a two-dimensional shape

2-D model the computer graphics equivalent of an orthographic projection created using a pencil and paper

U

Ultimate Tensile Strength (UTS) the maximum stress we can apply before damage

unidirectional having the same orientation

unit allows us to quantify a dimension—to state a number describing how much of that dimension exists in a specific situation. Units are defined by convention and related to an accepted standard.

V

values the principles we consider worthwhile that influence all of our decisions

variables the different factors that affect the performance of a design

variance To determine a variance (σ^2), the difference between each point and the mean is determined, and each difference is squared to keep all terms positive. This sum is then divided by one less than the number of data points.

venture capitalist a person who supplies money to finance the start-up of a new company

virtual reality using artificial intelligence to create a computer-generated environment

visible lines thick, solid lines that represent the outline of the object that can be seen in the current view

voltage a measure of the force with which electric current is pushed through wires

W

watt the SI unit of power

wireframe model the simplest 3-D model; only edges of the object are represented

witness lines thin lines that extend from the object to the dimension line to indicate which feature is associated with the dimension

work something produced or accomplished through effort or activity

written algorithms a narrative set of instructions required to solve a problem

Y

yield stress point on the stress/strain curve where a material enters a plastic region

Young's Modulus the slope of the line, λ

INDEX

Page numbers followed by t *or* f *refer to tables and figures on respective pages.*

F

N

O

P

Q

R